STUDENT'S SOLUTIONS MANUAL

JAMES LAPP

MATHEMATICS ALL AROUND

FOURTH EDITION

Thomas L. Pirnot

Kutztown University of Pennsylvania

Addison-Wesley
is an imprint of

PEARSON

Reproduced by Pearson Addison-Wesley from electronic files supplied by the author.

Publishing as Pearson Addison-Wesley, 75 Arlington Street, Boston, MA 02116.

ISBN-13: 978-0-321-57584-5
ISBN-10: 0-321-57584-9

2 3 4 5 6 CRS 12 11 10 09

Addison-Wesley
is an imprint of

PEARSON

www.pearsonhighered.com

CONTENTS

Chapter 1 Problem Solving: Strategies and Principles 1
1.1 Problem Solving 1
1.2 Inductive and Deductive Reasoning 5
1.3 Estimation 8
Chapter Review Exercises 9
Chapter Test 10

Chapter 2 Set Theory: Using Mathematics to Classify Objects 13
2.1 The Language of Sets 13
2.2 Comparing Sets 14
2.3 Set Operations 17
2.4 Survey Problems 18
2.5 Looking Deeper: Infinite Sets 20
Chapter Review Exercises 22
Chapter Test 23

Chapter 3 Logic: The Study of What's True or False or Somewhere in Between 25
3.1 Statements, Connectives, and Quantifiers 25
3.2 Truth Tables 26
3.3 The Conditional and Biconditional 30
3.4 Verifying Arguments 33
3.5 Using Euler Diagrams to Verify Syllogisms 38
3.6 Looking Deeper: Fuzzy Logic 39
Chapter Review Exercises 40
Chapter Test 42

Chapter 4 Graph Theory (Networks): The Mathematics of Relationships 45
4.1 Graphs, Puzzles and Map Coloring 45
4.2 The Traveling Salesman Problem 48
4.3 Directed Graphs 52
4.4 Looking Deeper: Scheduling Projects Using Pert 55
Chapter Review Exercises 61
Chapter Test 63

Chapter 5 Numeration Systems: Does It Matter How We Name Numbers? 65
5.1 The Evolution of Numeration Systems 65
5.2 Place Value Systems 68
5.3 Calculating in Other Bases 73
5.4 Looking Deeper: Modular Systems 77
Chapter Review Exercises 81
Chapter Test 83

Chapter 6 Number Theory and the Real Number System: Understanding the Numbers All Around Us 85
6.1 Number Theory 85
6.2 The Integers 87
6.3 The Rational Numbers 90
6.4 The Real Number System 95
6.5 Exponents and Scientific Notation 98
6.6 Looking Deeper: Sequences 100
Chapter Review Exercises 104
Chapter Test 106

Chapter 7 Algebraic Models: How Do We Approximate Reality? 109
7.1 Linear Equations 109
7.2 Modeling with Linear Equations 114
7.3 Modeling with Quadratic Equations 117
7.4 Exponential Equations and Growth 123
7.5 Proportions and Variation 127
7.6 Functions 132
7.7 Looking Deeper: Dynamical Systems 134
Chapter Review Exercises 136
Chapter Test 138

Chapter 8 Modeling with Systems of Linear Equations and Inequalities: What's the Best Way to Do It? 141
8.1 Systems of Linear Equations 141
8.2 Systems of Linear Inequalities 146
8.3 Looking Deeper: Linear Programming 155
Chapter Review Exercises 166
Chapter Test 167

Chapter 9 Consumer Mathematics: The Mathematics of Everyday Life 169
9.1 Percent Change And Taxes 169
9.2 Interest 172
9.3 Consumer Loans 175
9.4 Annuities 178
9.5 Amortization 183
9.6 Looking Deeper: Annual Percentage Rate 189
Chapter Review Exercises 191
Chapter Test 194

Chapter 10 Geometry: Ancient and Modern Mathematics Embrace 197
10.1 Lines, Angles, and Circles 197
10.2 Polygons 199
10.3 Perimeter and Area 202
10.4 Volume and Surface Area 209
10.5 The Metric System and Dimensional Analysis 213
10.6 Geometric Symmetry and Tessellations 216
10.7 Looking Deeper: Fractals 218
Chapter Review Exercises 220
Chapter Test 222

Chapter 11 Apportionment: How Do We Measure Fairness? 225
11.1 Understanding Apportionment 225
11.2 The Huntington–Hill Apportionment Principle 228
11.3 Applications of the Apportionment Principle 232
11.4 Other Paradoxes and Apportionment Methods 235
11.5 Looking Deeper: Fair Division 244
Chapter Review Exercises 246
Chapter Test 247

Chapter 12 Voting: Using Mathematics to Make Choices 251
12.1 Voting Methods 251
12.2 Defects in Voting Methods 254
12.3 Weighted Voting Systems 258
12.4 Looking Deeper: The Shapley–Shubik Index 265
Chapter Review Exercises 271
Chapter Test 273

Chapter 13 Counting: Just How Many Are There 277
13.1 Introduction to Counting Methods 277
13.2 The Fundamental Counting Principle 283
13.3 Permutations and Combinations 285
13.4 Looking Deeper: Counting and Gambling 288
Chapter Review Exercises 289
Chapter Test 290

Chapter 14 Probability: What Are the Chances 291
14.1 The Basics of Probability Theory 291
14.2 Complements and Unions of Events 294
14.3 Conditional Probability and Intersections of Events 297
14.4 Expected Value 303
14.5 Looking Deeper: Binomial Experiments 305
Chapter Review Exercises 308
Chapter Test 309

Chapter 15 Descriptive Statistics: What a Data Set Tells Us 311
15.1 Organizing and Visualizing Data 311
15.2 Measures of Central Tendency 313
15.3 Measures of Dispersion 317
15.4 The Normal Distribution 324
15.5 Looking Deeper: Linear Correlation 327
Chapter Review Exercises 331
Chapter Test 332

Chapter 1
Problem Solving: Strategies and Principles

Section 1.1 Problem Solving

1. Understand the problem, devise a plan, carry out your plan, check your answer.

3. Answers will vary.

5. *How to Solve It*

7. Drawings may vary.

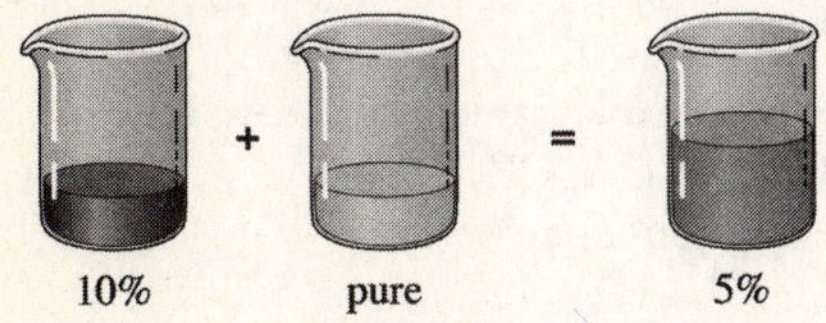

9. Drawings may vary.

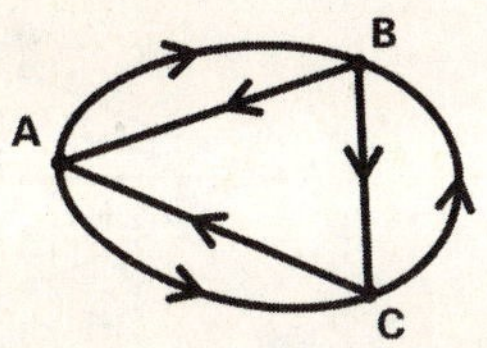

11. Answers may vary.
Let w be the width of the solar panel and l or $3w$ be the length.

13. Answers may vary.
Let M be Meredith, C be Christina, I be Izzie, A be Alex, and D be Derek.

15. Answers may vary.
Let s be the dollar amount invested in stocks and b be the dollar amount invested in bonds.

17. Answers (order) may vary.

Penny	Nickel
Heads	Heads
Heads	Tails
Tails	Heads
Tails	Tails

Combinations would be *HH*, *HT*, *TH*, and *TT*.

19. $2\times2\times2\times2=16$

21. Answers (order) may vary.
There are six distinct segments:

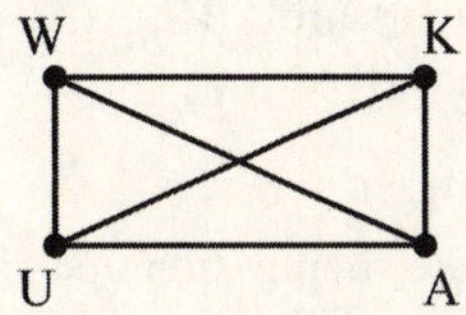

Pairs would be *WK*, *WA*, *WU*, *KA*, *KU*, and *AU*.

23. (1,1), (1,2), (1,3), (1,4), (2,1), (2,2), (2,3), (2,4), (3,1), (3,2), (3,3), (3,4), (4,1), (4,2), (4,3), (4,4)

25. r_3 is the set of people who are good singers and appeared on "American Idol". r_4 is the set of people who have appeared on "American Idol" and are not good singers.

27. 7, 14, 21, 28, **35, 42, 49, 56, 63,...**

29. Answers may vary.
ab, ac, ad, ae, bc, bd, be, ***bf, cd, ce, cf, cg,...***

31. 1, 1, 2, 3, 5, 8, 13, **21, 34, 55, 89, 144,...**

33. Answers may vary.
Three people are being honored for their work in reducing pollution. In how many ways can we line up these people for a picture?

Let the people be labeled A, B, and C.

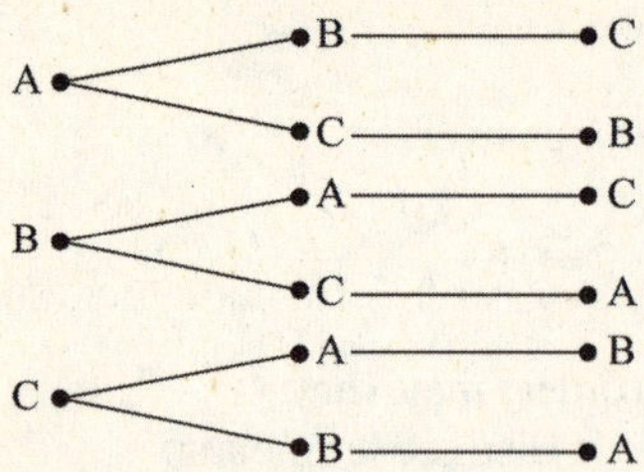

The possible orders are ABC, ACB, BAC, BCA, CAB, and CBA.
6 different ways

Four people are being honored for their work in reducing pollution. In how many ways can we line up these people for a picture?

Let the people be labeled A, B, C, and D.

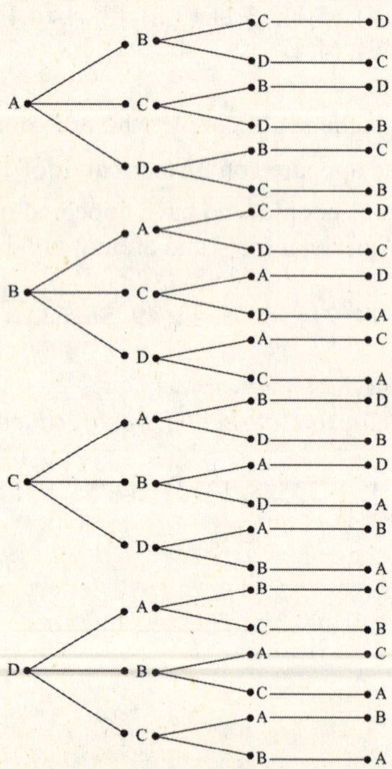

The possible orders are ABCD, ABDC, ACBD, ACDB, ADBC, ADCB, BACD, BADC, BCAD, BCDA, BDAC, BDCA, CABD, CADB, CBAD, CBDA, CDAB, CDBA, DABC, DACB, DBAC, DBCA, DCAB, and DCBA.
24 different ways

35. Answers may vary.
Using the first three letters of the alphabet, how many two-letter codes can we form if we are allowed to use the same letter twice?

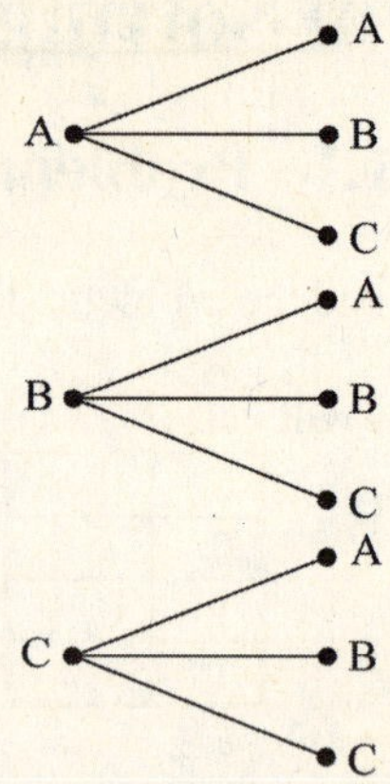

The possible codes are AA, AB, AC, BA, BB, BC, CA, CB, and CC.
9 different codes

Using the first five letters of the alphabet, how many two-letter codes can we form if we are allowed to use the same letter twice?

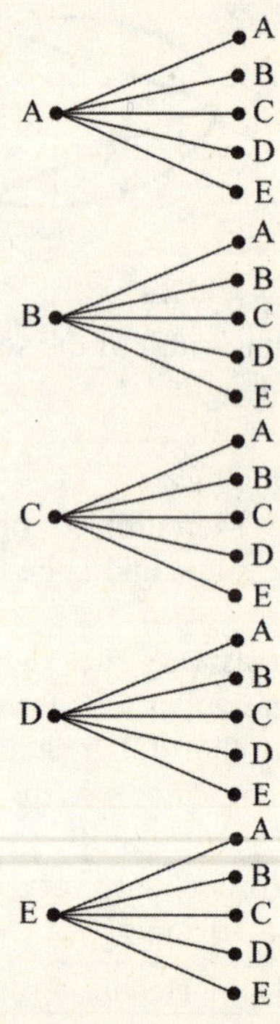

The possible codes are AA, AB, AC, AD, AE, BA, BB, BC, BD, BE, CA, CB, CC, CD, CE, DA, DB, DC, DD, DE, EA, EB, EC, ED, and EE.
25 different codes

37. Answers may vary.
An electric blue sports car comes with two options: air conditioning and a CD player. You may buy the car with any combination of the options (including none). How many different choices do you have?

Let A be air conditioning and C be CD player. If a feature is not included, it is indicated by a "0". If it is included, it is indicated by a "1".

A	C
0	0
0	1
1	0
1	1

4 different choices

An electric blue sports car comes with three options: air conditioning, CD player, and air bags. You may buy the car with any combination of the options (including none). How many different choices do you have?

Let A be air conditioning, C be CD player, and B be air bags. If a feature is not included, it is indicated by a "0". If it is included, it is indicated by a "1".

A	C	B
0	0	0
0	0	1
0	1	0
0	1	1
1	0	0
1	0	1
1	1	0
1	1	1

8 different choices

39. False, counterexamples may vary.
September has only 30 days.

41. False, counterexamples may vary.

$$\frac{1}{2}+\frac{3}{4}=\frac{2}{4}+\frac{3}{4}=\frac{5}{4}$$

$$\frac{1+3}{2+4}=\frac{4}{6}=\frac{2}{3},\quad \frac{5}{4}\neq\frac{2}{3}$$

43. False, A is the grandfather of C.

45. False, counterexamples may vary.
$-9<-5$ but $(-9)^2>(-5)^2$ since $81>25$.

47. False, counterexamples may vary.
If a square has a side of length 3 units, its area is 9 square units. If the length is doubled to 6 units, the area is 36 square units, not 18 square units.

49. Explanations may vary.
These two sequences do not give the same results. The question here is equivalent to asking if the algebraic statement, $x^2+5=(x+5)^2$, is true.

If we let $x=1$, we have a counterexample.

$$1^2+5\stackrel{?}{=}(1+5)^2$$

$$1+5\stackrel{?}{=}6^2$$

$$6\neq 36$$

Hence, the statement is false.

51. Explanations may vary.
These two sequences do give the same results. The question here is equivalent to asking if the algebraic statement, $\frac{x+y}{3}=\frac{x}{3}+\frac{y}{3}$, is true.

If you think of dividing by 3 equivalent to multiplying by $\frac{1}{3}$, then you can use the distributive property to prove this statement.

$$\frac{x+y}{3}=\frac{1}{3}(x+y)=\frac{1}{3}x+\frac{1}{3}y=\frac{x}{3}+\frac{y}{3}$$

53. Answers may vary.
5 is a number; $\{5\}$ is a number with braces around it. Moreover, 5 is a singly listed element, while $\{5\}$ is a set that contains the single element 5.

55. Answers may vary.
One is uppercase and the other is lowercase. Moreover, U usually denotes the universal set, and the lower case letters are usually elements that appear in some set.

57. Answers may vary.
{ } are different from ( ). Moreover, {1,2} is the set that contains only the elements 1 and 2. (1,2) is the interval that contains all real numbers between 1 and 2 (not including 1 and 2 themselves).

Note: Another interpretation of (1,2) is being a point in a plane. However, with {1,2} in this same problem, the interpretation should be relative to sets.

59. The symbol on the right is the number zero; the symbol on the left is not a number.

61. – 63. No solution provided.

65. Answers may vary.
Let O be the age of the older building and Y be the age of the younger building.

Guesses for O and Y	Good points	Weak Points
100, 221	Sum is 321.	The older is more than twice the younger.
110, 211	Sum is 321.	The older is less than twice the older.
107, 214	All conditions satisfied. We have the solution.	

67. Answers may vary.
Let S be the number of hours Janine worked in the sporting store and P be the number of hours Janine gave piano lessons.

Guesses for S and P	Good points	Weak Points
11 and 4	Sum is 15.	Amount earned is less than $137.25.
8 and 7	Sum is 15.	Amount earned is more than $137.25.
9 and 6	All conditions satisfied. We have the solution	

69. Answers may vary.
Let L be the number of times LaDainian carried the football, S be the number of times Steven carried the football, and J be the number of times Larry carried the football.

Guesses for E, B and R	Good points	Weak Points
342, 340, and 410	L is 29 more than S and 68 less than J.	Sum is less than 1,110.
352, 350, and 420	L is 29 more than S and 68 less than J.	Sum is more than 1,110.
348, 346, and 416	All conditions satisfied. We have the solution.	

71. Answers may vary.
Let A be the amount invested at 8%and B be the amount invested at 6%.

Guesses for A and B	Good points	Weak Points
3000 and 5000	Sum is $8000.	Return is less than $550.
4000 and 4000	Sum is $8000.	Return is more than $550.
3500 and 4500	All conditions satisfied. We have the solution	

73. Answers may vary.
Let A be the number of administrators, S be the number of students, and F be the number of faculty members.

Guesses for A, S and F	Good points	Weak Points
10, 2, and 7	A is 5 times S and F is 5 more than S.	There are less than 26 people.
20, 4, and 9	A is 5 times S and F is 5 more than S.	There are more than 26 people.
15, 3, and 8	All conditions satisfied. We have the solution.	

There are 3 students.

75. Good names make it easier to remember what the unknowns represent.

77. Examining simpler examples may help you see a pattern that can be applied to the original problem.

79. – 83. No solution provided.

85. $2^1 = 2$; $2^2 = 4$; $2^3 = 8$; $2^4 = 16$; $2^5 = 32\ldots$, following this pattern, 2^{100} will end with a 6.

87. $6\times6\times6 = 216$

89. LCHPL, LCPHL, LHCPL, LHPCL, LPCHL, LPHCL

91. (3, 5), (5, 7), (11, 13), (17, 19), (29, 31), **(41, 43), (59, 61), (71, 73)**; pairs of sequential primes that differ by 2.

93. There are a total of 55 rectangles.

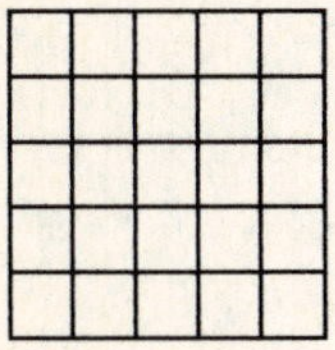

1×1	25
2×2	16
3×3	9
4×4	4
5×5	1

Section 1.2 Inductive and Deductive Reasoning

1. Example 2; the test for divisibility by nine.

3. The reasoning uses accepted facts and general principles.

5. Answers will vary.

7. inductive

9. deductive

11. inductive

13. deductive

15. inductive

17. 16

19. 96

21. $\frac{1}{64}$

23. 21

25.

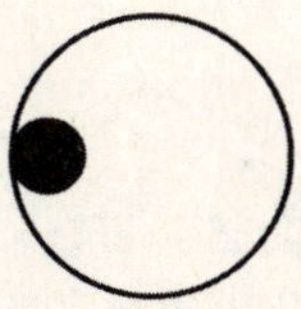

27.

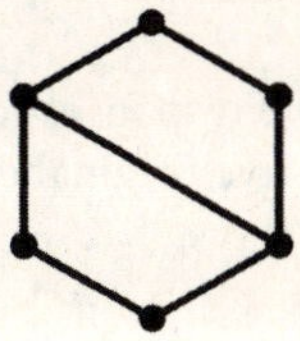

29. Hint: Think of prime numbers.

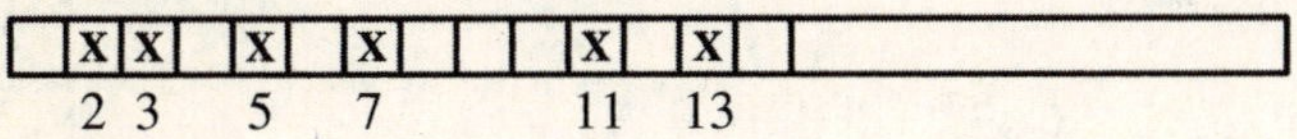

31. $3+13$ or $5+11$

33. $3+17$ or $7+13$

35. $1+4+9+16=30$

37. a) The total of all the numbers in the square is $1+2+3+...+16=136$.

    b) The total of the numbers for each row, column, and diagonal would be $\frac{136}{4}=34$.

    c) One can deduce the missing numbers to yield the following.

7	6	12	9
10	11	5	8
13	16	2	3
4	1	15	14

In Exercises 39 – 41, only the answers to Part b, the expected conjecture, and Part c, the algebraic justification, are provided.

39. In this trick, you will always get the result three.

    a) Call the number n.

    b) $3n$

    c) $3n+9$

    d) $\dfrac{3n+9}{3}=\dfrac{3(n+3)}{3}=n+3$

    e) $(n+3)-n=n+3-n=3$

41. In this trick, you will always get a result that is twice the number that you started with.

    a) Call the number n.

    b) $8n$

    c) $8n+12$

    d) $\dfrac{8n+12}{4}=\dfrac{4(2n+3)}{4}=2n+3$

    e) $(2n+3)-3=2n+3-3=2n$

43.

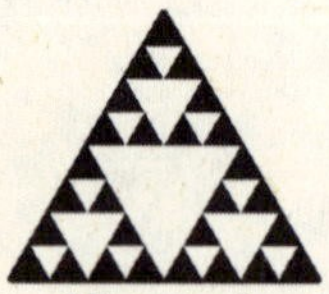

(d) step 3

45. Answers may vary. Possible answers include: Inductive reasoning is the process of drawing a general conclusion by observing a pattern in specific instances. Deductive reasoning uses accepted facts and general principles to arrive at a specific conclusion.

47. Answers may vary. Possible responses include that Sharifa would begin by visiting one of the six branches and then visit the five original cities in 120 ways. The total number of ways she could make her visits is $6\times 120=720$ ways.

49. 36644633; the letters represent the numbers in reverse order.

51. 986763; reverse the numbers and delete on number from any pair of numbers.

53. No solution provided.

55. $1+2+3+4+5=\dfrac{5\times 6}{2}$, $1+2+3+4+5+6=\dfrac{6\times 7}{2}$

57. $1+3+5+7+9=25$, $1+3+5+7+9+11=36$

59. There are a total of 20 squares.
2 3×3 squares; 6 2×2 squares; 12 1×1 squares

61. a) There are a total of 60 rectangles.

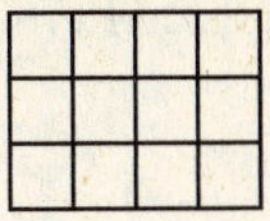

1×1	12
1×2	9
1×3	6
1×4	3
2×1	8
2×2	6
2×3	4
2×4	2
3×1	4
3×2	3
3×3	2
3×4	1

b) There are a total of 210 rectangles.

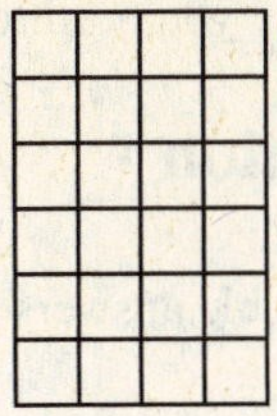

1×1	24
1×2	18
1×3	12
1×4	6
2×1	20
2×2	15
2×3	10
2×4	5
3×1	16
3×2	12
3×3	8
3×4	4

4×1	12
4×2	9
4×3	6
4×4	3
5×1	8
5×2	6
5×3	4
5×4	2
6×1	4
6×2	3
6×3	2
6×4	1

63. The base is a 6×4 rectangle. If you consider the diagram below as the base, we have $6\times4=24$ baseballs. In order to build the next level, we are looking for the number of places in which four baseballs (squares) meet. There are $5\times3=15$ such places.

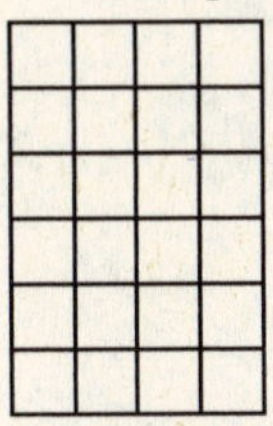

For the next level we would have $4\times2=8$ meeting places.

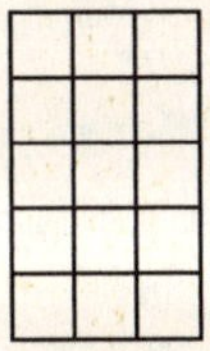

For the last level we would have $3\times1=3$ meeting places.

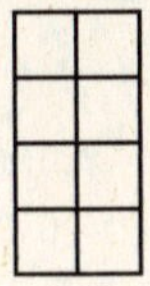

This yields a total of $6\times4+5\times3+4\times2+3\times1=24+15+8+3=50$ baseballs.

65. No solution provided.

67. If you expand the expression as follows,

$$(2n+5)\cdot 50+1759-1991=100n+250+1759-1991=100n+2009-1991=100n+18,$$

you see that the $1759+250$ gives a multiple of 100 plus 2009 – 1991, which is your age. If you have already had your birthday, you need to add the extra year, which is why we would then add 1760.

Section 1.3 Estimation

1. Answers may vary. Possible answers include 3279 rounding up to 3300.

3. No solution provided.

5. Eratosthenes made a remarkably accurate estimate of the circumference of the earth.

7. 37,300

9. 8,300

11. Answers may vary.
$20+40+190+40=290$

13. Answers may vary.
$35-15=20$

15. Answers may vary.
$5\times 16=80$

17. Answers may vary.
$18\div 3=6$

19. Answers may vary.
$0.1\times 800=80$

21. Answers may vary.
$9\%\times 1{,}000=0.09\times 1{,}000=90$

23. Answers may vary.
$4\times 5\times 6=120$ miles
The estimate is larger than the exact answer.
Exact answer is 111 miles.

25. Answers may vary.
$325\div 50=6.5$ more hours, 7:30PM
The estimate is earlier than the actual time.
Actual answer is 7:50PM

27. Answers may vary.
$\$80.00\times 15\%=\$80.00\times 0.15=\$12.00$
The estimate is less than the exact answer.
Exact answer is \$12.37.

29. Answers may vary.
$(3\times \$3.00)+(4\times \$1.50)+\$3.00=$
$\$9.00+\$6.00+\$3.00=\$18.00$
The estimate is larger than the exact answer.
Exact answer is \$17.20.

31. Answers may vary.
It seems safe. Alicia probably weighs less than 200 pounds, so that leaves $2{,}300-200=2{,}100$ pounds for the 21 students. They probably each weigh less than 100 pounds.

33. Answers may vary.
$\$40{,}000\times 4\%=\$40{,}000\times 0.04=\$1{,}600$
The estimate is larger than the exact answer.
Exact answer is \$1324.40.

35. Answers may vary.
Her total expenses are about \$100 per month; $100\div 7\approx 14$ $12\times \$14=\$168$
The estimate is larger than the actual answer.
Actual answer is \$163 (you round to nearest dollar on deductions).

37. Dentistry
$$\frac{1803-700}{700}\approx 158\% \text{ increase}$$

39. Between 1980 and 1990
1980 to 1990: 15406 – 10754 = 4652
1990 to 2000: 17514 – 15406 = 2108

41. Category 2.

43. Response 2.

45. Estimated answers may vary.
Exact answer is 20.6 million.

47. Estimated answers may vary.
Exact answer is 16 million.

49. $2,407 \times 0.36 = 866.52$
\$866.52 billion

51. $2,407 \times 0.15 = 361.05$
\$361.05 billion

53. $705,361 \times 0.302 \approx 213,019$
The actual number of immigrants was 213,019.

55. $130,661 \div 705,361 \approx 0.185$
18.5%

57. Determining whether your estimate is too large or too small.

59. Answers may vary. Possible answers include not always having a calculator available.

61. Skill with mental manipulations will help you make quick, reliable estimates.

63. No solution provided.

65. For 33: $\dfrac{1600-1324.34}{1324.34} \approx 20.8\%$

For 35: $\dfrac{168-163}{163} \approx 3.1\%$

67. Answers may vary.
The amount of lawn that needs to be fertilized is represented by the size of the lot, less the non-grassy areas such as the garden, driveway and house.

$$96 \cdot 169 - 96 \cdot 30 - 65 \cdot 28 - 18 \cdot 65 =$$
$$16,224 - 2,880 - 1,820 - 1,170 = 10,354$$

If you divide the grassy area into rectangles, you get 10,354 square feet. They need slightly over two bags of fertilizer.

69. No solution provided.

71. a) $\dfrac{7169-3486}{24} = 153.46$ b) $7169 + 153.46 \times 10 = 8704$

Chapter Review Exercises

1. Understand the problem, devise a plan, carry out your plan, check your answer.

3. 10.

Let A, C, L, R, and T represent Amber, Chris, Lawrence, "Thirteen," and Travis.

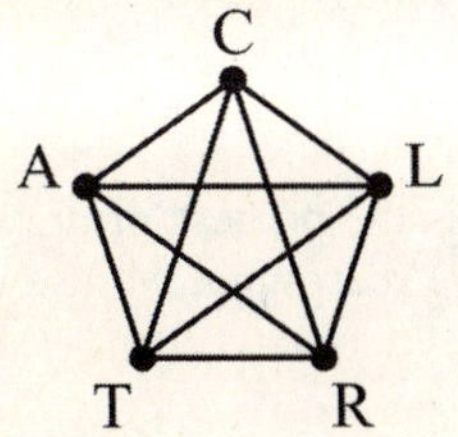

The pairs are AC, AL, AR, AT, CL, CR, CT, LR, LT, and RT.

5. Answers may vary.
Let P be the number of hours Picaboo worked as a stock person and I be the number of hours she worked as a ski instructor.

Guesses for S and I	Good points	Weak Points
9 and 11	Sum is 20.	Amount earned is less than \$141.20.
7 and 13	Sum is 20.	Amount earned is more than \$141.20.
8 and 12	All conditions satisfied. We have the solution.	

7. No solution provided.

9. a) deductive
   b) inductive

13. In this exercise, only the answer to Part b, the expected conjecture, and Part c, the algebraic justification, are provided.

    In this trick, you will always get twice the number you started with.
    a) Call the number n.
    b) $8n$
    c) $8n+12$
    d) $\frac{8n+12}{4}=\frac{4(2n+3)}{4}=2n+3$
    e) $(2n+3)-3=2n+3-3=2n$

15. a) $210-60=150$
    b) $6\times 15=90$

17. Estimated answers may vary.
    a) The exact value is $51,700.
    b) The exact value is $10,500.

Chapter Test

1. Answers will vary.

3. Answers may vary.
   Let W be the number of Wii's sold, X be the number of Xbox 360's sold, and P be the number of PS3's sold.

Guesses for P, X, and W	Good points	Weak Points
130, 275, and 401	Xbox's are 15 more than twice PS3's and Wii's are 4 less than Xbox's and PS3's combined.	Total number is less than 812.
135, 285, and 416	Xbox's are 15 more than twice PS3's and Wii's are 4 less than Xbox's and PS3's combined.	Total number is more than 812.
131, 277, and 404	All conditions satisfied. We have the solution.	

5. a) 36,000
   b) 36,500

7. Answers may vary. Possible answers include: Inductive reasoning is the process of drawing a general conclusion by observing a pattern in specific instances. In deductive reasoning, we use accepted facts and general principles to arrive at a specific conclusion.

9. Mathematical ideas can be understood verbally, graphically, and through examples.

11. cde, cdf, cdg

13. $7+53$, $13+47$, $19+41$, or $29+31$

15. In this exercise, only the answer to Part b, the expected conjecture, and Part c, the algebraic justification, are provided.

In this trick, you will always get twice the number you started with.

a) Call the number n.
b) $4n$
c) $4n+40$
d) $\frac{4n+40}{2}=\frac{2(2n+20)}{2}=2n+20$
e) $(2n+20)-20=2n+20-20=2n$

Chapter 2

Set Theory: Using Mathematics to Classify Objects

Section 2.1 The Language of Sets

1. Answers may vary. Possible answers include $\{x : x$ is an even counting number$\}$ and $\{2, 4, 6, 8, \ldots\}$

3. Answers may vary. Possible answers include $3 \in \{1,3,5,7,8\}$ and $4 \notin \{1,3,5,7,8\}$.

5. In learning new ideas, babies are classifying objects according to common characteristics similar to what we do in set theory.

7. $\{10,11,12,13,14,15\}$

9. $\{17, 18, 19, 20, 21, 22, 23, 24, 25\}$

11. $\{4, 8, 12, 16, 20, 24, 28\}$

13. {Sunday, Monday, Tuesday, Wednesday, Thursday, Friday, Saturday}

15. $\varnothing$

17. $\varnothing$

19. Answers may vary. Possible answers include $\{x : x$ is a multiple of 3 between 3 and 12 inclusive$\}$.

21. Answers may vary. Possible answers include $\{y : y$ is a letter in the word *music*$\}$.

23. $\{-1, -2, -3, \ldots\}$

25. $\varnothing$

27. Answers may vary. Possible answers include $\{101, 102, 103, \ldots\}$.

29. Answers may vary. Possible answers include $\{x : x$ is an even natural number between 1 and 101$\}$.

31. well defined

33. not well defined

35. not well defined

37. well defined

39. $\notin$

41. $\in$

43. $\in$

45. $\in$

47. $\notin$

49. $\in$

51. 6

53. 0

55. 4

57. 2 elements; $\{1, 2\}$, $\{1, 2, 3\}$

59. 1 element; $\{\{\varnothing\}\}$

61. finite

63. infinite

65. Answers may vary. Possible answers include 4.5.

67. Answers may vary. Possible answers include Sony.

69. Answers may vary. Possible answers include George W. Bush.

71. Answers may vary. Possible answers include Sunday.

73. $\{x : x$ is a humanities elective$\}$

75. {History012, History223, Geography115, Anthropology111}

77. {AZ, GA, LA, MN, NJ, TX, VA}

79. $\{x : x$ had an average price of gasoline above 290$\}$

81. {Amazon, Apple, CNET, eBay, Fox, Time Warner, Wikipedia}

83. $\{x : x$ had less than 155 million visitors$\}$

85. when the set is too large or to complicated to list all the elements

87. $\varnothing$ is the empty set, it contains no elements. $\{\varnothing\}$ is not empty, it contains 1 element, $\varnothing$.

89. – 93. No solution provided.

95. If the barber shaves himself, then he (the barber) does not shave himself. If the barber does not shave himself, then he (the barber) does shave himself.
Conclusion: This is a paradox.

97. If $S \in S$, then $S \notin S$. If $S \notin S$, then $S \in S$.
Conclusion: This is a paradox.

Section 2.2 Comparing Sets

1. The sets contain exactly the same elements.

3. Every element added to a set doubles the number of subsets.

5. These two sets are equal. They have the same elements arranged in a different order.

7. These two sets are not equal. The second set contains (infinitely many) elements that don't appear in the first set.

9. These two sets are equal. They are both $\{1, 3, 5, \ldots, 99\}$.

11. These two sets are equal. It is understood that the second set contains all twelve months of the year.

13. These two sets are equal. Common sense dictates that nobody born before 1800 should be living.

15. true; All the elements of the first set are understood to be elements of the second set.

17. false; The first set is the same as the second set, so the first set cannot be a <u>proper</u> subset.

19. false; The letter "y" is an element of the first set and not an element of the second set.

21. true; The null set is a subset of all sets.

23. false; The set that contains the null set (which is not an empty set) is not a subset of the set $\{0\}$. Each of the two sets (left and right) contains one element, which are not the same.

25. The first set is equivalent to the second set because they both have the same number of elements.

27. The first set is equivalent to the second set because they both have that same number of elements, namely 4.

29. The first set is not equivalent to the second set. The first set has 0 elements while the second set has 1 element.

31. The first set is equivalent to the second set. They both have 8 elements.

33. The first set is not equivalent to the second set. The first set has 366 elements while the second set has 365 elements.

35. $\{1, 2\}, \{1, 3\}, \{2, 3\}$

37. $\{1, 2, 3\}, \{1, 2, 4\}, \{1, 3, 4\}, \{2, 3, 4\}$

39. There are $2^5 = 32$ subsets and $2^5 - 1 = 31$ proper subsets.

41. {Carmen, Frank, Ivana}= *V*

43. The set of lowerclassmen = L; Note: the **bolded** value indicates the set with the largest cardinality.
$n(U)= 4$, $\mathbf{n(L)= 6}$, $n(S)= 2$, $n(V)= 3$,
$n(A)= 2$, $n(T)= 3$, and $n(D)=2$

45. $2^7 = 128$

47. $2^8 = 256$

49. 7

51. the order and repetition of the elements

53. Notations that looks similar may have very different meanings.

55. No solution provided.

57. 25 is not a power of 2.

59. Answers may vary.
Since Pete probably confused 5^2 with 2^5, then the set most likely has 5 elements.

61. Answers may vary.
At the first branching, the "yes" or "no" indicates that in forming a subset of $\{1, 2\}$, we will either take the 1 or omit it. The second branching in the tree indicate whether we are going to take the 2 as a member of the subset that we are forming. The tree shows all possible ways that we can decide to take 1 and 2 in forming a subset of $\{1, 2\}$. The top branch corresponds to the subset $\{1, 2\}$, the second branch corresponds to the subset $\{1\}$, the third branch corresponds to the subset $\{2\}$, and the bottom branch corresponds to the subset $\varnothing$.

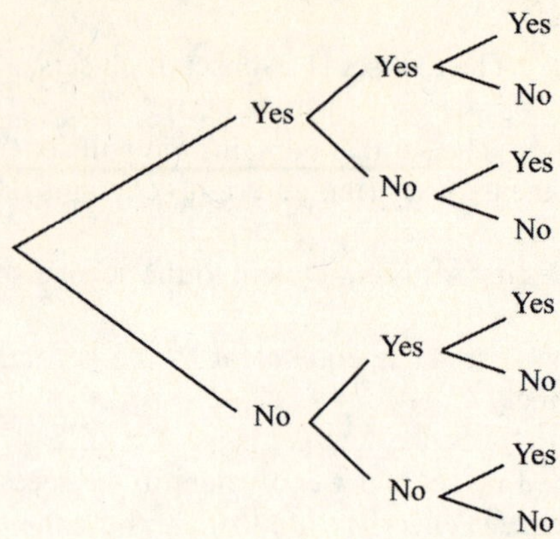

63. Answers will vary.

65. $k = 31$; $k = 32$ would take longer than 100 years.

67. No solution provided.

69. This line represents the counts of the number of subsets of a five-element set of sizes 0, 1, 2, 3, 4, and 5 elements.

71. The sixth, seventh, eighth, and ninth lines are:

1 6 15 20 15 6 1
1 7 21 35 35 21 7 1
1 8 28 56 70 56 28 8 1
1 9 36 **84** 126 126 84 36 9 1

So the newspaper editor can send three of them in 84 different ways.

1 → 1
1 1 → 2
1 2 1 → 4
1 3 3 1 → 8
1 4 6 4 1 → 16
1 5 10 10 5 1 → 32
1 6 15 20 15 6 1 → 64
1 7 21 35 35 21 7 1 → 128
1 8 28 56 70 56 28 8 1 → 256
1 9 36 84 126 126 84 36 9 1 → 512
1 10 45 120 210 252 210 120 45 10 1 → 1024

The sum across the rows is always a power of 2, specifically 2^n, where n is the number of the row that is being summed. Note: Recall we start counting these lines with 0, not 1.

75. Corresponding property: If $A \subseteq B$ and $B \subseteq C$, then $A \subseteq C$. If $x \in A$, then $x \in B$. Since $x \in B$, then $x \in C$. Therefore, if we have $x \in A$, we must also have $x \in C$. Note: We use capital letters for sets and lower case letters for elements.

77. No solution provided.

Section 2.3 Set Operations

1. $A \cup B = \{x : x \in A \text{ or } x \in B\}$

3. The complement of A is the set of all elements in the universal set that are not elements of A.

5. $(A \cup B)' = A' \cap B'$; We cannot reverse the order of taking the union of the sets and taking the intersection of the complements of the sets.
   $(A \cap B)' = A' \cup B'$; We cannot reverse the order of taking the intersection of the sets and taking the union of the complements of the sets

7. Words in everyday life that are the same as mathematical terminology have similar meanings.

9. $A \cap B = \{1, 3, 5\}$

11. $B \cup C = \{1, 2, 3, 4, 5, 6, 7, 8\}$

13. $A \cup \varnothing = \{1, 3, 5, 7, 9\} = A$

15. $A \cup U = \{1, 2, 3, 4, 5, 6, 7, 8, 9, 10\} = U$

17. $A \cap (B \cup C) = A \cap \{1, 2, 3, 4, 5, 6, 7, 8\} = \{1, 3, 5, 7\}$

19. $(A - B) \cap (A - C) = \{7, 9\} \cap \{1, 3, 5, 9\} = \{9\}$

21. $M \cap E =$ {potato chip, bread, pizza}

23. $E - M =$ {apple, fish, banana}

25. $M' \cap G' =$ {fish}

27. $A - (B \cup C)$

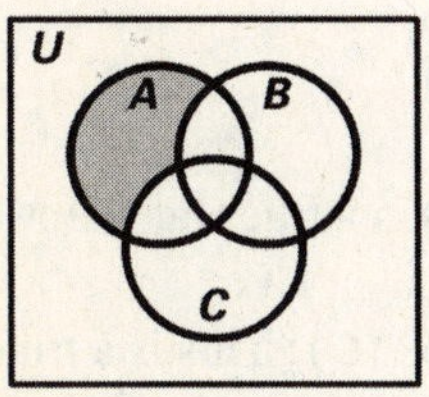

29. $(A \cap B) - C$

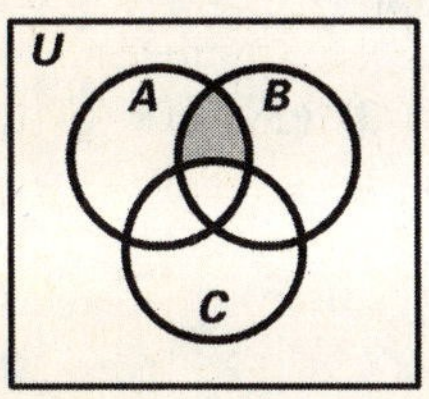

31. $A \cup (B - C)$

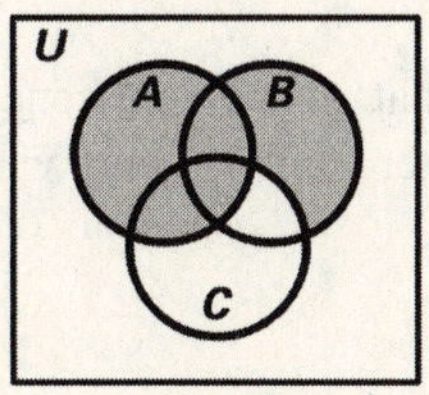

33. $(A \cup (B \cup C))'$

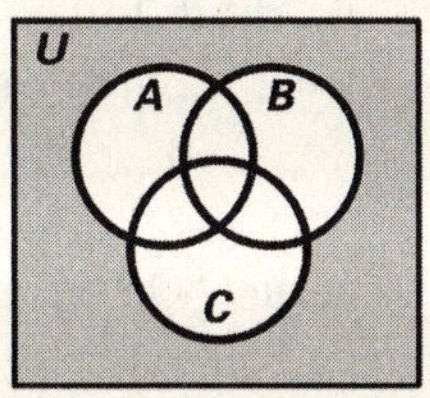

35. $B - A$

37. $(A \cup B)'$

39. $A \cap B \cap C$

41. $(A \cup C) - B$

43. 30

45. 28

47. 20

49. 27

51. A

53. B

55. $P \cap C$ = the set of cars whose price is above \$19,000 and is compact = $\{d, g\}$

57. $W \cap G'$ = the set of cars that have a warranty of at least three years and doesn't have a good safety rating = $\{c, d, f, g\}$

59. $P \cap (G \cup W)$ = the set of cars that have a price above \$19,000, and a good safety rating or a warrantee of at least three years = $\{b, d, g, h\} = P$

61. $P - (G \cup A)$ = the set of cars that have a price above \$19,000 and don't have a good safety rating and don't have an antitheft package = $\{d\}$

63. $P \cap (B \cup A) = P \cap \{m, mc, hc\} = \{m, mc, hc\}$

65. $P \cup C \cup B = \{m, mc, bc, c, hc\}$

67. $E \cap B$ = the set of movies that earned more than \$800 million and were made before 1960 = $\{c, d\}$

69. $O \cap B'$ = the set of movies that won an Oscar and were made after 1960 = $\{f, i\}$

71. $E \cap B \cap O$ = set of movies that earned more than \$800 million, were made before 1960, and won an Oscar = $\{c\}$

73. "Union" implies joining together . "Intersection" implies overlapping.

75. Answers may vary. Possible answers include confusing DeMorgan's laws with the distributive property.

77. false; It is possible that $A = B$ and hence $n(A) = n(B)$. Counterexamples may vary.

79. true

81. false
Sample counterexample:
Let $X = \{1,2,3,4\}$ and $Y = \{3,4,5,6\}$
$X - Y = \{1,2\}$, $n(X - Y) = 2$, $n(X) = 4$ and $n(Y) = 4$. $2 \neq 4 - 4$

83. "And" is intersection $(\cap)$, and "Or" is union $(\cup)$.

85. No solution provided.

87. $A \cap B = B \cap A$; This is a true statement.

89. $(A \cap B) \cap C = A \cap (B \cap C)$; This is a true statement.

91. Sample Counterexample:

$$2 + (3 \cdot 4) \stackrel{?}{=} (2 + 3) \cdot (2 + 4)$$

$$2 + 12 \stackrel{?}{=} 5 \cdot 6$$

$$14 \neq 30$$

Section 2.4 Survey Problems

1. $(A - B) \cup (A \cap B) \cup (B - A)$

3. The 14 in $A \cap C \cap R$ was subtracted from the 22 in $C \cap R$.

5. Many of his contemporaries scoffed at his ideas.

7. r_2, r_3

9. r_3

11. r_2

13. r_2, r_3, r_5, r_6

15. r_4, r_7

17. r_7; $\{r_6, r_7\} \cap \{r_1, r_4, r_7, r_8\} = \{r_7\}$

29. $n(A) = 18, n(B) = 15$, and $n(C) = 14$

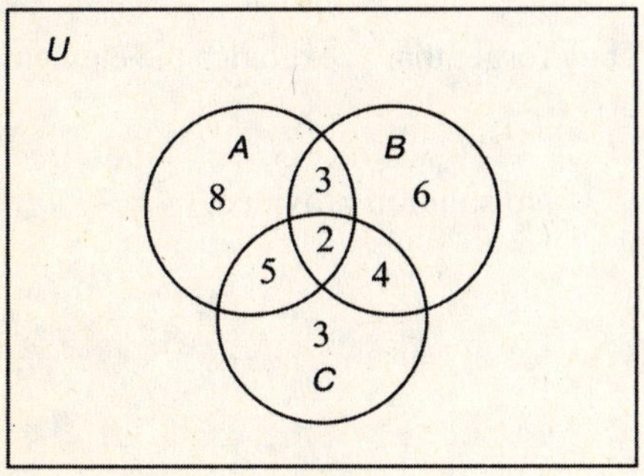

31. $n(A) = 5, n(B) = 14$, and $n(C) = 9$

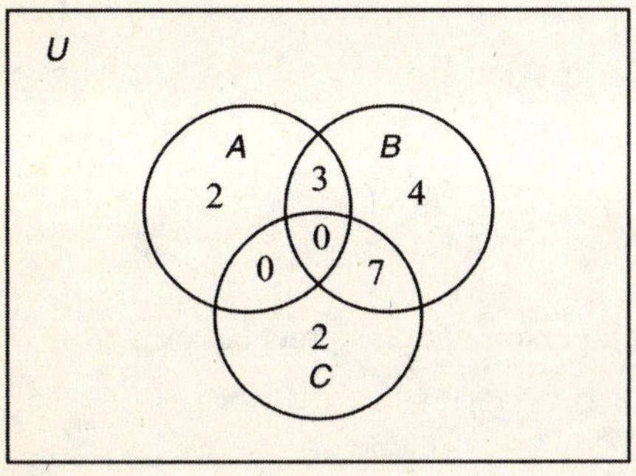

33. 59

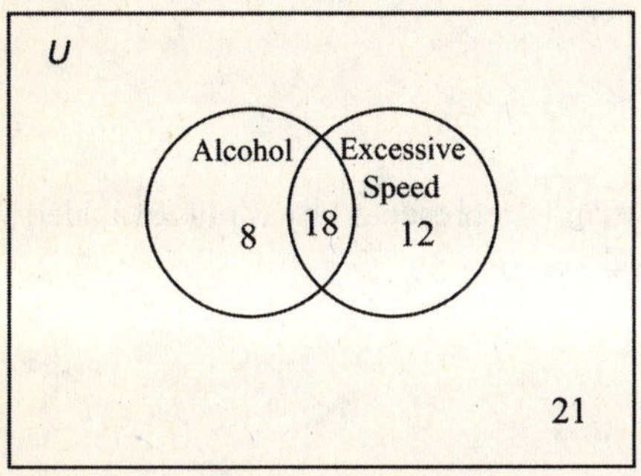

19. 25

21. 18

23. 16

25. 20

27. 19

35. 28

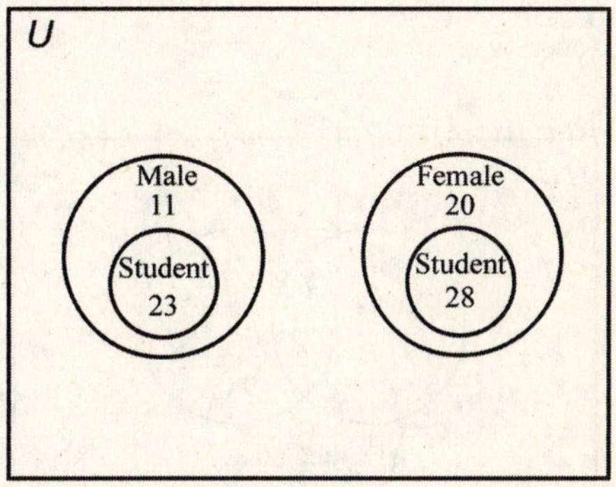

37. 30 attend the barbecue and 49 purchased a tour guide.

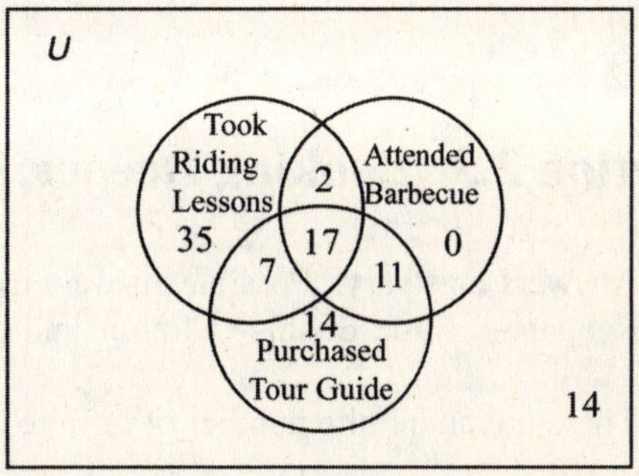

39. 23

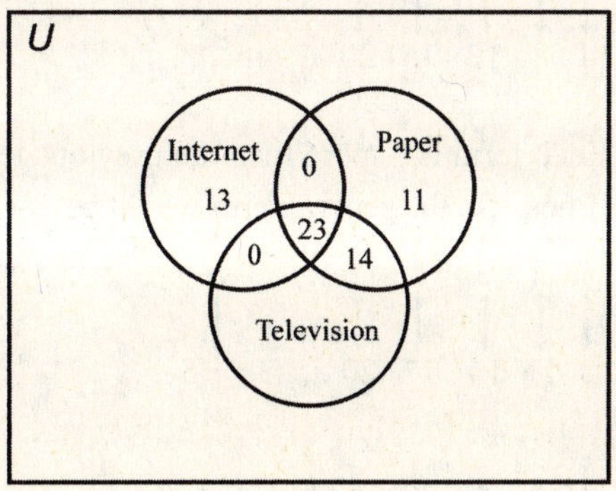

41.

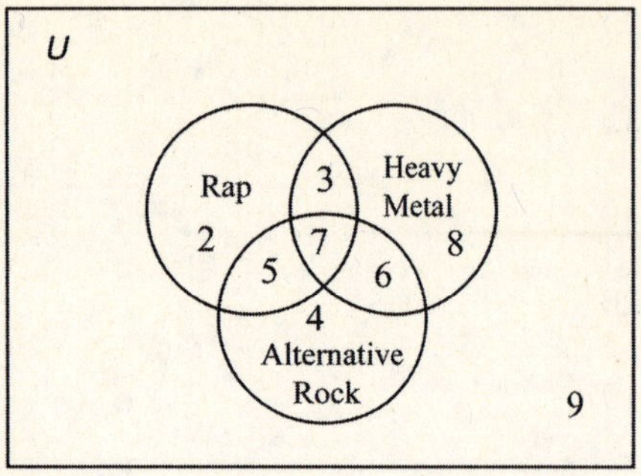

a) 44
b) 27
c) 8

43.

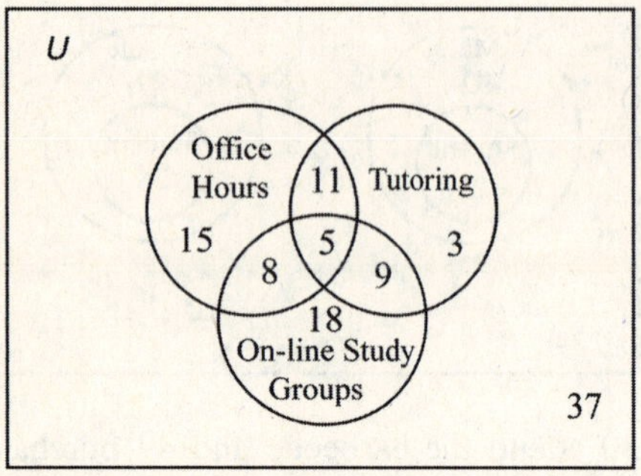

a) 106
b) 15
c) 40

45. $n(A \cup P) = 95 + 63 = 158$

47. $n(A - (M \cup S)) = 95 - (44 + 31)$
$= 95 - 75 = 20$

49. A^-

51. A^-, B^-, or AB^-,

53. $A \cap B \cap Rh$

55. They forget that r_2 excludes the elements in $A \cap B$.

57. – 59. No solution provided.

61. 8

Section 2.5 Looking Deeper: Infinite Sets

1. Answers may vary. Possible answers include matching each person with his predecessor, Reagan — Bush, Bush — Clinton, and Chaney — Gore

3. There are an infinite number of infinite cardinal numbers.

5.
$$\begin{array}{cccccccc} 1 & 2 & 3 & 4 & 5 \ldots & n & \ldots \\ \updownarrow & \updownarrow & \updownarrow & \updownarrow & \updownarrow & \updownarrow & \\ 4 & 8 & 12 & 16 & 20 \ldots & 4n & \ldots \end{array}$$

7. Each term is 3 more than the previous term. The general term is represented by 8 plus 3, added $n-1$ times. So the general term would be $8 + (n-1)\cdot 3 = 8 + 3n - 3 = 3n + 5$.

$$\begin{array}{cccccccc} 1 & 2 & 3 & 4 & 5 \ldots & n & \ldots \\ \updownarrow & \updownarrow & \updownarrow & \updownarrow & \updownarrow & \updownarrow & \\ 8 & 11 & 14 & 17 & 20 \ldots & 3n+5 & \ldots \end{array}$$

9.
$$\begin{array}{cccccccc} 1 & 2 & 3 & 4 & 5 \ldots & n & \ldots \\ \updownarrow & \updownarrow & \updownarrow & \updownarrow & \updownarrow & \updownarrow & \\ 2 & 4 & 8 & 16 & 32 \ldots & 2^n & \ldots \end{array}$$

11.
$$\begin{array}{cccccccc} 1 & 2 & 3 & 4 & 5 \ldots & n & \ldots \\ \updownarrow & \updownarrow & \updownarrow & \updownarrow & \updownarrow & \updownarrow & \\ 1 & 1/2 & 1/3 & 1/4 & 1/5 \ldots & 1/n & \ldots \end{array}$$

13. $3 \cdot 1 = 3$; $3 \cdot 2 = 6$; $3 \cdot 3 = 9$; $3 \cdot 4 = 12$; $3 \cdot 5 = 15$

$$\begin{array}{cccccccc} 1 & 2 & 3 & 4 & 5 & \ldots & n & \ldots \\ \updownarrow & \updownarrow & \updownarrow & \updownarrow & \updownarrow & & \updownarrow & \\ 3 & 6 & 9 & 12 & 15 & \ldots & 3n & \ldots \end{array}$$

15. $3 \cdot 1 - 2 = 1$; $3 \cdot 2 - 2 = 4$; $3 \cdot 3 - 2 = 7$; $3 \cdot 4 - 2 = 10$; $3 \cdot 5 - 2 = 13$

$$\begin{array}{cccccccc} 1 & 2 & 3 & 4 & 5 & \ldots & n & \ldots \\ \updownarrow & \updownarrow & \updownarrow & \updownarrow & \updownarrow & & \updownarrow & \\ 1 & 4 & 7 & 10 & 13 & \ldots & 3n-2 & \ldots \end{array}$$

17. Match $\{2, 4, 6, 8, 10, \ldots\}$ with $\{4, 6, 8, 10, 12, \ldots\}$; in general, match $2n$ with $2(n+1) = 2n+2$.

19. Since each term of $\{7, 10, 13, 16, 19, \ldots\}$ is 3 more than the previous, the general term is $7 + 3(n-1) = 7 + 3n - 3 = 3n + 4$. Match $\{7, 10, 13, 16, 19, \ldots\}$ with $\{10, 13, 16, 19, 22, \ldots\}$; in general, match $3n+4$ with $3(n+1)+4 = 3n+3+4 = 3n+7$.

21. Match $\{2, 4, 8, 16, 32, \ldots\}$ with $\{4, 8, 16, 32, 64, \ldots\}$; in general, match 2^n with 2^{n+1}.

23. Match $\{1, \frac{1}{2}, \frac{1}{3}, \frac{1}{4}, \frac{1}{5}, \ldots\}$ with $\{\frac{1}{2}, \frac{1}{3}, \frac{1}{4}, \frac{1}{5}, \frac{1}{6}, \ldots\}$; in general, match $\frac{1}{n}$ with $\frac{1}{n+1}$.

25. Match $\{\frac{1}{2}, \frac{1}{4}, \frac{1}{6}, \frac{1}{8}, \frac{1}{10}, \ldots\}$ with $\{\frac{1}{4}, \frac{1}{6}, \frac{1}{8}, \frac{1}{10}, \frac{1}{12}, \ldots\}$; in general, match $\frac{1}{2n}$ with $\frac{1}{2(n+1)} = \frac{1}{2n+2}$.

For Exercises 27 – 29 refer to the following diagram.

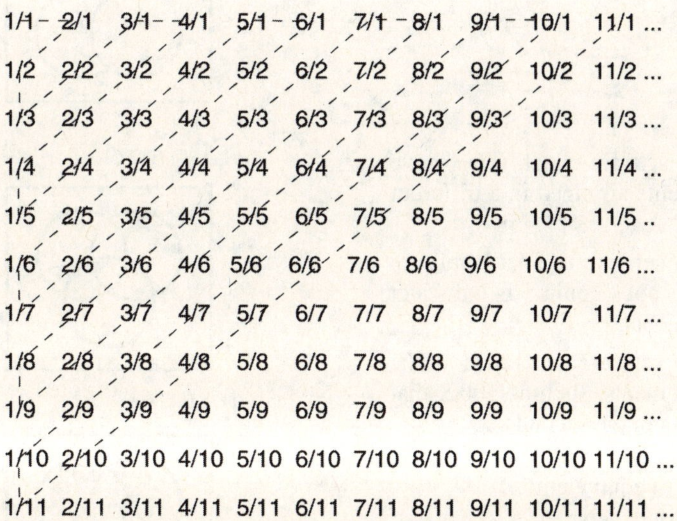

27. 6; We skip 2/2, 2/4, 3/3, and 4/2.

29. 25; We skip 2/2, 2/4, 3/3, 4/2, 2/6, 4/4, 6/2, and 6/3.

31. It was shown that the set of positive rational numbers has a one-to-one correspondence with the natural numbers.

33. {1,2,3,4,5} does not have a one-to-one correspondence with any of its 31 proper subsets.

35. They had already been matched as $1, \frac{1}{2}, 1,$ and 2.

37. No solution provided.

39. 6

1 ↔ 4	1 ↔ 4	1 ↔ 5	1 ↔ 5	1 ↔ 6	1 ↔ 6
2 ↔ 5	2 ↔ 6	2 ↔ 4	2 ↔ 6	2 ↔ 4	2 ↔ 5
3 ↔ 6	3 ↔ 5	3 ↔ 6	3 ↔ 4	3 ↔ 5	3 ↔ 4

41. Answers may vary.
If we take the union of $\{1\}$, which has cardinal number 1, and $\{2, 3, 4, ...\}$, which has cardinal number $\aleph_0$, we get $\{1, 2, 3, 4, ...\}$, which has cardinal number $\aleph_0$.

Chapter Review Exercises

1. a) Answers may vary. Possible answers include: $\{x : x$ is an even natural number between 1 and 19$\}$.
   b) $\{x : x$ is one of the fifty states in the U.S.A.$\}$
   c) $\{s,t,u,p,e,n,d,o\}$.
   d) $\varnothing$

3. 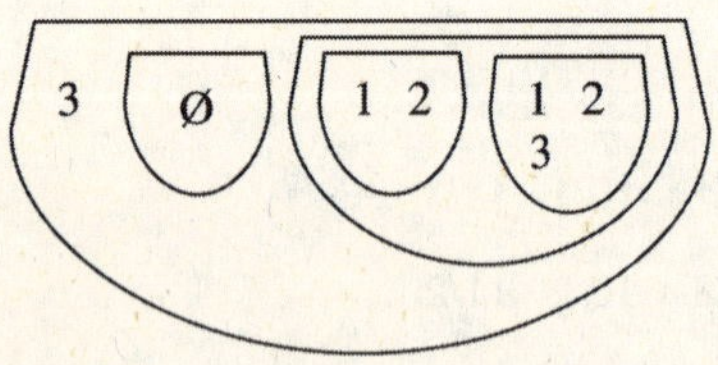

5. a) Yes, these two sets are equal. They have the same elements arranged in a different order.
   b) Yes, these two sets are equal. Duplicate elements do not count as distinct (different) elements.
   c) No, these two sets are not equal. The first set has elements, such as 1002, that are not elements in the second set.

7. a) not equivalent; b) equivalent; c) equivalent; d) equivalent

9. a) $A \cap B = \{5,7,9\}$
   b) $B \cup C = \{2,3,4,5,7,8.9.10\}$
   c) $C' = \{1,4,6,7,10\}$
   d) $A - C = \{7\}$

11. a) $A \cup B$

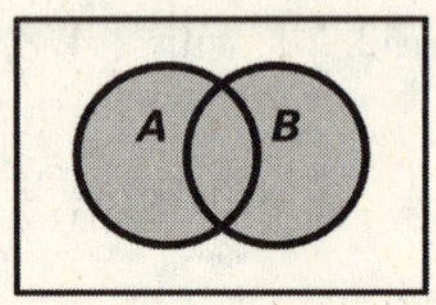

b) $B \cap C$

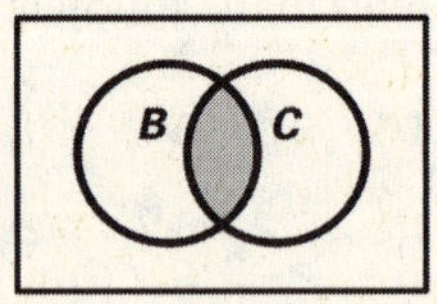

c) $A' \cap B \cap C$

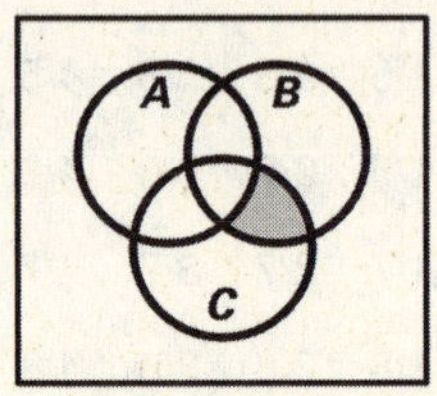

d) $(B \cup C) - A$

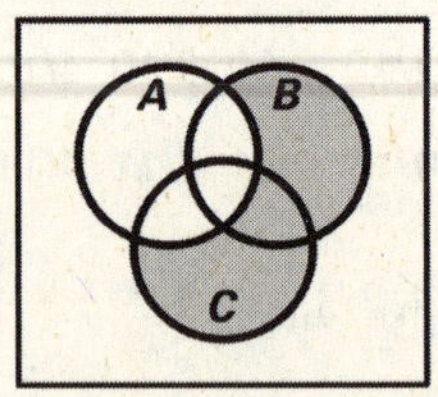

13. a) Any three of the following four are valid responses.
   1) Closure
   2) Commutativity; $A \cup B = B \cup A$
   3) Associativity; $(A \cup B) \cup C = A \cup (B \cup C)$
   4) Identity; $A \cap U = A$

   b) Any three of the following four are valid responses.
   1) Closure
   2) Commutativity; $A \cap B = B \cap A$
   3) Associativity; $(A \cap B) \cap C = A \cap (B \cap C)$
   4) Identity; $A \cup \varnothing = A$

   c) Union distributes over intersection $A \cup (B \cap C) = (A \cup B) \cap (A \cup C)$ and intersection distributes over union; $A \cap (B \cup C) = (A \cap B) \cup (A \cap C)$

15. 83 said academics. 58 said both academics and social life.

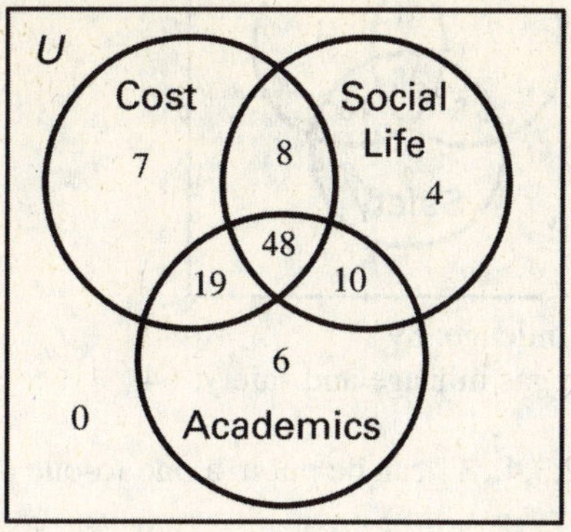

17. {1,2,3,4,...} can be put in a one-to-one correspondence with {2,4,6,8,...}

19. We chose a digit that was different from the third decimal place of the third number in the list.

Chapter Test

1. a) {x: x is a counting number greater than 100}
   b) {January,February,March,...,December}
   c) $\varnothing$

3. a) not equivalent
   b) equivalent;
   c) equivalent

5. $\varnothing$ contains no elements, but $\{\varnothing\}$ contains one element, namely $\varnothing$.

7. a) $2^8 = 256$

9. a) true; every element of the first set is a member of the second.
   b) true; every element of the first set is a member of the second
   c) false; $\varnothing$ is not one of the numbers 1,2,3.

11. a) $A \cap C$

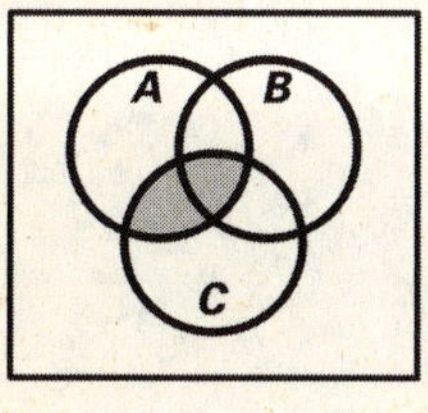

b) $(A \cup C) - B'$

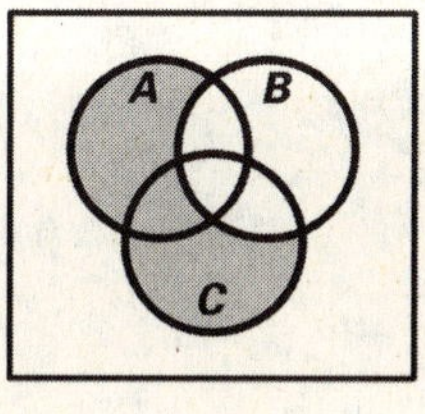

13. 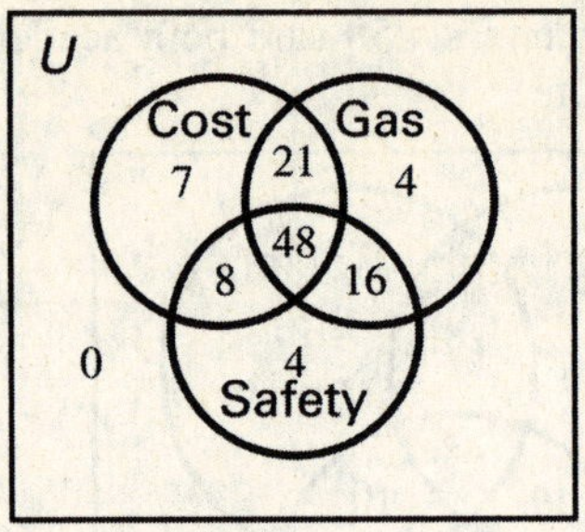

gas mileage: 89
both gas mileage and safety: 64

15. {1,2,3,4,...} can be put in a one-to-one correspondence with {2,4,6,8,...}

17. We chose a digit that was different from the third decimal place of the third number in the list.

Chapter 3

Logic: The Study of What's True or False or Somewhere in Between

Section 3.1 Statements, Connectives, and Quantifiers

1. not (negation), and (conjunction), or (disjunction), if … then (conditional), if and only if (biconditional)

3. Universal quantifiers state that all objects have a certain property.

5. "At least one is not."

7. To discover religious truths and lead more perfect lives.

9. statement

11. not a statement

13. not a statement

15. statement

17. not a statement

19. compound; conjunction, negation, disjunction, conditional

21. compound; disjunction

23. simple

25. compound; conditional, disjunction

27. compound; conditional

29. $g \vee \sim c$

31. $\sim g \rightarrow (\sim c \wedge \sim a)$

33. $g \leftrightarrow \sim a$

35. The radial tires are included or the sunroof is not extra.

37. It is not true that the sunroof is extra and the radial tires are included.

39. If the radial tires are included, then the sunroof is extra or power windows are not optional.

41. There exists at least one snake that is not poisonous.

43. All personal items are covered by this insurance policy.

45. All scientists believe that it is not true that an asteroid collision led to the extinction of the dinosaurs.

47. Some modern art is not difficult to understand.

49. false; Nadia is not a commuter.

51. false; All sophomores are athletes.

53. false; Lennox is a freshman that is on a scholarship.

55. – 65. Answers for these exercises have not been provided. These statements are complex and sophisticated. You will probably find that you, your classmates, and your instructor do not always agree as to which connectives are present. Nevertheless, it is interesting to try to determine the form of these statements.

67. true

69. false; There exists a whole number that is not a natural number.

71. true

73. In the first expression, the conjunction is taken before the negation; in the second expression, negations are taken before the conjunction.

75. To show that "not all objects have a property," it must be shown that "at least one does not have the property."

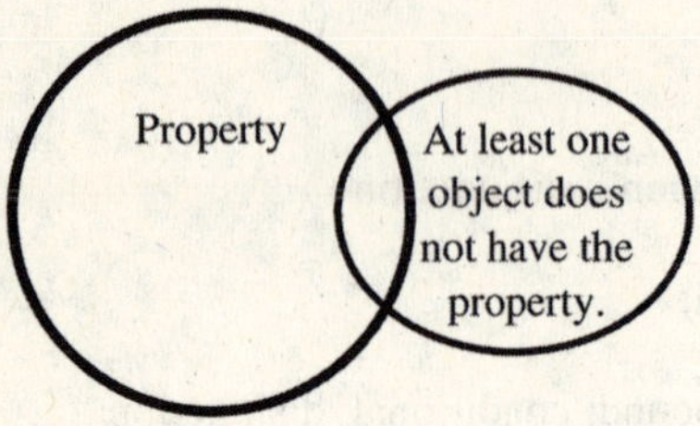

77. No solution provided.

79. a) true b) true

81. – 89. No solution provided.

Section 3.2 Truth Tables

1. The "and" is only true when both the components are true; the "or" is false only when both the components are false.

3. There is a difference between computing the "or" first before negating and negating before computing the "or". The situation is similar for "and."

5. Kurt Gödel

7.

		3 2 1
p	q	$\sim(p \vee \sim q)$

9.

		4 3 2 1
p	q	$p \wedge \sim(p \vee \sim q)$

11. T

13. T

15. F

17. T

19. $2^5 = 32$

21. Yes; $16 = 2^4$

23. No; 9 cannot be represented as a power of 2.

25.

		2 1
p	q	$p \wedge (\sim q)$
T	T	T **F** F
T	F	T **T** T
F	T	F **F** F
F	F	F **F** T

27.

		3 1 2
p	q	$\sim(\sim p \wedge q)$
T	T	**T** F FT
T	F	**T** F FF
F	T	**F** T TT
F	F	**T** T FF

29.

p	q	$\sim(p \vee q)$	$\wedge$	$\sim(p \wedge q)$
		2 1	5	4 3
T	T	F TTT	**F**	F TTT
T	F	F TTF	**F**	T TFF
F	T	F FTT	**F**	T FFT
F	F	T FFF	**T**	T FFF

31.

p	q	r	$(p \vee r)$	$\wedge$	$(p \wedge \sim q)$
			1	4	3 2
T	T	T	T T T	**F**	T F F
T	T	F	T T F	**F**	T F F
T	F	T	T T T	**T**	T T T
T	F	F	T T F	**T**	T T T
F	T	T	F T T	**F**	F F F
F	T	F	F F F	**F**	F F F
F	F	T	F T T	**F**	F F T
F	F	F	F F F	**F**	F F T

33.

p	q	r	$(p \vee q)$	$\wedge$	$(p \vee r)$
			1	3	2
T	T	T	T T T	**T**	T T T
T	T	F	T T T	**T**	T T F
T	F	T	T T F	**T**	T T T
T	F	F	T T F	**T**	T T F
F	T	T	F T T	**T**	F T T
F	T	F	F T T	**F**	F F F
F	F	T	F F F	**F**	F T T
F	F	F	F F F	**F**	F F F

35.

p	q	r	$\sim(p \vee \sim q)$	$\wedge$	r
			3 2 1	4	
T	T	T	F T T F	**F**	T
T	T	F	F T T F	**F**	F
T	F	T	F T T T	**F**	T
T	F	F	F T T T	**F**	F
F	T	T	T F F F	**T**	T
F	T	F	T F F F	**F**	F
F	F	T	F F T T	**F**	T
F	F	F	F F T T	**F**	F

37. exclusive or

39. inclusive or

41. p: Bill is tall.
q: Bill is thin
It is not true that Bill is tall and thin.
$\sim(p \wedge q) \Leftrightarrow \ \sim p \vee \sim q$
Negation: Bill is not tall or Bill is not thin.

43. p: Christian will apply for a loan.
q: Christian will apply for work study.
It is not true that Christian will apply for either a loan or work study.
$\sim(p \vee q) \ \Leftrightarrow \ \sim p \wedge \sim q$
Negation: Christian will not apply for a loan and he will not apply for work study.

45. p: Ken qualifies for a rebate.
q: Ken qualifies for a reduced interest rate.
It is not true that Ken qualifies for a rebate or a reduced interest rate.
$\sim(p \vee q) \ \Leftrightarrow \ \sim p \wedge \sim q$
Negation: Ken does not qualify for a rebate and he does not qualify for a reduced interest rate.

47. p: The number x is equal to five.
q: The number s is odd.
It is not true that The number x is not equal to five and s is not odd.
$\sim(\sim p \wedge \sim q) \Leftrightarrow p \vee q$
Negation: The number x is equal to five or s is odd.

49. Yes; $\sim(p \wedge \sim q)$ is logically equivalent to $\sim p \vee q$ by DeMorgan's laws.

51. No; $\sim(p \vee \sim q) \wedge \sim(p \vee q)$ is not logically equivalent to $p \vee (p \wedge q)$.

		3 2 1 6 5 4
p	q	$\sim(p \vee \sim q) \wedge \sim(p \vee q)$
T	T	F T T F **F** F TTT
T	F	F T T T **F** F TTF
F	T	T F F F **F** F FTT
F	F	F F T T **F** T FFF

		2 1
p	q	$p \vee (p \wedge q)$
T	T	T **T** TTT
T	F	T **T** TFF
F	T	F **F** FFT
F	F	F **F** FFF

53. Yes; $\sim(p \vee \sim q) \wedge \sim(p \vee q)$ is logically equivalent to $(\sim p \wedge q) \wedge (\sim p \wedge \sim q)$ by DeMorgan's laws.

55. Yes; $p \vee (\sim q \wedge r)$ is logically equivalent to $(p \vee (\sim q)) \wedge (p \vee r)$.

			3 1 2
p	q	r	$p \vee (\sim q \wedge r)$
T	T	T	T **T** F FT
T	T	F	T **T** F FF
T	F	T	T **T** T TT
T	F	F	T **T** T FF
F	T	T	F **F** F FT
F	T	F	F **F** F FF
F	F	T	F **T** T TT
F	F	F	F **F** T FF

			2 1 4 3
p	q	r	$(p \vee (\sim q)) \wedge (p \vee r)$
T	T	T	T T F **T** TTT
T	T	F	T T F **T** TTF
T	F	T	T T T **T** TTT
T	F	F	T T T **T** TTF
F	T	T	F F F **F** FTT
F	T	F	F F F **F** FFF
F	F	T	F T T **T** FTT
F	F	F	F T T **F** FFF

57. true

p	q	$\sim p$	$\vee$	$\sim q$
T	F	F	**T**	T

59. false

p	q	r	$(\sim p \vee q)$	$\wedge$	r
T	F	F	F F F	**F**	F

61. true

p	q	r	p	$\vee$	$(\sim q \wedge r)$
T	F	F	T	**T**	T F F

63. p: The earned income did reduce the tax you owe.
q: The earned income gave you a refund.
$\sim p \vee \sim q \Leftrightarrow \sim(p \wedge q)$
It is not true that the earned income did reduce the tax you owe and gave you a refund.

65. p: You are single.
q: You are the head of a household.
$\sim p \wedge \sim q \Leftrightarrow \sim(p \vee q)$
It is not true that you are single or the head of a household.

67. $(p \vee q) \vee (p \vee r)$ is logically equivalent to $p \vee (q \vee r)$.

1 3 2

p	q	r	$(p \vee q) \vee (p \vee r)$		
T	T	T	TTT	**T**	TTT
T	T	F	TTT	**T**	TTF
T	F	T	TTF	**T**	TTT
T	F	F	TTF	**T**	TTF
F	T	T	FTT	**T**	FTT
F	T	F	FTT	**T**	FFF
F	F	T	FFF	**T**	FTT
F	F	F	FFF	**F**	FFF

2 1

p	q	r	$p \vee (q \vee r)$		
T	T	T	T	**T**	TTT
T	T	F	T	**T**	TTF
T	F	T	T	**T**	FTT
T	F	F	T	**T**	FFF
F	T	T	F	**T**	TTT
F	T	F	F	**T**	TTF
F	F	T	F	**T**	FTT
F	F	F	F	**F**	FFF

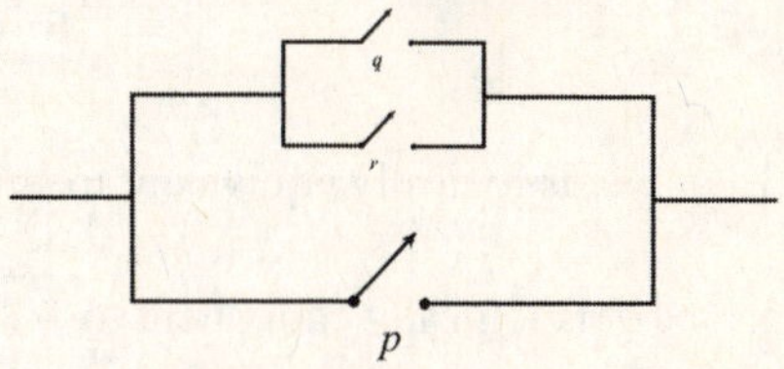

69. $(p \wedge q) \vee \sim q$ is logically equivalent to $p \vee \sim q$.

1 3 2

p	q	$(p \wedge q) \vee \sim q$		
T	T	TTT	**T**	F
T	F	TFF	**T**	T
F	T	FFT	**F**	F
F	F	FFF	**T**	T

2 1

p	q	$p \vee \sim q$		
T	T	T	**T**	F
T	F	T	**T**	T
F	T	F	**F**	F
F	F	F	**T**	T

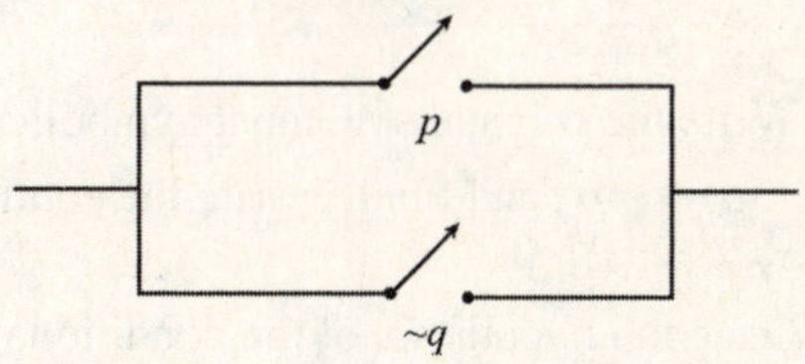

71. line one

73. Answers may vary. One possibility is "paper or plastic?"

75. Answers may vary. Possible responses include that we want the truth table to cover every possibility for the truth values for the three variables. There are eight different ways to assign truth values to the three variables.

77. – 79. No solution provided.

81. A "T" would imply that the element is in the subset, and an "F" would imply that it is not an element of the subset.

p	q	r	Set of Variables That Have True Values
T	T	T	$\{p, q, r\}$
T	T	F	$\{p, q\}$
T	F	T	$\{p, r\}$
T	F	F	$\{p\}$
F	T	T	$\{q, r\}$
F	T	F	$\{q\}$
F	F	T	$\{r\}$
F	F	F	$\varnothing$

83. $p \wedge q$ is logically equivalent to $\sim(\sim p \vee \sim q)$ by DeMorgan's laws.

85. $p \mid q$ is logically equivalent to $\sim(p \wedge q)$.

		2 1	
p	q	$\sim(p \wedge q)$	$p \mid q$
T	T	**F** T T T	**F**
T	F	**T** T F F	**T**
F	T	**T** F F T	**T**
F	F	**T** F F F	**T**

87.

		1 3 2	
p	q	$(p \mid p) \mid (q \mid q)$	$p \vee q$
T	T	F **T** F	**T**
T	F	F **T** T	**T**
F	T	T **T** F	**T**
F	F	T **F** T	**F**

Section 3.3 The Conditional and Biconditional

1. The only way a statement of the form $p \rightarrow q$ can be false is the p is true and q is false.

3. Write the original statement symbolically in the form $p \rightarrow q$, then write the contrapositive in the form $\sim q \rightarrow \sim p$, and finally write the contrapositive in words.

5. Once the hypothesis of the conditional of the contract is false, the deal is no longer in effect.

7. true

		2 1 4 3
p	q	$\sim(p \vee q) \rightarrow \sim p$
T	F	F TTF **T** F

9. true

			1 3 2
p	q	r	$(p \wedge q) \rightarrow (q \vee r)$
T	F	T	TFF **T** FTT

11. true

			1 3 2 4
p	q	r	$(\sim p \vee \sim q) \rightarrow r$
T	F	T	FT T **T** T

13. false

			3 1 2 5 4
p	q	r	$\sim(\sim p \wedge q) \rightarrow \sim r$
T	F	T	T FFF **F** F

15.

		2 1
p	q	$p \rightarrow \sim q$
T	T	T **F** F
T	F	T **T** T
F	T	F **T** F
F	F	F **T** T
		F **T** F
		T **T** T
		T **F** F

17.

		2 1
p	q	$\sim (p \rightarrow q)$
T	T	**F** TTT
T	F	**T** TFF
F	T	**F** FTT
F	F	**F** FTF

19.

			1 4 3 2
p	q	r	$(p \vee r) \rightarrow (p \wedge \sim q)$
T	T	T	TTT **F** TFF
T	T	F	TTF **F** TFF
T	F	T	TTT **T** TTT
T	F	F	TTF **T** TTT
F	T	T	FTT **F** FFF
F	T	F	FFF **T** FFF
F	F	T	FTT **F** FFT
F	F	F	FFF **T** FFT

21.

			2 1 5 4 3
p	q	r	$\sim (p \vee r) \rightarrow \sim (p \wedge q)$
T	T	T	F TTT **T** F TTT
T	T	F	F TTF **T** F TTT
T	F	T	F TTT **T** T TFF
T	F	F	F TTF **T** T TFF
F	T	T	F FTT **T** T FFT
F	T	F	T FFF **T** T FFT
F	F	T	F FTT **T** T FFF
F	F	F	T FFF **T** T FFF

23.

			1 3 2
p	q	r	$(p \vee q) \leftrightarrow (p \vee r)$
T	T	T	TTT **T** TTT
T	T	F	TTT **T** TTF
T	F	T	TTF **T** TTT
T	F	F	TTF **T** TTF
F	T	T	FTT **T** FTT
F	T	F	FTT **F** FFF
F	F	T	FFF **F** FTT
F	F	F	FFF **T** FFF

25.

		1 2 5 3 4
p	q	$(\sim p \rightarrow q) \leftrightarrow (\sim q \rightarrow p)$
T	T	FTT **T** FTT
T	F	FTF **T** TTT
F	T	TTT **T** FTF
F	F	TFF **T** TFF

27. If it pours, then it rains.

29. If you do not buy all-weather radial tires, then they will not last for 80,000 miles.

31. If the sides of a geometric figure are not all equal in length, then it is not an equilateral triangle.

33. If x does not evenly divide 6, then x does not evenly divide 9.

35. contrapositive

37. original conditional

39. original conditional

41. Converse: $q \rightarrow \sim p$
Inverse: $p \rightarrow \sim q$
Contrapositive: $\sim q \rightarrow p$

43. Converse: $\sim (q \wedge r) \rightarrow (\sim p)$
Inverse: $p \rightarrow (q \wedge r)$
Contrapositive: $(q \wedge r) \rightarrow p$

45. The second statement is the contrapositive of the first. They are logically equivalent.

47. The second statement is the converse of the first. They are not logically equivalent.

49. If I finish my workout, then I'll take a break.

51. If you qualify for this deduction, then you complete Form 3093.

53. If you receive a free cell phone, then you sign up before March 1.

55. If you remain accident free for three years, then you get a reduction on your auto insurance.

57. If you can be claimed by someone else as a dependent, then your gross income is not more than $2,250.

59. If you decrease the amount being withheld from your pay, then the amount you overpaid is large.

61. true; F → T

63. true; F → F

65. $(p \leftrightarrow q) \Leftrightarrow (p \rightarrow q) \wedge (q \rightarrow p)$

			1	3	2
p	q	$p \leftrightarrow q$	$(p \rightarrow q)$	$\wedge$	$(q \rightarrow p)$
T	T	T **T** T	TTT	**T**	TTT
T	F	T **F** F	TFF	**F**	FTT
F	T	F **F** T	FTT	**F**	TFF
F	F	F **T** F	FTF	**T**	FTF

67. true

w	f	w	$\rightarrow$	f
T	T	T	**T**	T

69. false

s	w	i	s	$\rightarrow$	$(w \wedge i)$
T	T	F	T	**F**	TFF

71. Converse: switch clauses; inverse: negate both clauses; contrapositive: switch and negate both clauses.

73. $F \rightarrow T$; Inverse is $\sim F \rightarrow \sim T$ or $T \rightarrow F$.

75. $(p \rightarrow q) \Leftrightarrow (\sim p \vee q)$

		1	4	3 2
p	q	$(p \rightarrow q)$	$\leftrightarrow$	$(\sim p \vee q)$
T	T	TTT	**T**	FTT
T	F	TFF	**T**	FFF
F	T	FTT	**T**	TTT
F	F	FTF	**T**	TTF

77. $\sim (p \rightarrow q) \Leftrightarrow \sim (\sim p \vee q) \Leftrightarrow p \wedge \sim q$

79. Jamie has not been a member for at least ten years.

81. Using the contrapositive and DeMorgan's laws, $r \rightarrow (p \wedge q)$.

83. $p \vee\ (r \wedge (p \wedge q))$ is logically equivalent to p.

p	q	r	p	$\vee$ (3)	$(r \wedge (p \wedge q))$ (2, 1)
T	T	T	T	**T**	TTTTT
T	T	F	T	**T**	FFTTT
T	F	T	T	**T**	TFTFF
T	F	F	T	**T**	FFTFF
F	T	T	F	**F**	TFFFT
F	T	F	F	**F**	FFFFT
F	F	T	F	**F**	TFFFF
F	F	F	F	**F**	FFFFF

85. $(p \rightarrow q) \Leftrightarrow (\sim p \vee q)$

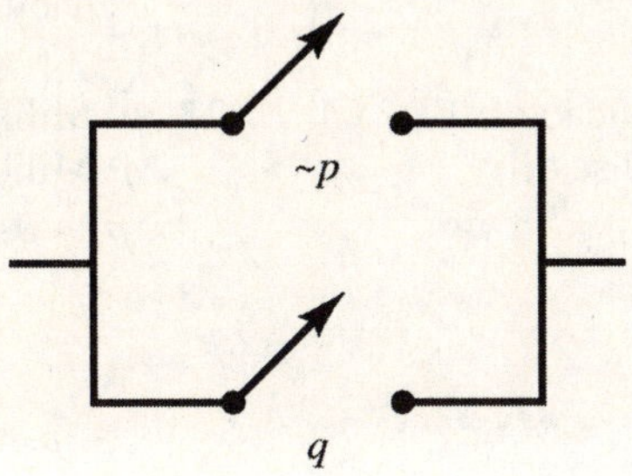

87. $(\sim p) \rightarrow \sim (q \wedge r) \Leftrightarrow p \vee \sim (q \wedge r) \Leftrightarrow\ p \vee (\sim q \vee \sim r)$

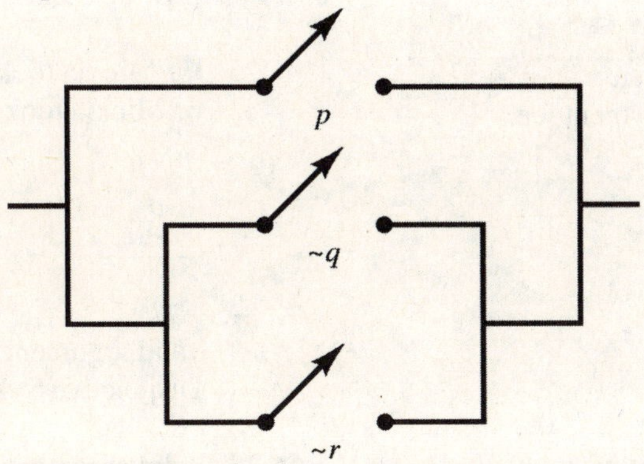

Section 3.4 Verifying Arguments

1. $(p_1 \wedge p_2) \rightarrow c$

3. The final column of the truth table had all trues.

5. Answers will vary.

7. *p:* The car has air bags.
q: The car is safe.

$p \rightarrow q$

p

$\therefore q$

valid argument
Law of Detachment

9. *p:* The movie is exciting.
q: The movie will gross a lot of money.

$p \rightarrow q$

q

$\therefore p$

invalid argument
Fallacy of the Converse

11. *p:* The laptop has the enhanced video card.
q: The computer has an optical disc drive.

$p \vee q$

$\sim p$

$\therefore q$

valid argument
Disjunctive Syllogism

13. *p:* You pay your tuition late.
q: You will pay a late payment fee.

$p \rightarrow q$

$\sim p$

$\therefore \sim q$

invalid argument
Fallacy of the Inverse

15. *p:* You watch *The Apprentice*.
q: You will succeed in business.
r: A skyscraper will be named after you.

$p \rightarrow q$

$q \rightarrow r$

$\therefore p \rightarrow r$

valid argument
Law of Syllogism

17. *p:* You buy the sports package.
q: You get the leather seats.

$\sim p \rightarrow \sim q$

q

$\therefore p$

valid argument
Law of Contraposition

19. *p:* Phillipe joins the basketball team.
q: Phillipe will be able to work part time.

$p \rightarrow \sim q$

$\sim p$

$\therefore q$

invalid argument
Fallacy of the Inverse

21. *p:* Milik buys a satellite dish.
q: Milik will get 128 TV channels.

$p \rightarrow q$

$\sim q$

$\therefore \sim p$

valid argument
Law of Contraposition

23. *p:* Minxia took the job in Boston.
q: Minxia took the job in San Diego.

$p \vee q$

$\sim p$

$\therefore q$

valid argument
Disjunctive Syllogism

25. valid argument

		3	2	1	5	4
p	q	$[p \wedge (q \rightarrow \sim p)] \rightarrow \sim q$				
T	T	T F T F F			T	F
T	F	T T F T F			T	T
F	T	F F T T T			T	F
F	F	F F F T T			T	T

27. invalid argument

		1 3 2 6 4 5
q	r	$[\sim r \wedge (r \rightarrow q)] \rightarrow (\sim q \wedge r)$
T	T	F F T T T **T** F F T
T	F	T T F T T **F** F F F
F	T	F F T F F **T** T T T
F	F	T T F T F **F** T F F

29. invalid argument

			1 2 5 4 3 8 6 7
p	q	r	$[(\sim q \rightarrow p) \wedge (r \rightarrow \sim q)] \rightarrow (\sim p \rightarrow r)$
T	T	T	F T T F T F F **T** F T T
T	T	F	F T T T F T F **T** F T F
T	F	T	T T T T T T T **T** F T T
T	F	F	T T T T F T T **T** F T F
F	T	T	F T F F T F F **T** T T T
F	T	F	F T F T F T F **F** T F F
F	F	T	T F F F T T T **T** T T T
F	F	F	T F F F F T T **T** T F F

31. invalid argument

			3 2 1 7 4 6 5 8
p	q	r	$\{p \wedge (q \rightarrow \sim p) \wedge [\sim q \rightarrow (r \vee p)]\} \rightarrow r$
T	T	T	T F TF F F F T TTT **T** T
T	T	F	T F TF F F F T FTT **T** F
T	F	T	T T FT F T T T TTT **T** T
T	F	F	T T FT F T T T FTT **F** F
F	T	T	F F TT T F F T TTF **T** T
F	T	F	F F TT T F F T FFF **T** F
F	F	T	F F FT T F T T TTF **T** T
F	F	F	F F FT T F T F FFF **T** F

33. valid argument

			3 2 1 5 4 6
p	q	r	$[r \wedge (r \rightarrow \sim q) \wedge (p \vee q)] \rightarrow p$
T	T	T	T F TFF F TTT **T** T
T	T	F	F F FTF F TTT **T** T
T	F	T	T T TTT T TTF **T** T
T	F	F	F F FTT F TTF **T** T
F	T	T	T F TFF F FTT **T** F
F	T	F	F F FTF F FTT **T** F
F	F	T	T T TTT F FFF **T** F
F	F	F	F F FTT F FTF **T** F

35. invalid argument

p	q	r	$[(q \to \sim p) \wedge (r \to \sim q)] \to (\sim p \to r)$
			2 1 5 4 3 8 6 7
T	T	T	T F F F T F F T F T T
T	T	F	T F F F F T F T F T F
T	F	T	F T F T T T T T F T T
T	F	F	F T F T F T T T F T F
F	T	T	T T T F T F F T T T T
F	T	F	T T T T F T F F T F F
F	F	T	F T T T T T T T T T T
F	F	F	F T T T F T T F T F F

37. invalid argument
p: The product has a lower price.
q: The product has quality.
r: The product is less reliable.

p	q	r	$\{(p \to \sim q) \wedge [(\sim p \vee \sim q) \to \sim r] \wedge p\} \to r$
			2 1 8 3 5 4 7 6 9 10
T	T	T	TFF F F FF T F FT T T
T	T	F	TFF F F FF T T FT T F
T	F	T	TTT F F TT F F FT T T
T	F	F	TTT T F TT T TTT F F
F	T	T	FTF F T TF F F FF T T
F	T	F	FTF T T TF T T FF T F
F	F	T	FTT F T TT F F FF T T
F	F	F	FTT T T TT T T FF T F

39. valid argument
p: You buy from a reputable breeder.
q: Your labradoodle requires shots.
r: It will cost you extra.

p	q	r	$\{(p \to \sim q) \wedge (\sim p \to r) \wedge \sim r\} \to \sim q$
			2 1 5 3 4 7 6 9 8
T	T	T	TF F F F T T FF T F
T	T	F	TF F F F T F FT T F
T	F	T	TT T T F T T FF T T
T	F	F	TT T T F T F TT T T
F	T	T	FT F T T T T FF T F
F	T	F	FT F F T F F FT T F
F	F	T	FT T T T T T FF T T
F	F	F	FT T F T F F FT T T

41. valid argument

p: Health care is improved.
q: Quality of life will be high
r: The incumbents will be reelected.

			1	3	2			4	6		9	8	11	10	
p	q	r	$\{(\sim p$	$\rightarrow$	$\sim q)$	$\wedge$	$[(p$	$\wedge$	$q)$	$\rightarrow$	$r]$	$\wedge$	$q\}$	$\rightarrow$	r
T	T	T	F	T	F	T	T	T	T	T	T	T	T	**T**	T
T	T	F	F	T	F	F	T	T	T	F	F	F	T	**T**	F
T	F	T	F	T	T	F	T	F	F	T	T	F	F	**T**	T
T	F	F	F	T	T	T	T	F	F	T	F	F	F	**T**	F
F	T	T	T	F	F	F	F	F	T	T	T	F	T	**T**	T
F	T	F	T	F	F	F	F	F	T	T	F	F	T	**T**	F
F	F	T	T	T	T	T	F	F	F	T	T	F	F	**T**	T
F	F	F	T	T	T	T	F	F	F	T	F	F	F	**T**	F

43. invalid argument

p: Jamie is fluent in Spanish.
q: Jamie will work in Madrid.
r: Jamie will visit Mexico.

			2	1	6	3 5 4	8			
p	q	r	$[p\wedge$	$(p\rightarrow q)$	$\wedge$	$(\sim r\vee\sim q)]$	$\rightarrow$	r
T	T	T	T T	TTT	F	FFF	**T**	T
T	T	F	T T	TTT	T	TTF	**F**	F
T	F	T	T F	TFF	F	FTT	**T**	T
T	F	F	T F	TFF	F	TTT	**T**	F
F	T	T	F F	FTT	F	FFF	**T**	T
F	T	F	F F	FTT	F	TTF	**T**	F
F	F	T	F F	FTF	F	FTT	**T**	T
F	F	F	F F	FTF	F	TTT	**T**	F

45. The statement *p* allows us to detach *q* from $p \rightarrow q$.

47. In the disjunction $p \vee q$, if we don't have *p*, then we must have *q*.

49. If we have the premise ~*p*, then we conclude ~*q*, which is the inverse of $p \rightarrow q$.

51. – 57. No solution provided.

59. $a \wedge b$

$b \rightarrow c$

$d \rightarrow \sim c$

$\therefore \sim d$

(1) If we assume that $a \wedge b$ is true, then *b* must be true.
(2) By the law of detachment if *b* and $b \rightarrow c$ are true, then *c* is true.
(3) Because $d \rightarrow \sim c$ is equivalent to its contrapositive, we know $c \rightarrow \sim d$ is true.
(4) By the law of detachment again, with *c* true and $c \rightarrow \sim d$ true we conclude ~*d* is true.

61. – 63. No solution provided.

Section 3.5 Using Euler Diagrams to Verify Syllogisms

1. A syllogism is valid if whenever all its premises are true, then the conclusion is also true.

3. Whenever you draw an Euler diagram, you will always draw some extra conditions that were not stated as premises.

5. Jan Lukasiewicz

7. valid

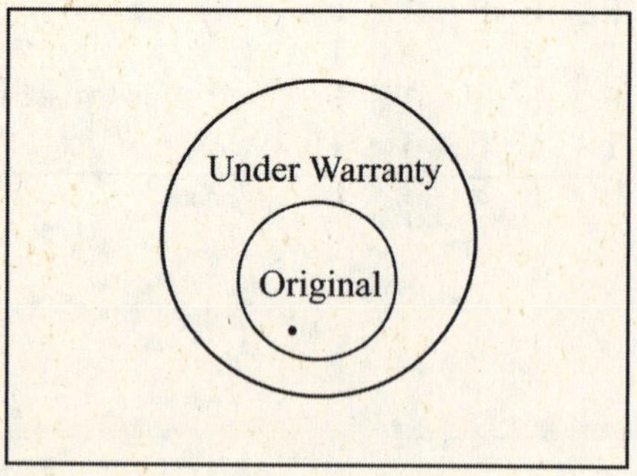

9. invalid

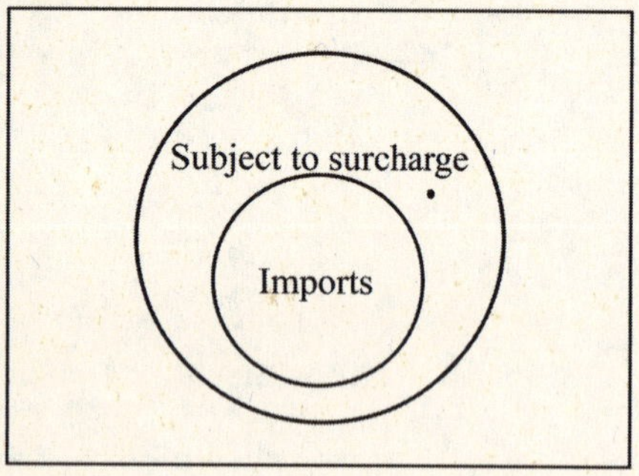

11. invalid

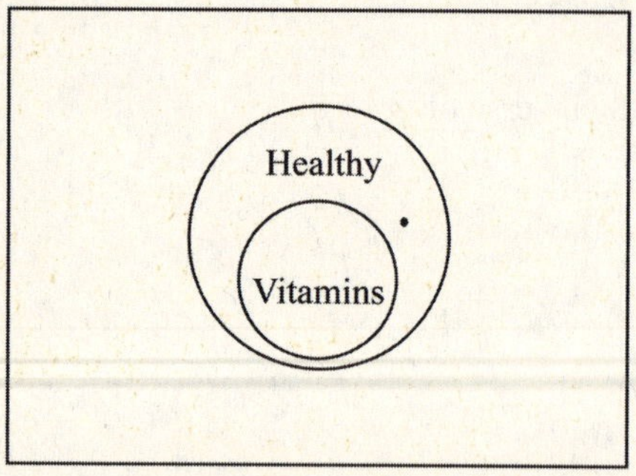

13. invalid

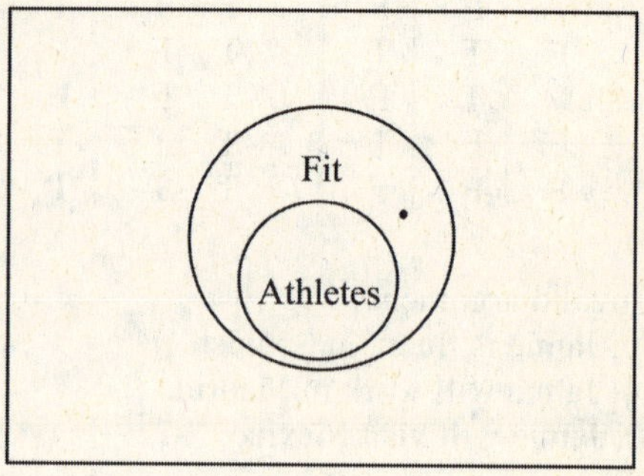

15. valid

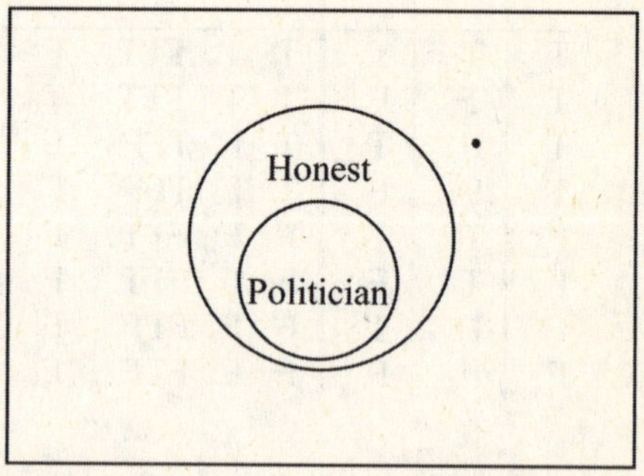

17. invalid

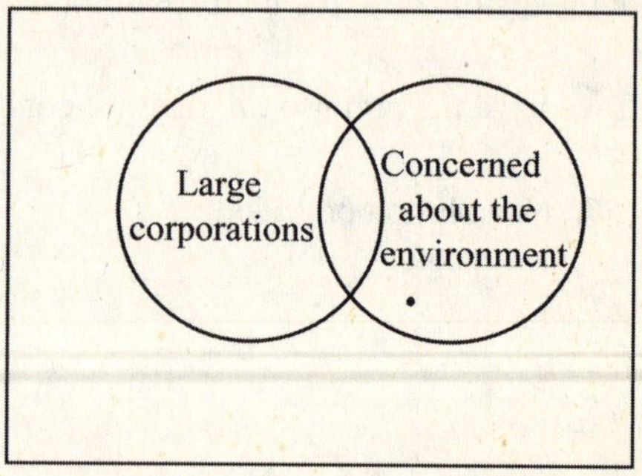

19. invalid

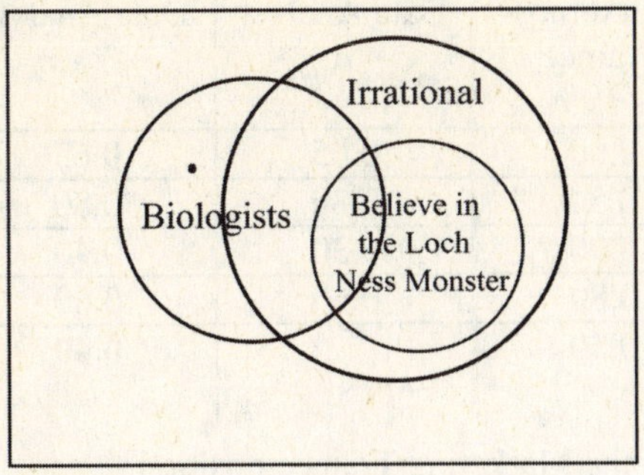

21. valid

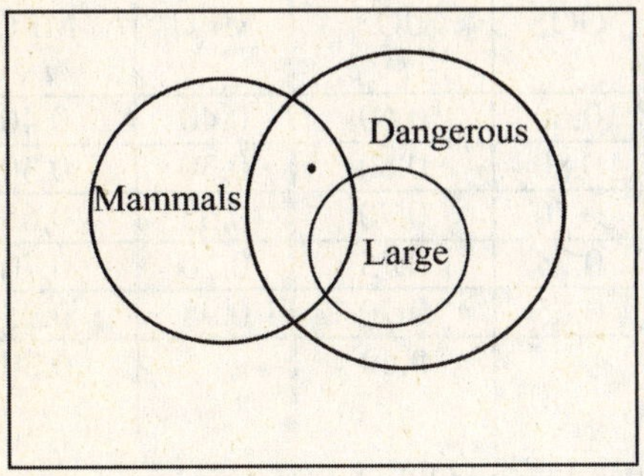

27. – 29. Many diagrams are possible.

23. Some taxes should be abolished.

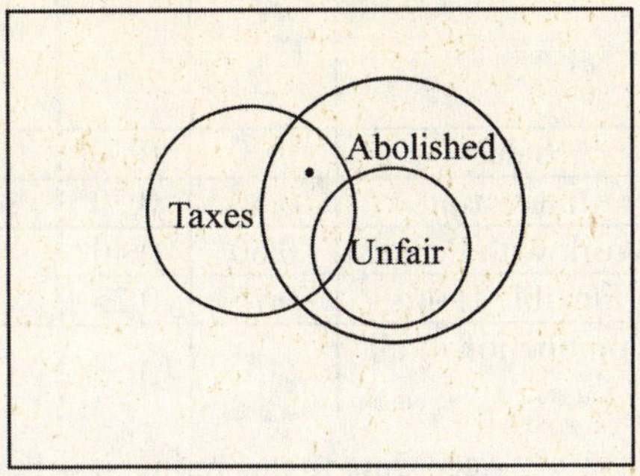

25. Some teams that wear red uniforms do not play in a domed stadium.

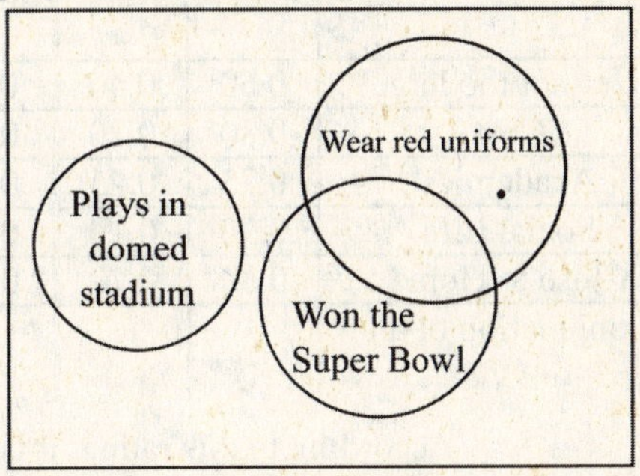

Note: The "Plays in domed stadium" and "Wear red uniforms" could be overlapping (like logic would suggest), but the conclusion would be the same.

31. – 37. No solution provided.

Section 3.6 Looking Deeper: Fuzzy Logic

1. There are an infinite number of values between 0 and 1, inclusive.

3. the minimum of the truth values for p and q

5. – 11. No solution provided.

13. $1-0.95=0.05$

15. $1-0.75=0.25$

17. $0.75 \vee 0.85 = 0.85$

19. $0.75 \wedge \sim 0.85 = 0.75 \wedge 0.15 = 0.15$

21. $0.27 \wedge \sim (0.64) = 0.27 \wedge 0.36 = 0.27$

23. $\sim (0.27 \vee \sim 0.64) = \ \sim (0.27 \vee 0.36) = \ \sim 0.36 = 0.64$

25. $\sim (0.27 \vee \sim 0.64) \wedge \sim 0.71 = \ \sim (0.27 \vee 0.36) \wedge 0.29 = \sim 0.36 \wedge 0.29 = \ 0.64 \wedge 0.29 = 0.29$

27. $0.64 \rightarrow (0.27 \vee 0.71) = \sim 0.64 \vee (0.27 \vee 0.71) = \ 0.36 \vee 0.71 = 0.71$

29.

	i	$\sim i$	Marketing Trainee c	Marketing Trainee $\sim i \vee c$	Data Analyst c	Data Analyst $\sim i \vee c$
Salary	0.70	0.30	0.60	0.60	0.65	0.65
Interesting	0.80	0.20	0.50	0.50	0.80	0.80
Work with People	0.60	0.40	0.90	0.90	0.30	0.40
Flexible Hours	0.75	0.25	0.80	0.80	0.60	0.60
Conjunction of all $\sim i \vee c$				**0.50**		**0.40**

According to this rating, it is better for you to take the Marketing Trainee position.

31.

	i	$\sim i$	LSU c	LSU $\sim i \vee c$	GOS c	GOS $\sim i \vee c$	MFU c	MFU $\sim i \vee c$
Size (not too large)	0.60	0.40	0.95	0.95	0.60	0.60	0.40	0.40
Cost	0.80	0.20	0.40	0.40	0.80	0.80	0.30	0.30
Academics	0.75	0.25	0.80	0.80	0.70	0.70	0.65	0.65
Social Life	0.90	0.10	0.55	0.55	0.75	0.75	0.70	0.70
Close to Home	0.30	0.70	0.70	0.70	0.65	0.70	0.95	0.95
Conjunction of all $\sim i \vee c$				**0.40**		**0.60**		**0.30**

According to this rating, it is better for her to attend Good Old State (GOS).

33. If we only allowed the values 1 (for true) and 0 (for false), then the rules for computing truth values in fuzzy logic are exactly the same as the rules for computing truth tables.

35. – 39. No solution provided.

Chapter Review Exercises

1. a) not a statement
   b) statement
   c) not a statement

3. a) It is not true that Antonio is fluent in Spanish and he has not lived in Spain for a semester.
   b) Antonio is not fluent in Spanish or he has not lived in Spain for a semester.

5. a) true

		2 1
p	q	$p \wedge (\sim q)$
T	F	T **T** T

b) false

			2 1
p	q	r	$r \vee (\sim p \wedge q)$
T	F	F	F **F** F F F

c) false

			3 3 4 1
p	q	r	$\sim(p \vee q) \wedge \sim r$
T	F	F	F TTF **F** T

7. a)

		3 2 1
p	q	$\sim(p \vee \sim q)$
T	T	**F** T T F
T	F	**F** T T T
F	T	**T** F F F
F	F	**F** F T T

b)

			3 2 1 5 4
p	q	r	$\sim(p \vee \sim q) \wedge \sim r$
T	T	T	F TTF **F** F
T	T	F	F TTF **F** T
T	F	T	F TTT **F** F
T	F	F	F TTT **F** T
F	T	T	T FFF **F** F
F	T	F	T FFF **T** T
F	F	T	F FTT **F** F
F	F	F	F FTT **F** T

9. a) $\sim(p \wedge \sim q)$ is logically equivalent to $\sim p \vee q$ by DeMorgan's laws.

b) $\sim(p \vee \sim q) \wedge \sim(p \vee q)$ is not logically equivalent to $p \vee (p \wedge q)$.

		3 2 1 6 5 4
p	q	$\sim(p \vee \sim q) \wedge \sim(p \vee q)$
T	T	F T T F **F** F TTT
T	F	F T T T **F** F TTF
F	T	T F F F **F** F FTT
F	F	F F T T **F** T FFF

		2 1
p	q	$p \vee (p \wedge q)$
T	T	T **T** TTT
T	F	T **T** TFF
F	T	F **F** FFT
F	F	F **F** FFF

11. a)

		1 2
p	q	$\sim p \to q$
T	T	F **T** T
T	F	F **T** F
F	T	T **T** T
F	F	T **F** F

b)

			2 1 5 4 3
p	q	r	$\sim(p \wedge r) \leftrightarrow \sim(p \vee q)$
T	T	T	F TTT **T** F TTT
T	T	F	T TFF **F** F TTT
T	F	T	F TTT **T** F TTF
T	F	F	T TFF **F** F TTF
F	T	T	T FFT **F** F FTT
F	T	F	T FFF **F** F FTT
F	F	T	T FFT **T** T FFF
F	F	F	T FFF **T** T FFF

13. a) If the Knicks get to the finals, then they beat the Lakers.

b) If you are an astronaut, then you have a pilot's license.

15. invalid argument

			1 3 2 6 4 5 7
p	q	r	$\{\sim p \wedge (q \to p) \wedge [(p \vee q) \to r]\} \to r$
T	T	T	F F TTT F TTT T T **T** T
T	T	F	F F TTT F TTT F F **T** F
T	F	T	F F FTT F TTF T T **T** T
T	F	F	F F FTT F TTF F F **T** F
F	T	T	T F TFF F FTT T T **T** T
F	T	F	T F TFF F FTT F F **T** F
F	F	T	T T FTF T FFF T T **T** T
F	F	F	T T FTF T FFF T F **F** F

17. invalid argument

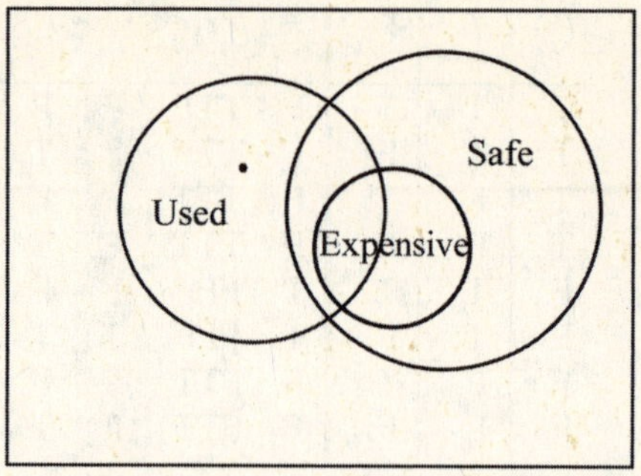

19. a) $0.47 \wedge \sim 0.82 = 0.47 \wedge 0.18 = 0.18$

b) $\sim(0.47 \vee \sim 0.82) = \sim(0.47 \vee 0.18) = \sim 0.47 = 0.53$

c) $0.47 \to \sim 0.82 = \sim 0.47 \vee \sim 0.82 = 0.53 \vee 0.18 = 0.53$

Chapter Test

1. a) statement
   b) not a statement

3. a) $p \vee \sim f$
   b) $\sim(\sim p \wedge f)$

5. $2^6 = 64$

7. a) $\sim(0.65 \vee 0.75) = \sim .075 = 0.25$

   b) $\sim 0.75 \vee \sim(0.65 \wedge 0.38) =$
   $0.25 \vee \sim 0.38 = 0.25 \vee 0.62 = 0.62$

9. a) If you go to Wikipedia, then you will get enough sources for your term paper.
   b) If you pay a fee, then Ticket Master will mail the concert tickets.

11. a) $\sim(p \vee \sim q)$ is logically equivalent to $\sim p \wedge q$ by DeMorgan's laws.
    b) $(\sim p \vee \sim q) \wedge (\sim p \vee q)$ is not logically equivalent to $\sim p \wedge q$.

		1 3 2 6 4 5
p	q	$(\sim p \vee \sim q) \wedge (\sim p \vee q)$
T	T	F F F **F** F T T
T	F	F T T **F** F F F
F	T	T T F **T** T T T
F	F	T T T **T** T T F

		1 2
p	q	$\sim p \wedge q$
T	T	F **F** T
T	F	F **F** F
F	T	T **T** T
F	F	T **F** F

13. a) true

		2 1 4 3
p	q	$(p \vee \sim q) \to \sim q$
T	F	T T T **T** T

b) true

			1 2 4 3
p	q	r	$(\sim p \wedge q) \to \sim r$
T	F	T	F F F **T** F

c) true

		2 1 5 4 3
p	q	$(p \vee \sim q) \leftrightarrow \sim(p \wedge q)$
T	F	T T T **T** T TFF

15. valid argument

p	q	r	$\{p \wedge [\sim q \rightarrow \sim r] \wedge [(q \vee r) \rightarrow \sim p]\}$	$\rightarrow$	$\sim r$
			4 1 3 2 8 5 7 6	9	
T	T	T	T T FTF F TTT F F	**T**	F
T	T	F	T T FTT F TTF F F	**T**	T
T	F	T	T F TFF F FTT F F	**T**	F
T	F	F	T T TTT T FFF T F	**T**	T
F	T	T	F F FTF F TTT T T	**T**	F
F	T	F	F F FTT F TTF T T	**T**	T
F	F	T	F F TFF F FTT T T	**T**	F
F	F	F	F F TTT F FFF T T	**T**	T

17. $\sim p \vee q$

19. invalid argument

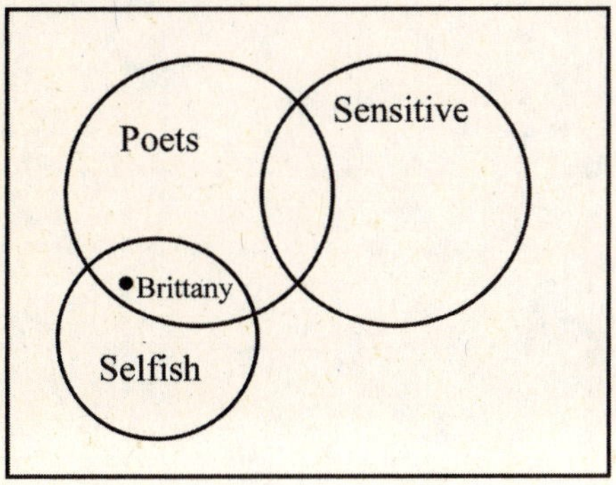

Chapter 4

Graph Theory (Networks): The Mathematics of Relationships

Section 4.1 Graphs, Puzzles and Map Coloring

1. An odd vertex is connected to an odd number of vertices; an even vertex is connected to an even number of edges.

3. If FC were traversed and removed, the graph would become disconnected and the remaining graph could not be traced.

5. Leonhard Euler and Arthur Cayley

7. Connected; A and B are odd vertices; C and D are even vertices.

9. Connected; A, B, E, and F are odd vertices; C and D are even vertices.

11. Not connected; All vertices are even.

13. Connected; A, B, E, and F are odd vertices; C and D are even vertices.

15. Graphs may vary.

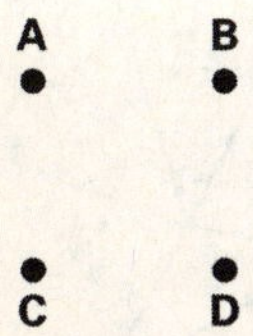

17. not possible

19. Graphs may vary.

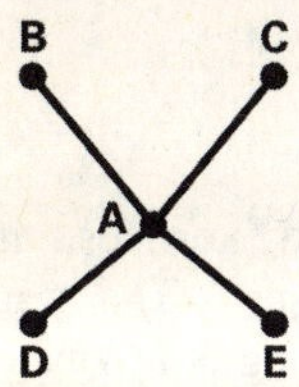

21. Graphs may vary.

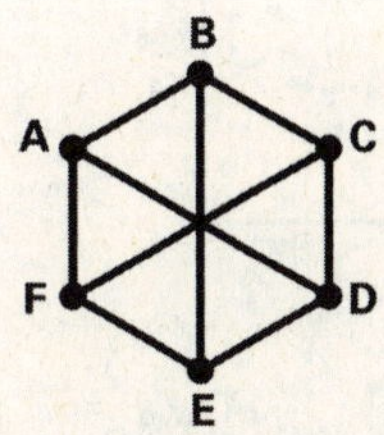

23. Yes, this graph can be traced.

25. No, this graph cannot be traced because it has four odd vertices.

27. No, this graph cannot be traced because it is not connected.

29. No, this graph cannot be traced because it has four odd vertices.

31. This is a Eulerian graph because all vertices are even. One possible Euler circuit is: ABFHI GCBEF GECDA

33. This is not an Eulerian graph.
Duplicate edges GH, GI, and FG.

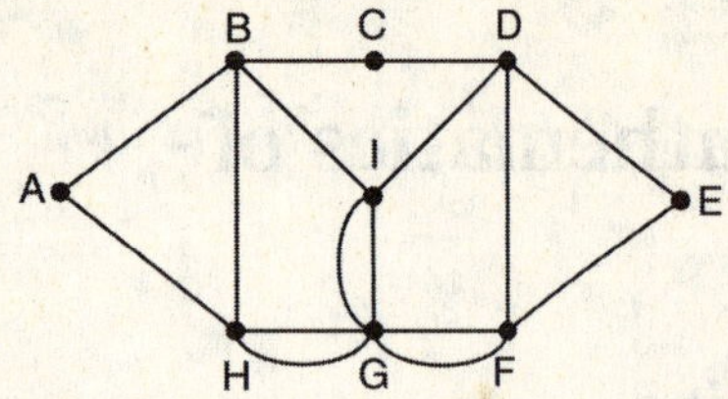

A possible path is ABCDE FDIGF GIBHG HA.

35. This is not an Eulerian graph.
Duplicate edges CF, DE, and EI.

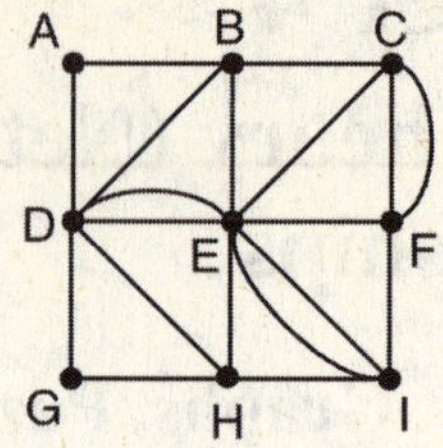

A possible path is ABDEB CEFCF IEIHE DHGDA.

37. No. If you consider the intersection of two streets as a vertex, then there are four odd vertices. The graph can therefore not be traced, and the taxi driver must travel over parts of the route more than once.

39. If you label the intersections that represent odd vertices as follows you can duplicate edges AB, DE, CF, and GH. Follow the possible path indicated in numerical order.

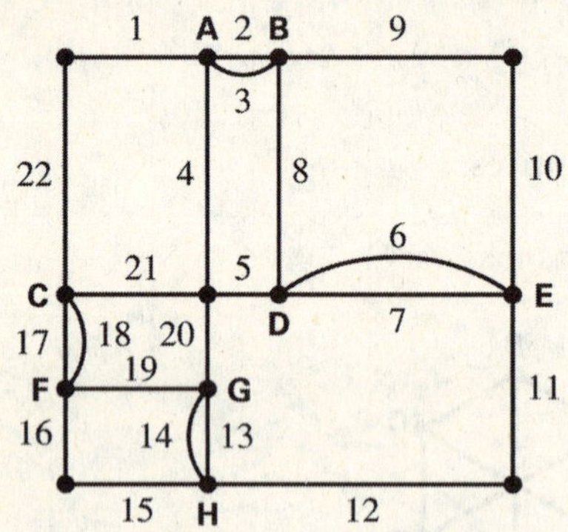

41.

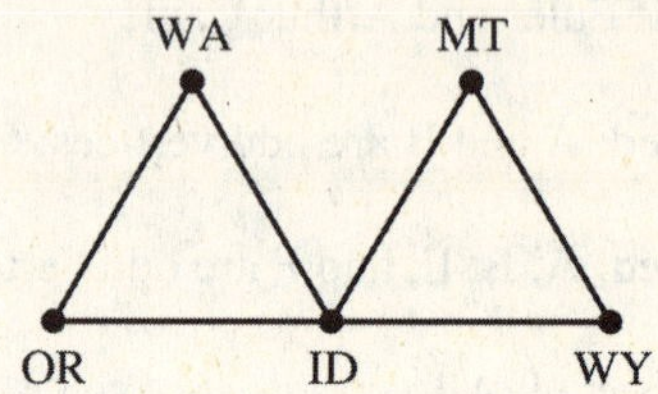

43.

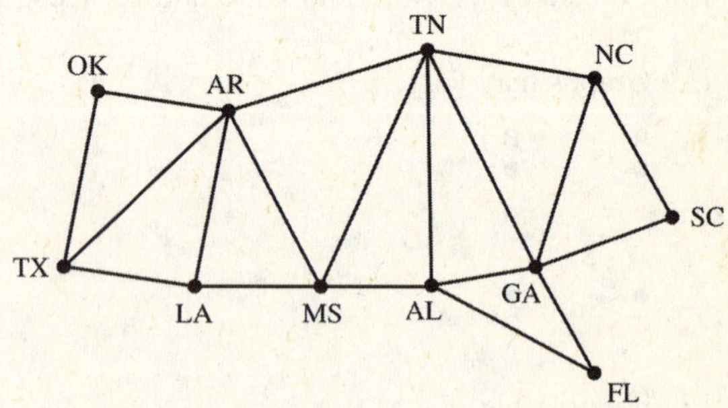

45. The trip is possible because the graph in Exercise 41 has no odd vertices.

47. The trip is not possible because the graph in Exercise 43 has more than two odd vertices.

49. It is not possible. The room can be graphed as follows:

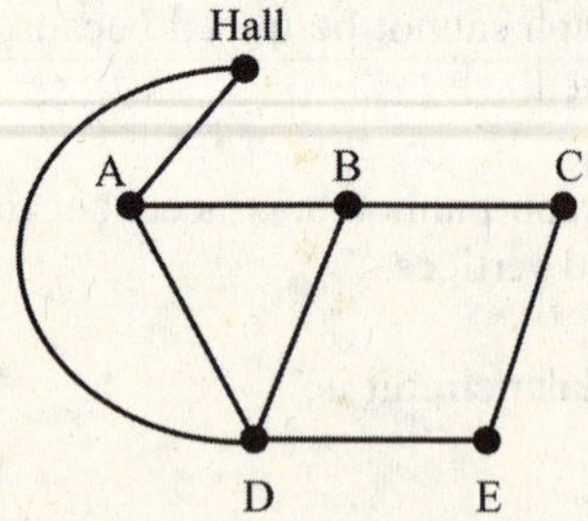

Since there are two odd vertices, it is possible to trace this graph. The tracing (locking of doors) must begin at one odd vertex and terminate at the other. Here this would imply that the guard starts in either Room B and ends in Room A or starts in Room A and ends in Room B. Either way, he would lock himself in the building because he cannot enter or exit through the hallway.

51. Answers may vary. The edges in the following graph represent family members that are unfriendly to each other.

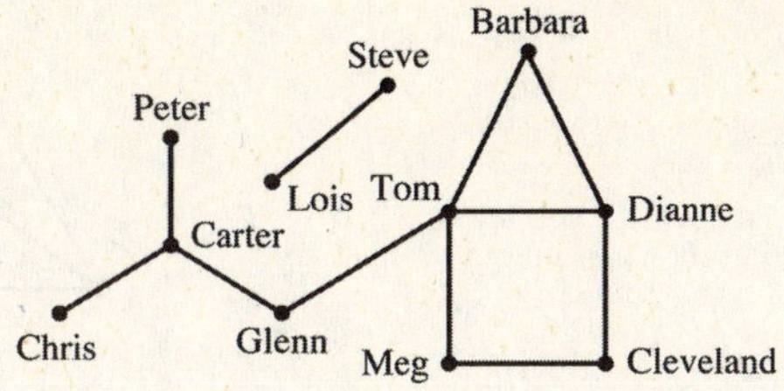

It is possible to color this graph with three colors (indicated by 1, 2, and 3 on the graph). This tells us that three tables will be satisfactory. A satisfactory seating arrangement would be:

Table 1) Peter, Chris, Glenn, and Dianne,

Table 2) Carter, Lois, Tom, and Cleveland,

Table 3) Barbara, Steve, and Meg.

53. Answers may vary.
The edges in the following graph represent committees that would have a conflict because they have common members.

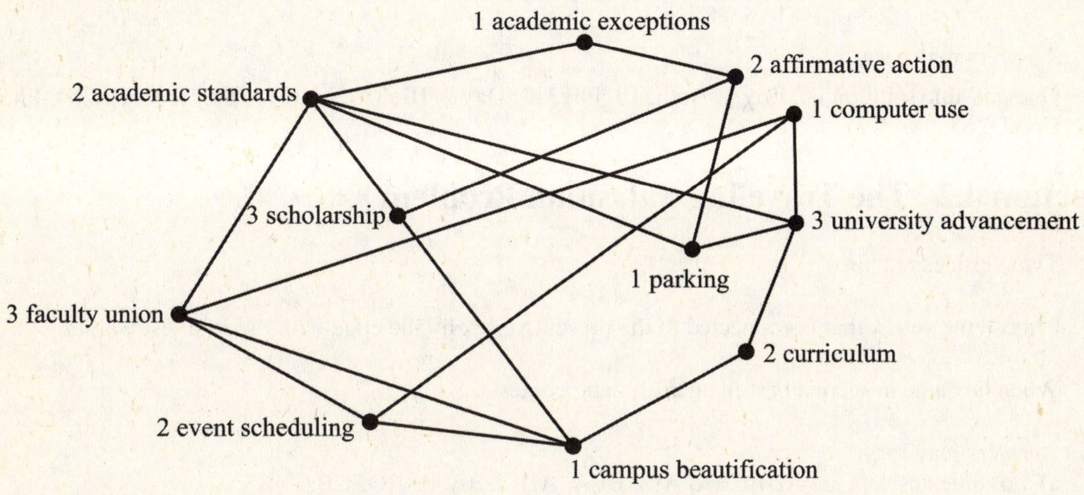

It is possible to color this graph with three colors (indicated by 1,2, and 3). This tells us that three times will be satisfactory to schedule meetings. A satisfactory schedule would be:

Time 1) academic exceptions, computer use, campus beautification, and parking,

Time 2) academic standards , affirmative action, event scheduling, and curriculum,

Time 3) university advancement, faculty union, and scholarship.

55. Every time the vertex is entered by one edge, it must be left by another.

57. No solution provided.

59. Examples and responses may vary.
Each edge connects to two vertices, so it contributes two of the total of all the degrees. The total of all degrees of all of the edges will be twice the number of edges.

61. – 67. No solution provided.

69. Answers may vary.

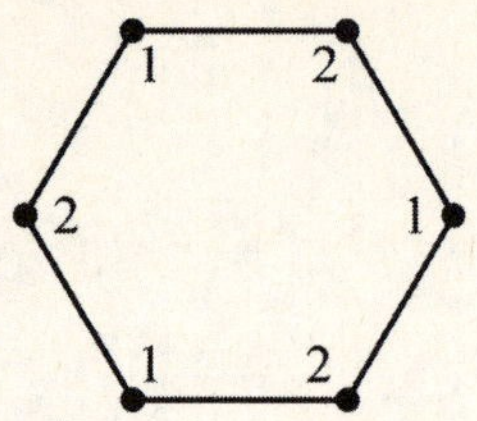

71. Answers may vary.

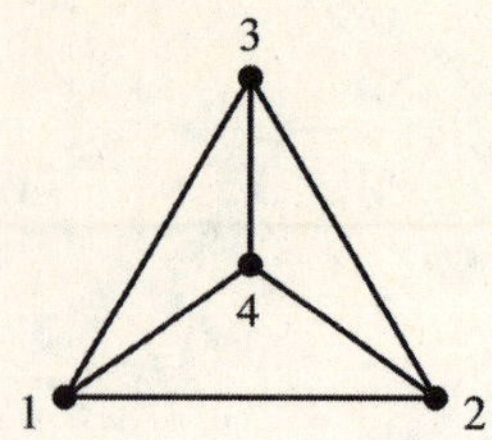

73. One possible sequence is C, F, E, C#, D, A, E, B.

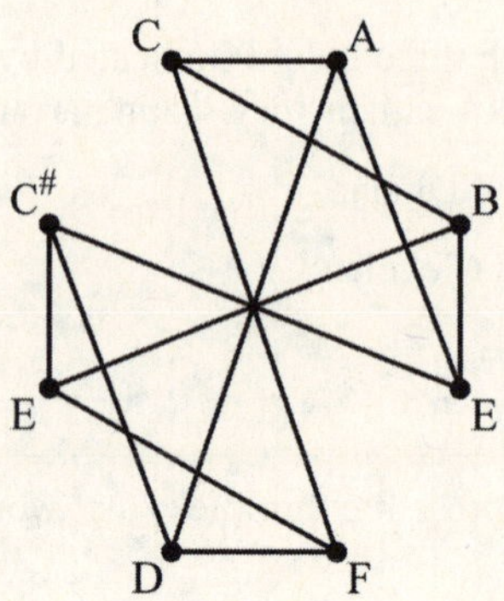

75. Answers may vary.
One possible solution is: Day 1: Anth215, Phy212; Day 2: Bio325, Mat311; Day 3: Chem264, Fin323, ComD265.

Section 4.2 The Traveling Salesman Problem

1. Two vertices remain.

3. Choose the vertex that is connected to the present vertex by the edge with the smallest weight.

5. When he came in second best in an arithmetic contest.

7. Answers may vary.
   a) Possible answers are ADBCEA, ADCBEA, ADCEBA, or ADECBA.
   b) Possible answers are EBCDAE or EBADCE.

9. Answers may vary.
   a) Possible answers are ADCBFEA, ADCEFBA, ADBFCEA, ADBCEFA, ADBCFEA, ADEFCBA, ADECBFA, or ADECFBA.
   b) Possible answers are EADCBFE, EABDCFE, EABFCDE, EADBFCE, EADBCFE, EAFBDCE, EAFCBDE, or EAFBCDE.

11. Possible circuits are ABCDEA, ABCEDA, ABDCEA, and ABECDA.

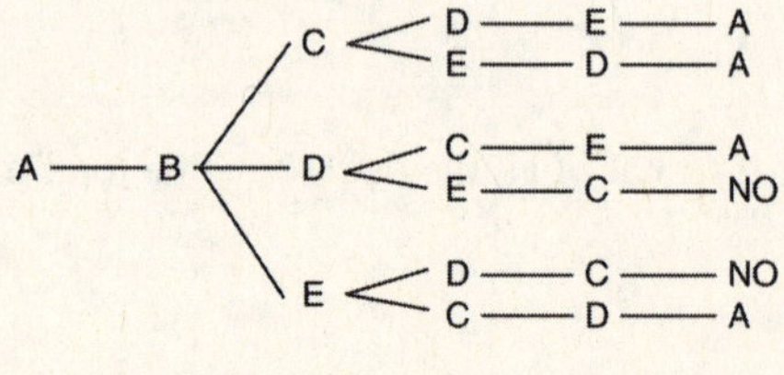

13. $(7-1)! = 6! = 720$

15. K_6

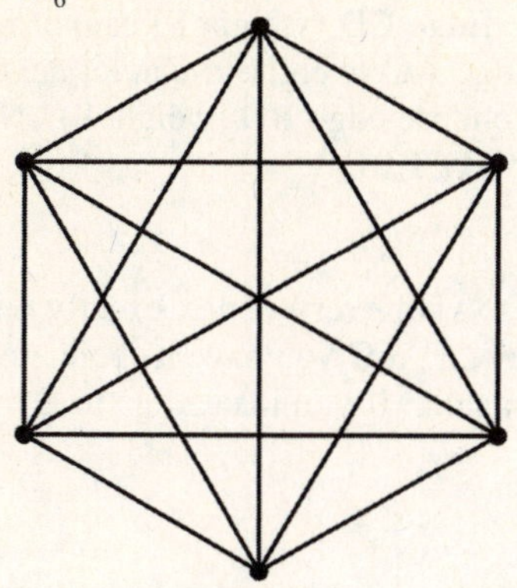

17. a) yes, $\frac{n}{2}=\frac{5}{2}=2.5$, and each vertex has degree of at least 3.

b) yes; $n=5$, and the sum of the degrees of all pairs of non-adjacent vertices is 6.

c) yes

19. a) no, $\frac{n}{2}=\frac{11}{2}=5.5$, and each vertex has degree of at most 4.

b) no; $n=11$, and the sum of the degrees of all pairs of non-adjacent vertices is at most 8.

c) yes

21. a) The weight of path AGEDCB would be $2+4+2+8+4=20$.
b) The weight of path ABCDGEF would be $3+4+8+1+4+3=23$.

23. Since Hamilton circuits occur in pairs, we need only calculate the weight of three Hamilton circuits.

Hamilton Circuit	Weight
ABCDA	3+4+5+3=15
ABDCA	3+6+5+1=15
ACBDA	**1+4+6+3=14**

Since 14 is the smallest number in the right hand column, the circuits ACBDA and ADBCA have minimal weight.

25. Since Hamilton circuits occur in pairs, we need only calculate the weight of three Hamilton circuits.

Hamilton Circuit	Weight
ABCDA	23+14+37+36=110
ABDCA	23+31+37+18=109
ACBDA	**18+14+31+36=99**

Since 99 is the smallest number in the right hand column, the circuits ACBDA and ADBCA have minimal weight.

27. ABDECA

29. ADCFEBA

31. Start by selecting DE which has the smallest weight (weight 2). Next choose edges AB and CD (weight 3). Edge BD (weight 4) cannot next be chosen because that will create a vertex that joins three edges. We can next choose edge AC (weight 5). Edge AE (weight 6) cannot next be chosen because that will create a vertex that joins three edges. Next choose edge BE (weight 6) to complete the circuit of ABEDCA or ACDEBA.

33. Start by selecting CF which has the smallest weight (weight 4). Next choose edge AD (weight 5). Next choose edge BE (weight 6). Next choose edge CE (weight 7). Edge CD (weight 8) cannot next be chosen because that will create a vertex that joins three edges. Edge AC (weight 9) cannot next be chosen because that will create a vertex that joins three edges. Next choose edge BD (weight 9). Next choose edge AF (weight 13) to complete the circuit of ADBECFA or AFCEBDA.

35. Answers may vary.
If you start at any vertex, you will see that you cannot follow edges to visit every vertex exactly once. For example, if you start at vertex A and go to B, you must then go to C. At C you have a choice to go to D or E. If you visit D, then you cannot go to E without visiting A again. If you instead go to E from C, then you cannot go to D without visiting A again.

37. Answers may vary.
If you start at any vertex, you will see that you cannot follow edges to visit every vertex exactly once. For example, if you start at vertex A you would have a choice to visit either B or E or F. If you choose to visit B then you go to C and then D. At D you have a choice to go to either F or E. If you go to F, you must go through A again in order to visit E. Also, if you go to E from D, you must visit A again prior to going to F. If back at A you decided to go to E instead of B initially, you would have similar problems. At E you would then go to D and here you would have the choice to go to F or C. If you go to F from D then you must visit A again prior to going to B. If at D you instead decided to go to C then B, you must visit A again prior to going to F. If back at A you decided to go to F instead of E or B initially, you would have again similar problems. At F you would then go to D and would be then forced to go to E then A would have to be revisited before reaching B and C.

39. a) This is a K_7 drawing. Since this graph has 7 vertices, there will be 6!=720 Hamilton circuits. We won't count reversals so $\frac{720}{2} = 360$ circuits need to be considered.

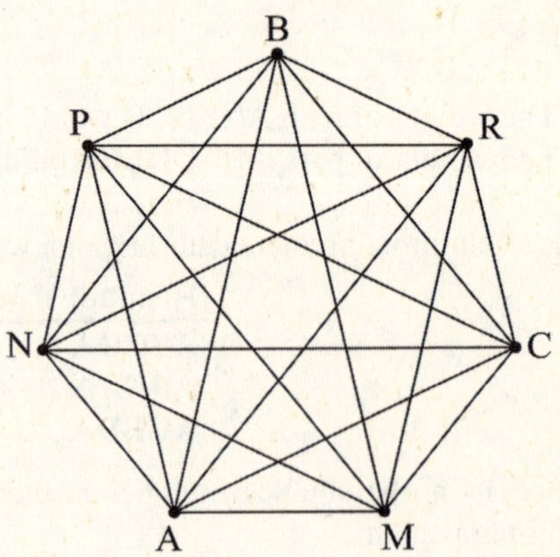

b) If you consider 360 circuits at one per minute, then it would take 360 minutes or 6 hours to do this problem by the brute force algorithm.

c) Using the nearest neighbor algorithm, the path Danielle should go is PNBRAMCP. The cost of this trip would be 210 + 180 + 290 + 90 + 170 + 240 + 420= $1,600.

d) Using the best edge algorithm, the edge with the smallest weight is AR with a weight of 90. The next edge of choice would be MR (weight 120). Edge AM (weight 170) cannot be next chosen because it will create a circuit. Edge BN (weight 180) would be chosen next. Edge NP (weight 210) would be chosen next followed by edge AP (weight 230). Edge PR (weight 240) cannot be chosen because it will create a vertex in which three edges meet. The next choice would be edge CM (weight 240). The remaining choice of BC (weight 380) would be made to complete the circuit. The path would therefore be PNBCMRAP or PARMCBNP. The cost would be 210 + 180 + 380 + 240 + 120 + 90 + 230 = $1,450.

41. Find a table to relate the distance between the stops and Papa John's Pizza.

	A	B	D	E	H	F	M
A	0	2	3	6	4	9	3
B	2	0	3	4	4	7	1
D	3	3	0	5	3	8	2
E	6	4	5	0	4	3	3
H	4	4	3	4	0	5	3
F	9	7	8	3	5	0	6
M	3	1	2	3	3	6	0

a) ABMDHEFA

b) Using the best edge algorithm, we can first choose edge BM because is has the lowest weight (weight 1). The next edges that can be chosen are AB and DM (weight 2). Of the edges of weight 3, you don't want to choose EM, AM, HM or BD because any of these will cause three edges to meet at one vertex. The next edge of weight 3 to choose would be EF. You now have to make a choice between edges AD and DH. They both have weight 3 and can't both be chosen because the combination of these two edges will cause three edges will meet at a vertex. AD cannot be chosen however because it will close a circuit that does not include all the stops. Therefore, if you next choose DH (weight 3), then the next to be chosen would be EH (weight 4). The next edge to choose would be AF (weight 9) to complete the circuit. The circuits that could be found would be ABMDHEFA or AFEHDMBA

43. a) Find a table to relate the distance between the stops and Fed Ex.

	X	A	B	C	D	E	F	P
X	0	9	8	9	8	5	2	7
A	9	0	3	2	3	4	7	6
B	8	3	0	5	4	3	6	5
C	9	2	5	0	1	6	7	4
D	8	3	4	1	0	5	6	3
E	5	4	3	6	5	0	3	4
F	2	7	6	7	6	3	0	5
P	7	6	5	4	3	4	5	0

Using nearest neighbor algorithm the path followed should be XFEBACDPX

b) Using the best edge algorithm, we can first choose edge CD because it has the lowest weight (weight 1). The next edges that can be chosen are AC and XF (weight 2). Of the edges of weight 3, you can choose EF, DP, AB and BE. If edge AD were chosen then it would cause three vertices to meet. The next edge to choose would be XP (weight 7) to complete the circuit. The circuits that could be found would be XFEBACDPX or XPDCABEFX.

45. A Hamilton circuit goes through every vertex but does not travel over every edge as an Euler circuit does.

47. They are faster than the brute force algorithm.

49. No solution provided.

51. The number of circuits will always be an even number except for the trivial case of K_2.

53. Answers may vary.

55. – 57. No solution provided.

59. A K_{20} figure would have 19!=121,645,100,408,832,000 circuits. We won't count reversals so $\frac{121,645,100,408,832,000}{2} = 60,822,550,204,416,000$ circuits need to be considered. If a computer can examine 1,000,000 Hamilton circuits per second, then it would take the computer 60,822,550,204.4 seconds. Now multiply 60 seconds per minute times 60 minutes per hour times 24 hours in a day times 365 days per year this gives us 31,536,000 seconds per year. $\frac{60,822,550,204.4}{31,536,000} \approx 1,928.67$. It would take over 1,928 years to complete.

61. It may be helpful to create a table for this size figure

	A	B	C	D	E	F
A	0	11	9	5	8	13
B	11	0	14	9	6	15
C	9	14	0	8	7	4
D	5	9	8	0	9	9
E	8	6	7	9	0	13
F	13	15	4	9	13	0

Using the nearest neighbor algorithm starting at A we have the circuit **ADCFEBA**. The weight of this circuit is 5 + 8 + 4 + 13 + 6 +11 = **47**.
Using the nearest neighbor algorithm starting at B we have the circuit **BECFDAB**. The weight of this circuit is 6 + 7 + 4 + 9 + 5 + 11 = **42**.
Using the nearest neighbor algorithm starting at C we have the circuit **CFDAEBC**. The weight of this circuit is 4 + 9 + 5 + 8 + 6 + 14 = **46**.
Using the nearest neighbor algorithm starting at D we have the circuit **DAEBCFD**. The weight of this circuit is 5 + 8 + 6 + 14 + 4 + 9 = **46**.
Using the nearest neighbor algorithm starting at E we have the circuit **EBDACFE**. The weight of this circuit is 6 + 9 + 5 + 9 + 4 + 13 = **46**.
The smallest weight among these is 42. There is essentially one circuit (not counting reversing the order) here that starts and stops at A. It is ABECFDA.

Section 4.3 Directed Graphs

1. because the rumors only travel in one direction

3. the total of the direct and two-stage influences of each committee member

5. Overnight shipping companies use graph theory to plan the routes packages travel for efficient delivery.

7. a) ABCE or ABCDE
b) ABC
c) ABCE
d) Not possible because you must first go from A to B. From B you cannot go directly to E.
e) Answers may vary. Possible answers include ABCEDA or ABCEBA.

9. a) BCE
b) Answers may vary. Possible answers included ABCFG or ACFG.
c) ACFGCE
d) Not possible because you cannot go to D. It is possible to go from D.
e) ABCFGC

11.

		To				
		A	B	C	D	E
From	A	1	1	1	0	0
	B	1	1	1	1	1
	C	1	1	0	2	2
	D	1	2	0	1	1
	E	2	1	1	1	1

13.

		To						
		A	B	C	D	E	F	G
From	A	0	1	2	0	1	1	0
	B	0	0	1	0	1	1	0
	C	0	0	0	0	1	1	1
	D	0	0	1	0	2	1	0
	E	0	0	0	0	0	0	0
	F	0	0	1	0	0	0	1
	G	0	0	1	0	1	1	0

15. a) Only Kevin; if anyone else starts the rumor, it would not spread to everyone. (Kevin would be excluded). b) Reverse the direction of the edge KD or RK.

17. The only possible sequence would be chief financial officer, then sales manager, then production manager, then company president, then marketing director. If the secretary starts with the sales manager, then he/she will not be able to obtain the signature of the chief financial officer. If the secretary starts with the production manager, then he/she will not be able to get the signatures of the sales manager and the chief financial officer. If the secretary starts with the company president, then he/she will not be able to get the signatures of the sales manager, chief financial officer, and the production manager. If the secretary starts with the marketing director, then he/she will not be able to get the signatures of the chief financial officer, sales manager, production manager, and the company president.

19.

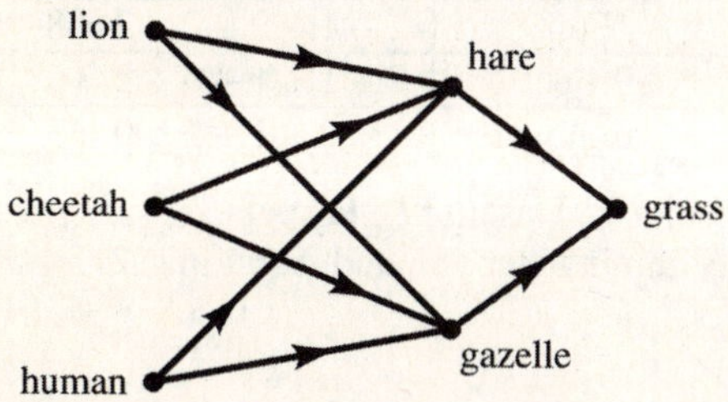

21.

		Georgia	Arkansas	LSU	Florida	Mississippi	Alabama
From	Georgia	0	0	0	0	0	0
	Arkansas	1	0	0	0	0	0
	LSU	2	1	0	0	0	0
	Florida	1	1	2	0	1	1
	Mississippi	1	1	1	1	0	1
	Alabama	0	0	1	1	1	0

Row	Sum of Entries
Georgia	$0+0+0+0+0+0=0$
Arkansas	$1+0+0+0+0+0=1$
LSU	$2+1+0+0+0+0=3$
Florida	$1+1+2+0+1+1=6$
Mississippi	$1+1+1+1+0+1=5$
Alabama	$0+0+1+1+1+0=3$

Florida (6), Mississippi (5), LSU and Alabama (3), Arkansas (1), and Georgia (3)

23.

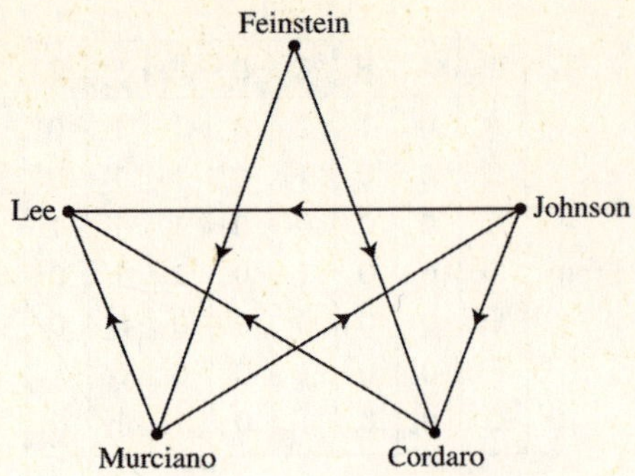

Row	Sum of Entries
Lee	0 + 0 + 0 + 0 + 0 = 0
Feinstein	2 + 0 + 1 + 1 + 1 = 5
Johnson	2 + 0 + 0 + 1 + 0 = 3
Cordaro	1 + 0 + 0 + 0 + 0 = 1
Murciano	2 + 0 + 1 + 1 + 0 = 4

Feinstein (5), Murciano (4), Johnson (3), Cordaro (1), and Murciano (0)

		To				
		Lee	Feinstein	Johnson	Cordaro	Murciano
	Lee	0	0	0	0	0
	Feinstein	2	0	1	1	1
From	Johnson	2	0	0	1	0
	Cordaro	1	0	0	0	0
	Murciano	2	0	1	1	0

25.

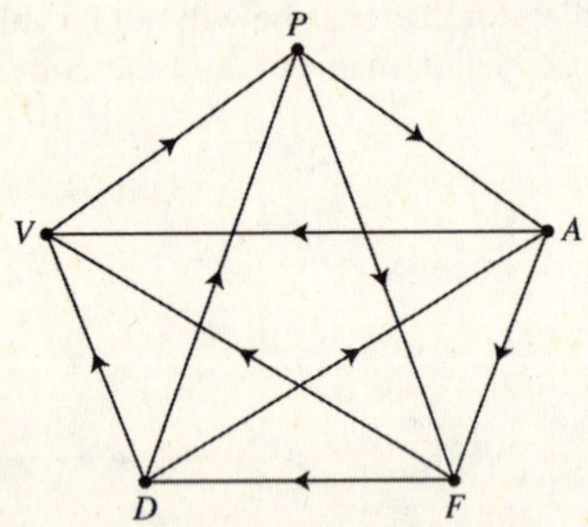

Row	Sum of Entries
Aquafina	0 + 0 + 0 + 1 + 1 = 2
Dasani	2 + 0 + 1 + 2 + 2 = 7
Fuji	2 + 1 + 0 + 2 + 3 = 8
Propel	2 + 1 + 1 + 0 + 2 = 6
Vitaminwater	1 + 0 + 1 + 1 + 0 = 3

Fuji (8), Dasani (7), Propel (6), Vitaminwater (3), and Aquafina (2)

		To				
		Aquafina	Dasani	Fuji	Propel	Vitaminwater
	Aquafina	0	0	0	1	1
	Dasani	2	0	1	2	2
From	Fuji	2	1	0	2	3
	Propel	2	1	1	0	2
	Vitaminwater	1	0	1	1	0

27. when the relationship goes in only one direction

29. I and B are in a block of vertices that do not have any directed edges leaving them. (Meaning J, D, or A could not be infected.)

31. The graph would have to have a directed edge or path from F to I and from F to J. We cannot do this by changing the direction of only one edge.

33. – 39. No solution provided.

Section 4.4 Looking Deeper: Scheduling Projects Using PERT

1. the time required, in months, to build the shell

3. a) Begin,A,B,E,I

Directed Path	Time to Complete
Begin,A,B,D,I	3+6+3+2=14 days
Begin,A,B,E,I	**3+6+5+2=16 days**
Begin,A,C,E,I	3+4+5+2=14 days
Begin,A,C,F,I	3+4+6+2=15 days

b) Begin,A,B,E

Directed Path	Time to Complete
Begin,A,B,E	**3+6+5=14 days**
Begin,A,C,E	3+4+5=12 days

c) On day 14

Directed Path	Time to Complete
Begin,A,C,F,H	**3+4+6=13 days**
Begin,A,C,G,H	3+4+3=10 days

d) After 16 days

Directed Path	Time to Complete
Begin,A,B,D,I	3+6+3+2=14 days
Begin,A,B,E,I	**3+6+5+2=16 days**
Begin,A,C,E,I	3+4+5+2=14 days
Begin,A,C,F,I	3+4+6+2=15 days

e) 17 days

Directed Path	Time to Complete
Begin,A,B,D,I,J,End	3+6+3+2+1=15 days
Begin,A,B,E,I,J,End	**3+6+5+2+1=17 days**
Begin,A,C,E,I,J,End	3+4+5+2+1=15 days
Begin,A,C,F,I,J,End	3+4+6+2+1=16 days
Begin,A,C,F,H,J,End	**3+4+6+3+1=17 days**
Begin,A,C,G,H,J,End	3+4+3+3+1=14 days

f) Begin,A,B,E,I,J,End or Begin,A,C,F,H,J,End

5. a) Begin,B,E,F,G

Directed Path	Time to Complete
Begin,A,D,G	2+4+2=8 days
Begin,A,E,G	2+2+2=6 days
Begin,B,E,G	3+2+2=7 days
Begin,B,E,F,G	**3+2+3+2=10 days**
Begin,C,F,G	3+3+2=8 days

b) Begin,B,E,F,H

Directed Path	Time to Complete
Begin,A,E,H	2+2+1=5 days
Begin,A,E,F,H	2+2+3+1=8 days
Begin,B,E,H	3+2+1=6 days
Begin,B,E,F,H	**3+2+3+1=9 days**
Begin,C,F,H	3+3+1=7 days

c) Begin,B,E,F,G,I,End

Directed Path	Time to Complete
Begin,A,D,G,I,End	2+4+2+4=12 days
Begin,A,E,G,I,End	2+2+2+4=10 days
Begin,A,E,H,I,End	2+2+1+4=9 days
Begin,A,E,F,G,I,End	2+2+3+2+4=13 days
Begin,A,E,F,H,I,End	2+2+3+1+4=12 days
Begin,B,E,G,I,,End	3+2+2+4=11 days
Begin,B,E,H,I,End	3+2+1+4=10 days
Begin,B,E,F,G,I,End	**3+2+3+2+4=14 days**
Begin,B,E,F,H,I,End	3+2+3+1+4=13 days
Begin,C,F,G,I,End	3+3+2+4=12 days
Begin,C,F,H,I,End	3+3+1+4=11 days

5. (continued)

d) Since it will take 3+2=5 days to complete the tasks that precede task F, task F should begin on day 6.

Directed Path	Time to Complete
Begin,A,E,F	2+2+3=7 days
Begin,B,E,F	**3+2+3=8 days**
Begin,C,F	3+3=6 days

e) From part a) we see that a critical path for G is Begin, B, E, F, G. Since it will take 3+2+3=8 days to complete the tasks that precede task G, task G should begin on day 9.

f) From part c) we determined that the critical path for "End" was Begin,B,E,F,G,I,End. This implies that it should take 14 days to complete the project.

7. Both tasks A and B can begin on day 1. We must however find the critical paths for tasks C,D,E,F,G,and H.

C: Since it will take 2 days to complete the tasks that precede task C, task C should begin on day 3.

Directed Path	Time to Complete
Begin,A,C	1+3=4 days
Begin,B,C	**2+3=5 days**

D: Since it will take 2 days to complete the tasks that precede task D, task D should begin on day 3.

Directed Path	Time to Complete
Begin,A,D	1+5=6 days
Begin,B,D	**2+5=7 days**

E: Since it will take 2+3=5 days to complete the tasks that precede task E, task E should begin on day 6.

Directed Path	Time to Complete
Begin,A,C,E	1+3+5=9 days
Begin,B,C,E	**2+3+5=10 days**

F: Since it will take 2+5=7 days to complete the tasks that precede task F, task F should begin on day 8.

Directed Path	Time to Complete
Begin,A,C,F	1+3+4=8 days
Begin,A,D,F	1+5+4=10 days
Begin,B,C,F	2+3+4=9 days
Begin,B,D,F	**2+5+4=11 days**

G: Since it will take 2+5=7 days to complete the tasks that precede task G, task G should begin on day 8.

Directed Path	Time to Complete
Begin,A,D,G	1+5+2=8 days
Begin,B,D,G	**2+5+2=9 days**

7. (continued)

H: Since it will take 2+5+4=11 days to complete the tasks that precede task H, task H should begin on day 8.

Directed Path	Time to Complete
Begin,A,C,E,H	1+3+5+6=15 days
Begin,A,C,F,H	1+3+4+6=14 days
Begin,A,D,F,H	1+5+4+6=16 days
Begin,A,D,G,H	1+5+2+6=14 days
Begin,B,C,E,H	2+3+5+6=16 days
Begin,B,C,F,H	2+3+4+6=15 days
Begin,B,D,F,H	**2+5+4+6=17 days**
Begin,B,D,G,H	2+5+2+6=15 days

9. Tasks A, B, and C can begin on day 1. We must however find the critical paths for tasks D,E,F,G,H, I, and J.

D: Since it will take 2 days to complete the tasks that precede task D, task D should begin on day 3.

Directed Path	Time to Complete
Begin,A,D	1+2=3 days
Begin,B,D	**2+2=4 days**

E: Since it will take 3 days to complete the tasks that precede task E, task E should begin on day 4.

Directed Path	Time to Complete
Begin,B,E	2+3=5 days
Begin,C,E	**3+3=6 days**

F: Since it will take 2+2=4 days to complete the tasks that precede task F, task F should begin on day 5.

Directed Path	Time to Complete
Begin,A,D,F	1+2+5=8 days
Begin,B,D,F	**2+2+5=9 days**

G: Since it will take 3+3=6 days to complete the tasks that precede task G, task G should begin on day 7.

Directed Path	Time to Complete
Begin,A,D,G	1+2+4=7 days
Begin,B,D,G	2+2+4=8 days
Begin,B,E,G	2+3+4=9 days
Begin,C,E,G	**3+3+4=10 days**

H: Since it will take 3+3=6 days to complete the tasks that precede task H, task H should begin on day 7.

Directed Path	Time to Complete
Begin,B,E,H	2+3+2=7 days
Begin,C,E,H	**3+3+2=8 days**

I: Since it will take 3+3+4=10 days to complete the tasks that precede task I, task I should begin on day 11.

Directed Path	Time to Complete
Begin,A,D,F,I	1+2+5+1=9 days
Begin,A,D,G,I	1+2+4+1=8 days
Begin,B,D,F,I	2+2+5+1=10 days
Begin,B,E,G,I	2+3+4+1=10 days
Begin,C,E,G,I	**3+3+4+1=11 days**

9. (continued)

J: Since it will take 3+3+4=10 days to complete the tasks that precede task J, task J should begin on day 11.

Directed Path	Time to Complete
Begin,A,D,G,J	1+2+4+3=10 days
Begin,B,D,G,J	2+2+4+3=11 days
Begin,B,E,G,J	2+3+4+3=12 days
Begin,B,E,H,J	2+3+2+3=10 days
Begin,C,E,G,J	**3+3+4+3=13 days**
Begin,C,E,H,J	3+3+2+3=11 days

11.

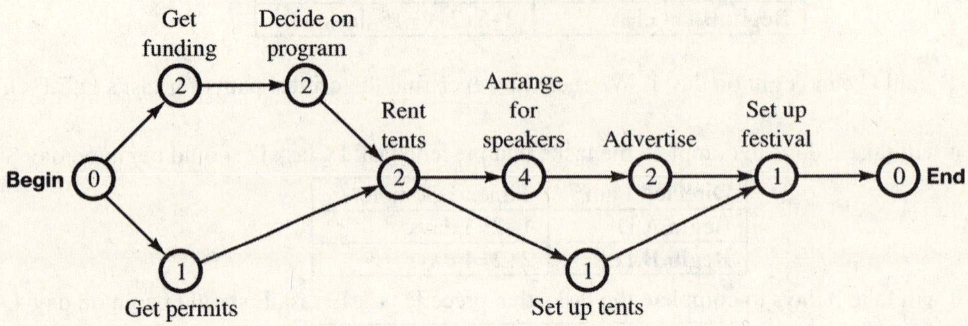

Funding and Permits can begin in week 1. Programs can begin in week 3. We must determine the critical paths for the rest of the tasks.

Rent tents: Since it will take 2+2=4 weeks to complete the tasks that precede renting tents, renting tents should begin on week 5.

Directed Path	Time to Complete
Begin,1,3,4	**2+2+2=6 weeks**
Begin,2,4	1+2=3 weeks

Arrange for speakers, etc.: Since it will take 2+2+2=6 weeks to complete the tasks that precede arranging for speakers, arranging for speakers should begin on week 7.

Directed Path	Time to Complete
Begin,1,3,4,5	**2+2+2+4=10 weeks**
Begin,2,4,5	1+2+4=7 weeks

Advertise: Since it will take 2+2+2+4=10 weeks to complete the tasks that precede arranging for advertising, arranging for advertising should begin on week 11.

Directed Path	Time to Complete
Begin,1,3,4,5,6	**2+2+2+4+2=12 weeks**
Begin,2,4,5,6	1+2+4+2=9 weeks

Set up tents, etc.: Since it will take 2+2+2=6 weeks to complete the tasks that precede setting up tents, setting up tents should begin on week 7.

Directed Path	Time to Complete
Begin,1,3,4,7	**2+2+2+1=7 weeks**
Begin,2,4,7	1+2+1=4 weeks

Set up festival: Since it will take 2+2+2+4+2=12 weeks to complete the tasks that precede setting up the festival, setting up the festival should begin on week 13.

Directed Path	Time to Complete
Begin,1,3,4,5,6,8	**2+2+2+4+2+1=13 weeks**
Begin,2,4,5,6,8	1+2+4+2+1=10 weeks
Begin,1,3,4,7,8	2+2+2+1+1=8 weeks
Begin,2,4,7,8	1+2+1+1=5 weeks

13.

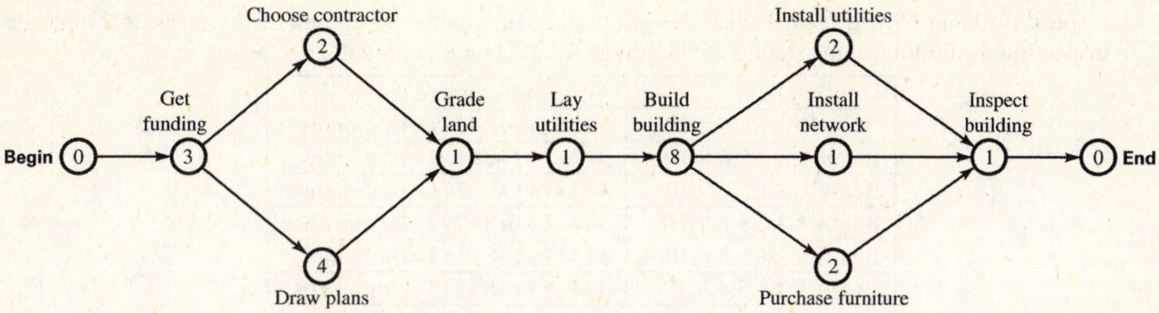

Getting funding can begin on month 1. Choosing a contractor and drawing the plans can begin on month 4. We must however determine the critical paths for the rest of the tasks.

Grade land: Since it will take 3+4=7 months to complete the tasks that precede grading the land, grading the land should begin on month 8.

Directed Path	Time to Complete
Begin,1,2,4	3+2+1=6 months
Begin,1,3,4	**3+4+1=8 months**

Lay underground utilities: Since it will take 3+4+1=8 months to complete the tasks that precede laying underground utilities, laying underground utilities should begin on month 9.

Directed Path	Time to Complete
Begin,1,2,4,5	3+2+1+1=7 months
Begin,1,3,4,5	**3+4+1+1=9 months**

Build buildings: Since it will take 3+4+1+1=9 months to complete the tasks that precede building buildings, building buildings should begin on month 10.

Directed Path	Time to Complete
Begin,1,2,4,5,6	3+2+1+1+8=15 months
Begin,1,3,4,5,6	**3+4+1+1+8=17 months**

Install utilities in building: Since it will take 3+4+1+1+8=17 months to complete the tasks that precede installing utilities in buildings, installing utilities in buildings should begin on month 14.

Directed Path	Time to Complete
Begin,1,2,4,5,6,7	3+2+1+1+8+2=17 months
Begin,1,3,4,5,6,7	**3+4+1+1+8+2=19 months**

Install computers: Since it will take 3+4+1+1+8=17 months to complete the tasks that precede installing computers, installing computers should begin on month 18.

Directed Path	Time to Complete
Begin,1,2,4,5,6,8	3+2+1+1+8+1=16 months
Begin,1,3,4,5,6,8	**3+4+1+1+8+1=18 months**

Purchase furniture: Since it will take 3+4+1+1+8=17 months to complete the tasks that precede purchase furniture, purchase furniture should begin on month 18.

Directed Path	Time to Complete
Begin,1,2,4,5,6,9	3+2+1+1+8+2=17 months
Begin,1,3,4,5,6,9	**3+4+1+1+8+2=19 months**

13. (continued)

Inspect Building: Since it will take 3+4+1+1+8+2=19 months to complete the tasks that precede inspecting the buildings, inspecting the buildings should begin on month 20.

Directed Path	Time to Complete
Begin,1,2,4,5,6,7,10	3+2+1+1+8+2+1=18 months
Begin,1,2,4,5,6,8,10	3+2+1+1+8+1+1=17 months
Begin,1,2,4,5,6,9,10	3+2+1+1+8+2+1=18 months
Begin,1,3,4,5,6,7,10	**3+4+1+1+8+2+1=20 months**
Begin,1,3,4,5,6,8,10	3+4+1+1+8+1+1=19 months
Begin,1,3,4,5,6,9,10	**3+4+1+1+8+2+1=20 months**

15.

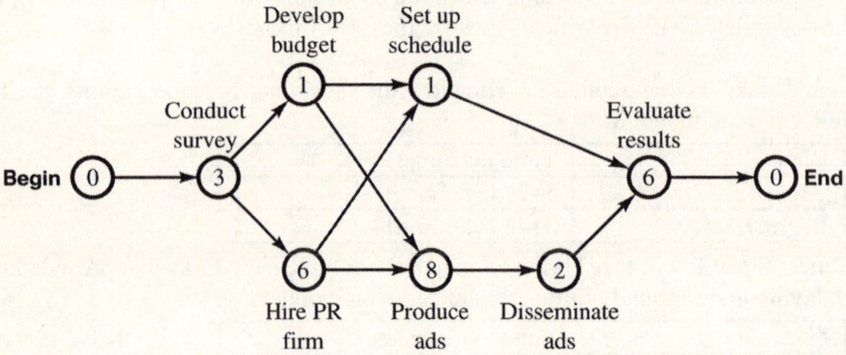

Conducting a survey can begin on month 1. Developing a budget and hiring a PR firm can begin on month 4. We must however determine the critical paths for the rest of the tasks.

Set up production schedule: Since it will take 3+6=9 months to complete the tasks that precede setting up the production schedule, setting up the production schedule should begin on month 10.

Directed Path	Time to Complete
Begin,1,2,4	3+1+1=5 months
Begin,1,3,4	**3+6+1=10 months**

Produce ads: Since it will take 3+6=9 months to complete the tasks that precede producing ads, producing ads should begin on month 10.

Directed Path	Time to Complete
Begin,1,2,5	3+1+8=12 months
Begin,1,3,5	**3+6+8=17 months**

Disseminate ads: Since it will take 3+6+8=17 months to complete the tasks that precede disseminating ads, disseminating ads should begin on month 18.

Directed Path	Time to Complete
Begin,1,2,5,6	3+1+8+2=14 months
Begin,1,3,5,6	**3+6+8+2=19 months**

Evaluate results: Since it will take 3+6+8+2=19 months to complete the tasks that precede evaluating results, evaluating results should begin on month 20.

Directed Path	Time to Complete
Begin,1,2,4,7	3+1+1+6=11 months
Begin,1,3,4,7	3+6+1+6=16 months
Begin,1,2,5,6,7	3+1+8+2+6=20 months
Begin,1,3,5,6,7	**3+6+8+2+6=25 months**

17. First, find a critical path for T. Then schedule T to begin after the sum of the times along its critical path.

19. No, because "Assemble shell" would no longer be on a critical path.

21. – 25. No solution provided.

Chapter Review Exercises

1. a) 12
   b) A, B, C, D, and E are even. F, G, H, and I are odd.
   c) Yes. The graph is connected.
   d) Yes. The bridges are FH, GH, and HI.

3. Graph (a) can be traced because it has only two odd vertices. Graph (b) cannot be traced because it has more than two odd vertices, namely vertices A, B, D, and E.

5. This is not an Eulerian graph. Duplicate edges BC, EI, HL, and NO so that all vertices are even. One possible route would be ABCDH GCBFG KLHLP ONJKO NMIEF JIEA.

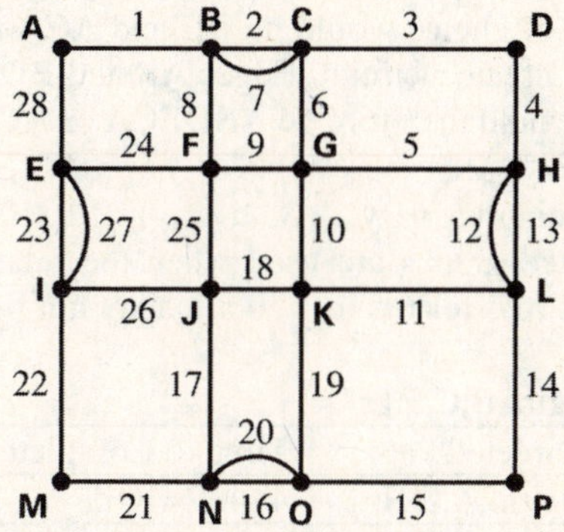

7. The edges in the following graph represent people that cannot be in the same canoe.

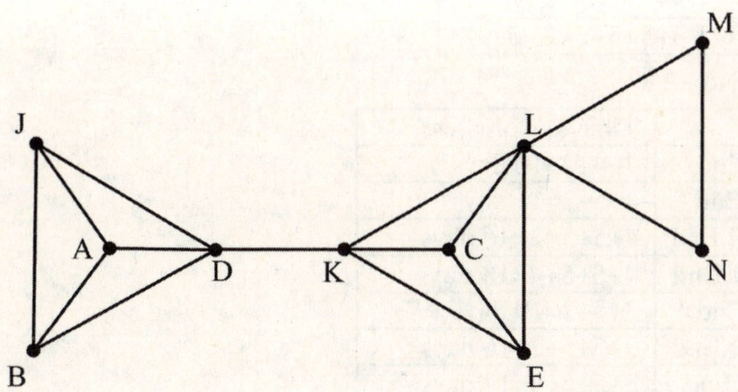

It is possible to color this graph with four colors (indicated by 1, 2, 3, and 4). This tells us that four canoes will be satisfactory. A satisfactory canoe assignment would be:

Canoe 1) Allison, Kami, and Nicole,

Canoe 2) Brandon, Colin, and Marshall,

Canoe 3) Donny and Erica,

Canoe 4) Jim and Lance.

9.

Hamilton Circuit	Weight
ABCEDA	4+8+1+2+3=18
ABDECA	**4+4+2+1+2=13**
ACEBDA	2+1+5+4+3=15
ACBEDA	2+8+5+2+3=20
ACEDBA	**2+1+2+4+4=13**
ADBECA	3+4+5+1+2=15
ADEBCA	3+2+5+8+2=20
ADECBA	3+2+1+8+4=18

Since 13 is the smallest number in the right hand column, the circuits ABDECA and ACEDBA have minimal weight.

11. Using the best edge algorithm, the edge with the smallest weight is EC with a weight of 1. The next edges of choice would be DE and AC(weight 2). Edge AD (weight 3) cannot be next chosen because it will create a circuit. Edges AB and BD (weight 4) would be chosen next to complete the circuit. The path would therefore be ABDECA or ACEDBA. The weight would be 4+4+2+1+2=13.

13. Answers may vary.
Directed graphs are used when the relationship modeled may apply in one direction. For example, object X is related to Y, but Y may not be related to X.

15. a) Begin, B,C,F,H

Directed Path	Time to Complete
Begin,A,E,H	4+4+2=10 days
Begin,A,C,H	4+5+2=11 days
Begin,B,C,H	3+5+2=10 days
Begin,B,C,H	3+5+2=10 days
Begin,B,D,F,H	3+3+5+2=13 days

b) Begin,A,C,F,I,End

Directed Path	Time to Complete
Begin,A,E,H,End	4+4+2=10 days
Begin,A,C,H,End	4+5+2=11 days
Begin,A,C,F,H,End	4+5+5+2=16 days
Begin,A,C,F,I,End	**4+5+5+4=18 days**
Begin,B,C,H,End	3+5+2=10 days
Begin,B,C,F,H,End	3+5+5+2=16 days
Begin,B,C,F,I,End	3+5+5+4=17 days
Begin,B,D,F,H,End	3+3+5+2=13 days
Begin,B,D,F,I,End	3+3+5+4=15 days
Begin,B,D,G,I,End	3+3+5+4=15 days

c) Since it will take 4+5=10 days to complete the tasks that precede task F, task F should begin on day 10.

Directed Path	Time to Complete
Begin,A,C,F	**4+5+5=14 days**
Begin,B,C,F	3+5+5=13 days
Begin,B,D,F	3+3+5=11 days

d) From part b) we determined that the critical path for "End" was Begin,A,C,F,I,End. This implies that it should take 18 days to complete the project.

Chapter Test

1. a) 9
b) E and F are odd; A, B, C, D, G, and H are even.
c) Yes, the graph is connected.
d) Yes, EF is a bridge.

3. Answers may vary. One possible circuit is ABDACEGIHGFEDFA.

5. This is not an Eulerian graph. Duplicate the edges as shown so that all vertices are even. One possible route is numbered on the diagram.

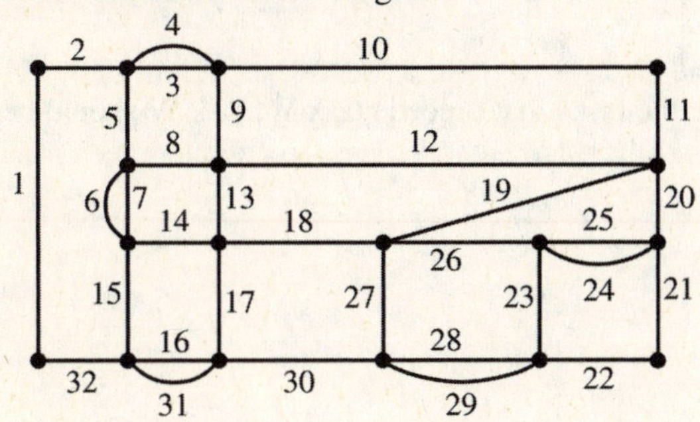

7.

Hamilton Circuit	Weight
ABCEDA	5+5+3+7+8=28
ABDECA	**5+2+7+3+2=19**
ACEBDA	**2+3+4+2+8=19**
ACBEDA	2+5+4+7+8=26
ACEDBA	2+3+7+2+5=19
ADBECA	8+2+4+3+2=19
ADEBCA	8+7+4+5+2=26
ADECBA	8+7+3+5+5=28

Since 19 is the smallest number in the right hand column, the circuits ABDECA and ACEBDA have minimal weight.

9. Using the best edge algorithm, the edges with the smallest weight are AC and BD (weight 2). The next edge of choice would be CE (weight 3). Edge BE (weight 4) cannot be next chosen because it will create a circuit. Edges AB and AD (weight 5) would be chosen next to complete the circuit. The path would therefore be ADBECA or ACEBDA. The weight would be 2+3+4+2+8=19.

11. a) Begin,X,P,B,Z

Directed Path	Time to Complete
Begin,P,B,Z	5+7+3=15 days
Begin,X,P,B,Z	**3+5+7+3=18 days**
Begin,X,Y,B,Z	3+4+7+3=17 days
Begin,X,Y,G,Z	3+4+5+3=15 days

11. (continued)

b) Begin,X,P,B,Z,End

Directed Path	Time to Complete
Begin,D,E,W,End	6+4+2=12 days
Begin,P,E,W,End	5+4+2=11 days
Begin,P,B,W,End	5+7+2=14 days
Begin,P,B,Z,End	5+7+3=15 days
Begin,X,P,E,W,End	3+5+4+2=14 days
Begin,X,P,B,W,End	3+5+7+2=17 days
Begin,X,P,B,Z,End	**3+5+7+3=18 days**
Begin,X,Y,B,W,End	3+4+7+2=16 days
Begin,X,Y,B,Z,End	3+4+7+3=17 days
Begin,X,Y,G,Z,End	3+4+5+3=15 days

c) Since it will take 3+5+7=15 days to complete the tasks that precede task W, task W should begin on day 16.

Directed Path	Time to Complete
Begin,D,E,W	6+4+2=12 days
Begin,P,E,W	5+4+2=11 days
Begin,P,B,W	5+7+2=14 days
Begin,X,P,B,W	**3+5+7+2=17 days**
Begin,X,Y,B,W	3+4+7+2=days

d) From part b) we determined that the critical path for "End" was Begin,X,P,B,Z,End. This implies that it should take 18 days to complete the project.

Chapter 5

Numeration Systems: Does It Matter How We Name Numbers?

Section 5.1 The Evolution of Numeration Systems

1. X, +, ∩, <

3. One scroll was rewritten as ten heel bones.

5. The Rhind papyrus

7. $300 + 20 + 4 = 324$

9. $20,000 + 1,000 + 300 + 20 + 4 = 21,324$

11. $2,000,000 + 100,000 + 20,000 + 1,000 + 200 + 10 = 2,121,210$

13. 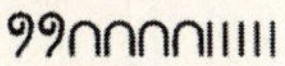

15. 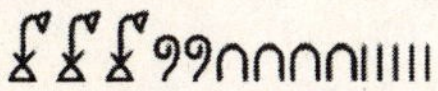

17. 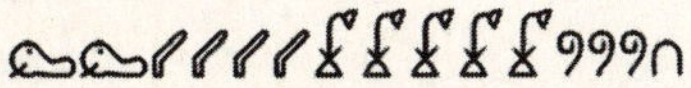

19.

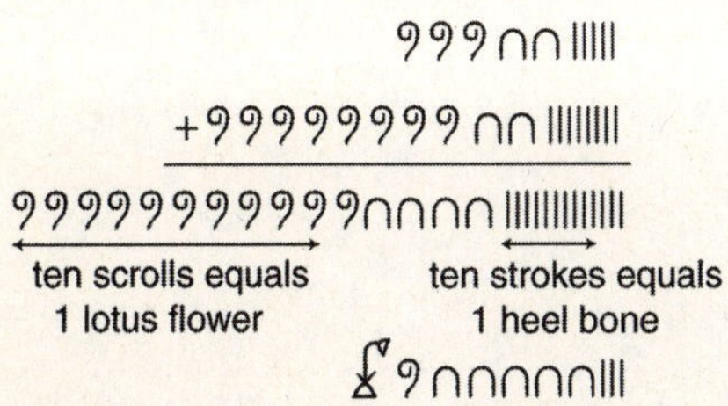

21.

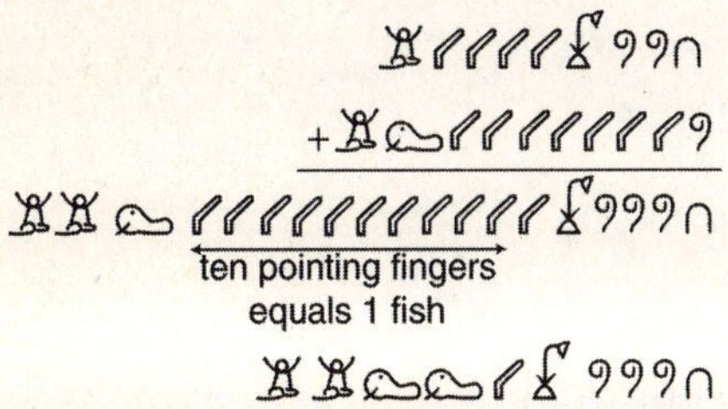

23.

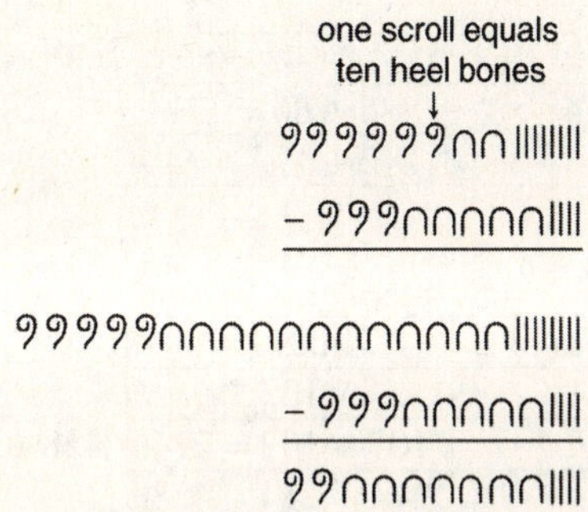

25.

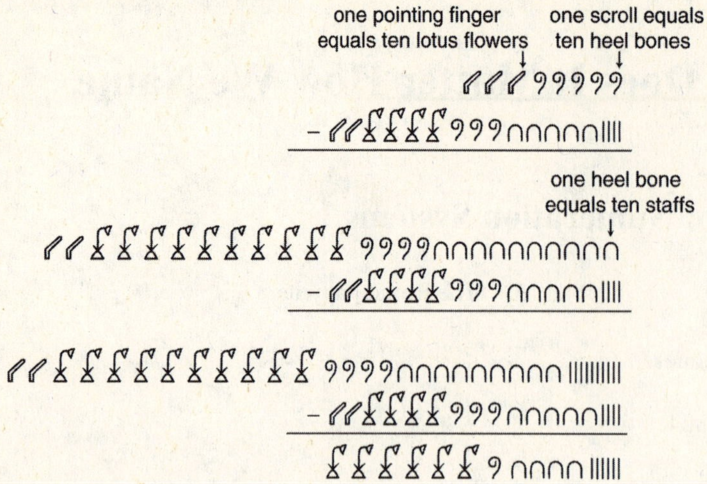

27. $14 = 2 + 4 + 8$

Powers of 2	Times 43
1	43
2	43 + 43 = 86
4	86 + 86 = 172
8	172 + 172 = 344

86 + 172 + 344 = 602

29. $12 = 4 + 8$

Powers of 2	Times 107
1	107
2	107 + 107 = 214
4	214 + 214 = 428
8	428 + 428 = 856

428 + 856 = 1,284

31. $35 = 1 + 2 + 32$

Powers of 2	Times 121
1	121
2	121 + 121 = 242
4	242 + 242 = 484
8	484 + 484 = 968
16	968 + 968 = 1,936
32	1,936 + 1,936 = 3,872

121 + 242 + 3,872 = 4,235

33. $50 + (10 - 1) = 50 + 9 = 59$

35. $500 + 50 + 10 + (5 - 1) =$
$500 + 60 + 4 = 564$

37. $1{,}000 + (1{,}000 - 100) + 50 + 10 + 3 = 1{,}000 + 900 + 60 + 3 = 1{,}963$

39. $5 \times 1{,}000 + 2{,}000 + 500 + (50 - 10) + (5 - 1) = 5{,}000 + 2{,}000 + 500 + 40 + 4 =$
$7{,}000 + 500 + 40 + 4 = 7{,}544$

41. $500 \times 100 + 200 + 50 + 10 + 2 = 50{,}000 + 200 + 60 + 2 = 50{,}262$

43. $5 \times 1{,}000 \times 100 + 1{,}000 + (500 - 100) + 20 = 500{,}000 + 1{,}000 + 400 + 20 = 501{,}420$

45. XLIV

47. CCLXXVIII

49. CDXLIV

51. $\overline{\text{IV}}$DCCXCV

53. $\overline{\text{LXXXIX}}$CDXXIII

55. $4 \times 100 + 3 \times 10 + 6 = 400 + 30 + 6 = 436$

57. $5 \times 1{,}000 + 6 \times 10 + 7 = 5{,}000 + 60 + 7 = 5{,}067$

59. $9 \times 1{,}000 + 9 \times 100 + 9 \times 10 + 9 = 9{,}000 + 900 + 90 + 9 = 9{,}999$

61. 四百九十五

63. 二千八百五

65. 九千八百四十六

67. $2 \times 1{,}000 + 6 \times 100 + 5 \times 10 = 2{,}000 + 600 + 50 = 2{,}650$ B.C.

69.

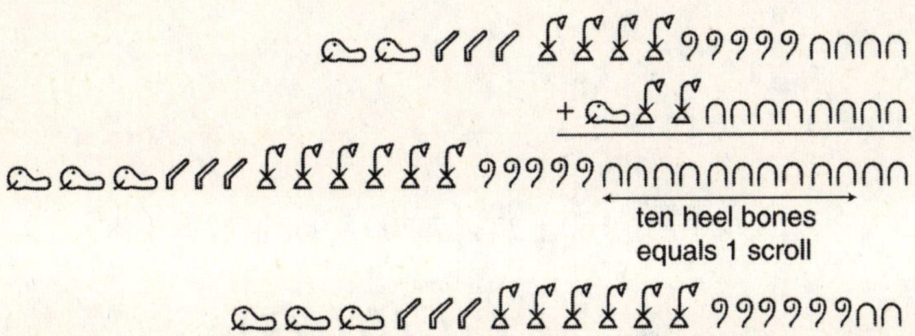

$$\begin{array}{r} \overset{1}{234{,}540} \\ +102{,}080 \\ \hline 336{,}620 \end{array}$$

71. MCMXXXIX

73. MCMXCIV

75. 一千五百二十三；一千二十九

77. A number tells "how many." A numeral is a symbol for writing a number.

79. To omit some power of ten, simply don't use the symbol for that power.

81. – 85. No solution provided.

87. 9,999,999

89. In the solution to this exercise, the symbol √. is being used instead of √.

a) ∝∝∝ ∇⊗⊗⊗⊗√.√. ; ∝∝∝∝∝∝∝ ∇∇√.√.√.

b)

∝∝∝ ∇⊗⊗⊗⊗√.√.
\+ ∝∝∝∝∝∝∝ ∇∇√.√.√.
∝∝∝∝∝∝∝∝∝∝ ∇∇∇⊗⊗⊗⊗√.√.√.√.√.

regroup ≈ ∇∇∇⊗⊗⊗⊗√.√.√.√.√.

c)

∝∝∝∝∝∝∝ ∇∇√.√.√.
− ∝∝∝ ∇⊗⊗⊗⊗√.√.

regroup

∝∝∝∝∝∝∝ ∇⊗⊗⊗⊗⊗⊗⊗⊗⊗⊗√.√.√.
− ∝∝∝ ∇⊗⊗⊗⊗√.√.
∝∝∝∝ ⊗⊗⊗⊗⊗⊗ √.

Section 5.2 Place Value Systems

1. Babylonian, Chinese, Mayan, Hindu–Arabic.

3. The Mayan system had a zero.

5. 20

7. 32

9. $(10+2)\times 60+(10+1)=12\times 60+11=720+11=731$

11. $2\times 60^2+(20-4)\times 60+(20+7)=2\times 3{,}600+16\times 60+27=7{,}200+960+27=8{,}187$

13. 

15. 

17. $8{,}235=2\times 60^2+17\times 60+15$  or
$8{,}235=2\times 60^2+(20-3)\times 60+15$ 

19. $18{,}397=5\times 60^2+6\times 60+37$  or
$18{,}397=5\times 60^2+6\times 60+(40-3)$ 

21. $123{,}485=34\times 60^2+18\times 60+5$  or
$123{,}485=34\times 60^2+(20-2)\times 60+5$ 

23. $188{,}289=52\times 60^2+18\times 60+9$  or
$188{,}289=52\times 60^2+(20-2)\times 60+9$ 

25. 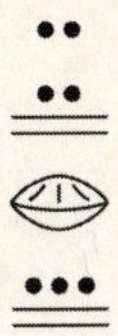

$2 \times 20 \times 18 \times 20 = 14{,}400$

$12 \times 20 \times 18 = 4{,}320$

$0 \times 20 = 0$

$13 = 13$

18,733

27. 

$2 \times 20 \times 18 \times 20^2 = 288{,}000$

$0 \times 20 \times 18 \times 20 = 0$

$6 \times 20 \times 18 = 2{,}160$

$0 \times 20 = 0$

$13 = 13$

290,173

29. 

31. 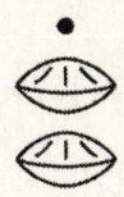

33. five hundred

35. five thousand

37. $2\times10^4 + 5\times10^3 + 3\times10^2 + 8\times10^1 + 9\times10^0$

39. $2\times10^5 + 7\times10^4 + 8\times10^3 + 0\times10^2 + 6\times10^1 + 3\times10^0$

41. $1\times10^6 + 2\times10^5 + 0\times10^4 + 0\times10^3 + 0\times10^2 + 4\times10^1 + 5\times10^0$

43. 5,368

45. 370,082

47. 3,070,502

49.
$$\begin{array}{rl} 2{,}863 = & 2\times10^3 + 8\times10^2 + 6\times10^1 + 3\times10^0 \\ +\ 425 = + & \qquad\quad 4\times10^2 + 2\times10^1 + 5\times10^0 \\ \hline & 2\times10^3 + 12\times10^2 + 8\times10^1 + 8\times10^0 \end{array}$$

$$\begin{aligned} 2\times10^3 + 12\times10^2 + 8\times10^1 + 8\times10^0 &= 2\times10^3 + (10+2)\times10^2 + 8\times10^1 + 8\times10^0 \\ &= 2\times10^3 + 10\times10^2 + 2\times10^2 + 8\times10^1 + 8\times10^0 \\ &= 2\times10^3 + 10^3 + 2\times10^2 + 8\times10^1 + 8\times10^0 \\ &= (2+1)\times10^3 + 2\times10^2 + 8\times10^1 + 8\times10^0 \\ &= 3\times10^3 + 2\times10^2 + 8\times10^1 + 8\times10^0 \\ &= 3{,}288 \end{aligned}$$

51.
$$\begin{array}{rl} 3{,}482 = & 3\times10^3 + 4\times10^2 + 8\times10^1 + 2\times10^0 \\ +2{,}756 = + & 2\times10^3 + 7\times10^2 + 5\times10^1 + 6\times10^0 \\ \hline & 5\times10^3 + 11\times10^2 + 13\times10^1 + 8\times10^0 \end{array}$$

$$\begin{aligned} 5\times10^3 + 11\times10^2 + 13\times10^1 + 8\times10^0 &= 5\times10^3 + (10+1)\times10^2 + (10+3)\times10^1 + 8\times10^0 \\ &= 5\times10^3 + 10\times10^2 + 1\times10^2 + 10\times10^1 + 3\times10^1 + 8\times10^0 \\ &= 5\times10^3 + 10^3 + 1\times10^2 + 10^2 + 3\times10^1 + 8\times10^0 \\ &= (5+1)\times10^3 + (1+1)\times10^2 + 3\times10^1 + 8\times10^0 \\ &= 6\times10^3 + 2\times10^2 + 3\times10^1 + 8\times10^0 \\ &= 6{,}238 \end{aligned}$$

53.
$$\begin{array}{rl} 926 = & 9\times10^2+2\times10^1+6\times10^0 \\ \underline{-784} = & \underline{-\ 7\times10^2+8\times10^1+4\times10^0} \end{array}$$

$$\begin{aligned} 9\times10^2+2\times10^1+6\times10^0 &= (8+1)\times10^2+2\times10^1+6\times10^0 \\ &= 8\times10^2+1\times10^2+2\times10^1+6\times10^0 \\ &= 8\times10^2+10\times10^1+2\times10^1+6\times10^0 \\ &= 8\times10^2+(10+2)\times10^1+6\times10^0 \\ &= 8\times10^2+12\times10^1+6\times10^0 \end{aligned}$$

$$\begin{array}{rl} 926 = & 8\times10^2+12\times10^1+6\times10^0 \\ \underline{-784} = & \underline{-\ 7\times10^2+8\times10^1+4\times10^0} \\ & 1\times10^2+4\times10^1+2\times10^0 \end{array}$$

or 142

55.
$$\begin{array}{rl} 5{,}238 = & 5\times10^3+2\times10^2+3\times10^1+8\times10^0 \\ \underline{-1{,}583} = & \underline{-\ 1\times10^3+5\times10^2+8\times10^1+3\times10^0} \end{array}$$

$$\begin{aligned} 5\times10^3+2\times10^2+3\times10^1+8\times10^0 &= 5\times10^3+(1+1)\times10^2+3\times10^1+8\times10^0 \\ &= 5\times10^3+1\times10^2+1\times10^2+3\times10^1+8\times10^0 \\ &= 5\times10^3+1\times10^2+10\times10^1+3\times10^1+8\times10^0 \\ &= 5\times10^3+1\times10^2+(10+3)\times10^1+8\times10^0 \\ &= 5\times10^3+1\times10^2+13\times10^1+8\times10^0 \end{aligned}$$

$$\begin{aligned} 5\times10^3+1\times10^2+13\times10^1+8\times10^0 &= (4+1)\times10^3+1\times10^2+13\times10^1+8\times10^0 \\ &= 4\times10^3+1\times10^3+1\times10^2+13\times10^1+8\times10^0 \\ &= 4\times10^3+10\times10^2+1\times10^2+13\times10^1+8\times10^0 \\ &= 4\times10^3+(10+1)\times10^2+13\times10^1+8\times10^0 \\ &= 4\times10^3+11\times10^2+13\times10^1+8\times10^0 \end{aligned}$$

$$\begin{array}{rl} 5{,}238 = & 4\times10^3+11\times10^2+13\times10^1+8\times10^0 \\ \underline{-1{,}583} = & \underline{-\ 1\times10^3+5\times10^2+8\times10^1+3\times10^0} \\ & 3\times10^3+6\times10^2+5\times10^1+5\times10^0 \end{array}$$

or 3,655

Note: In part b of the solutions to Exercises 57 – 63, the smaller number is placed below the larger in the conventional method.

57. a)

b)
$$\begin{array}{r} 235 \\ \times\ 4 \\ \hline 20 \\ 120 \\ \underline{800} \\ 940 \end{array}$$

59. a)

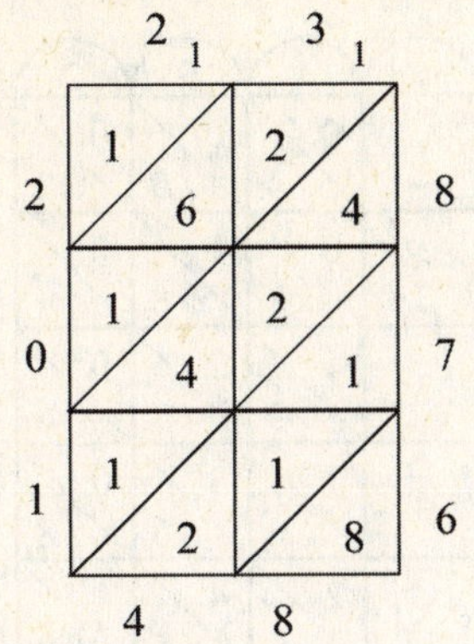

b)

$$\begin{array}{r} 876 \\ \times\ 23 \\ \hline 18 \\ {}_{1}210 \\ \overset{1}{2}400 \\ 120 \\ {}_{1}1400 \\ \overset{1}{1}6000 \\ \hline 20{,}148 \end{array}$$

18, 210, 2400: 876×3

120, 1400, 16000: 876×20

61. a)

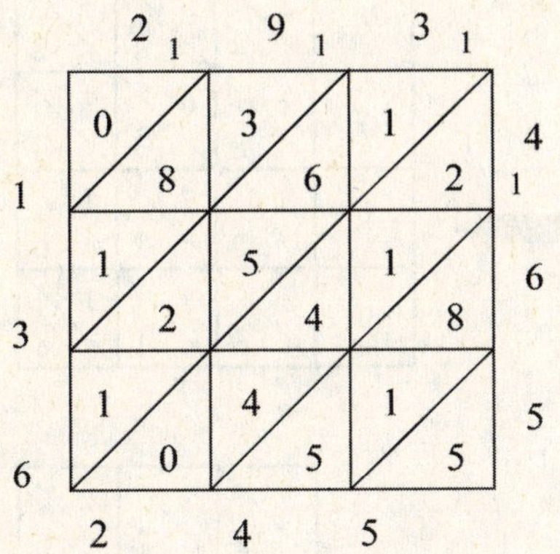

b)

$$\begin{array}{r} 465 \\ \times\ 293 \\ \hline {}_{1}15 \\ 180 \\ \overset{1}{1}200 \\ 450 \\ {}_{1}5400 \\ \overset{1}{3}6000 \\ 1000 \\ 12000 \\ 800000 \\ \hline 136{,}245 \end{array}$$

15, 180, 1200: 465×3

450, 5400, 36000: 465×90

1000, 12000, 800000: 465×200

63. $ad = 24$, $cd = 20$, $ae = 42$, and $be = 56 \Rightarrow ad = 6 \cdot 4$, $cd = 5 \cdot 4$, $ae = 6 \cdot 7$, and $be = 8 \cdot 7$

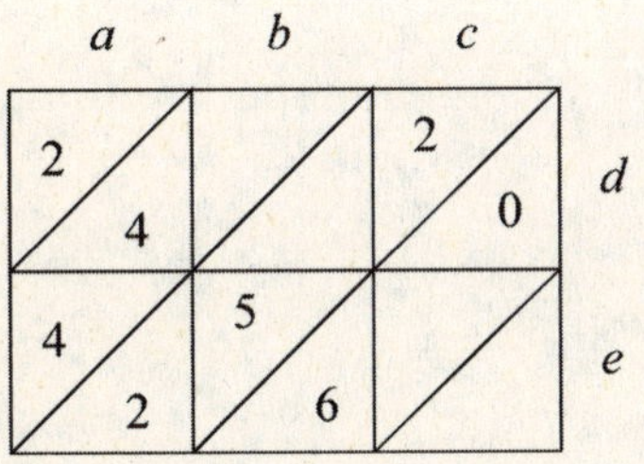

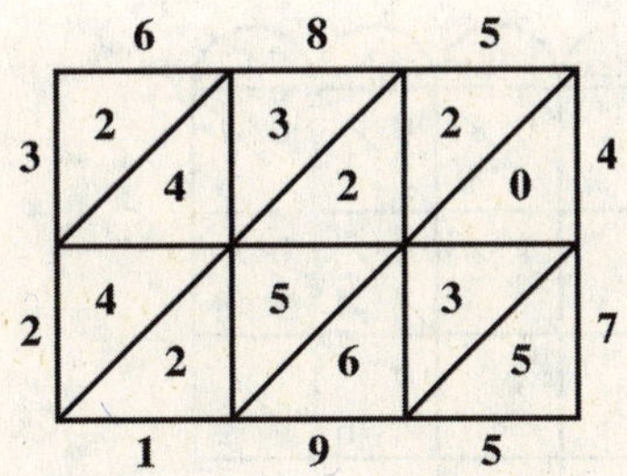

$685 \times 47 = 32{,}195$

65. 1,884

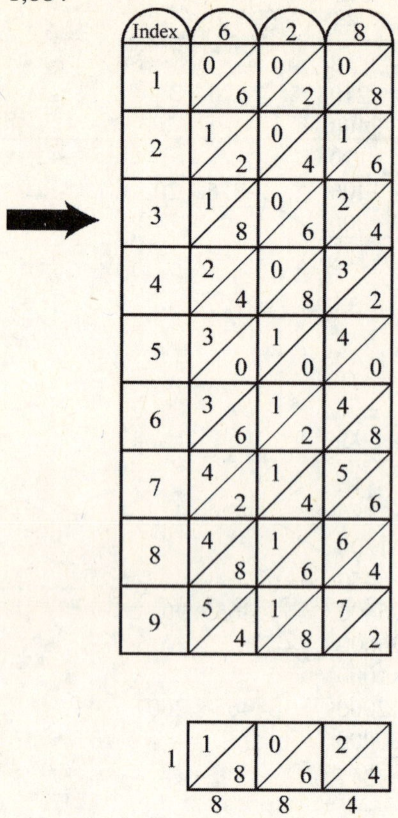

67. 3,936

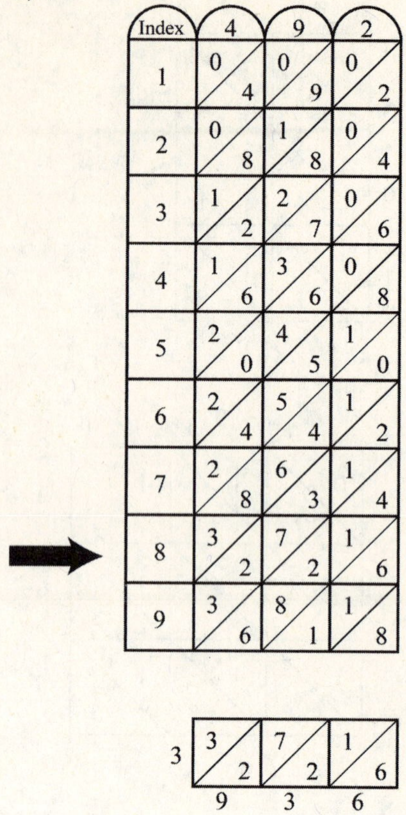

69. 5,544

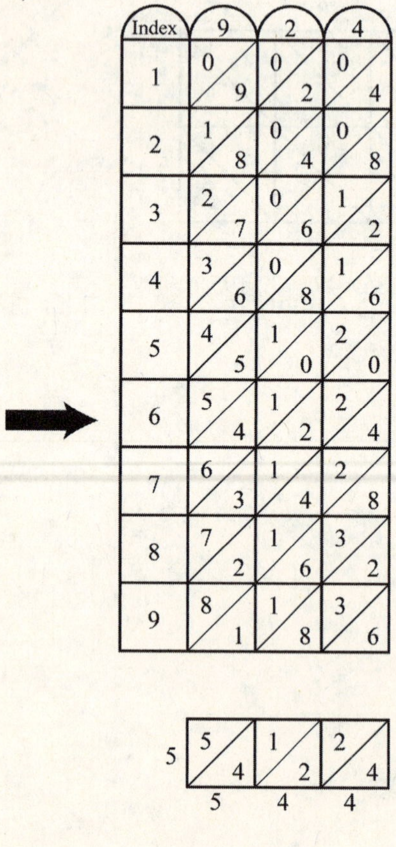

71. Answers may vary.

If you think of 324 as $300+20+4=3\cdot 100+2\cdot 10+4$ then you can use indices of 3, 2, and 4 for 615. You would position the rectangles to consider the place values of the 3, 2, and 4. Dashed boxes are included for ease in adding. The result is 199,260.

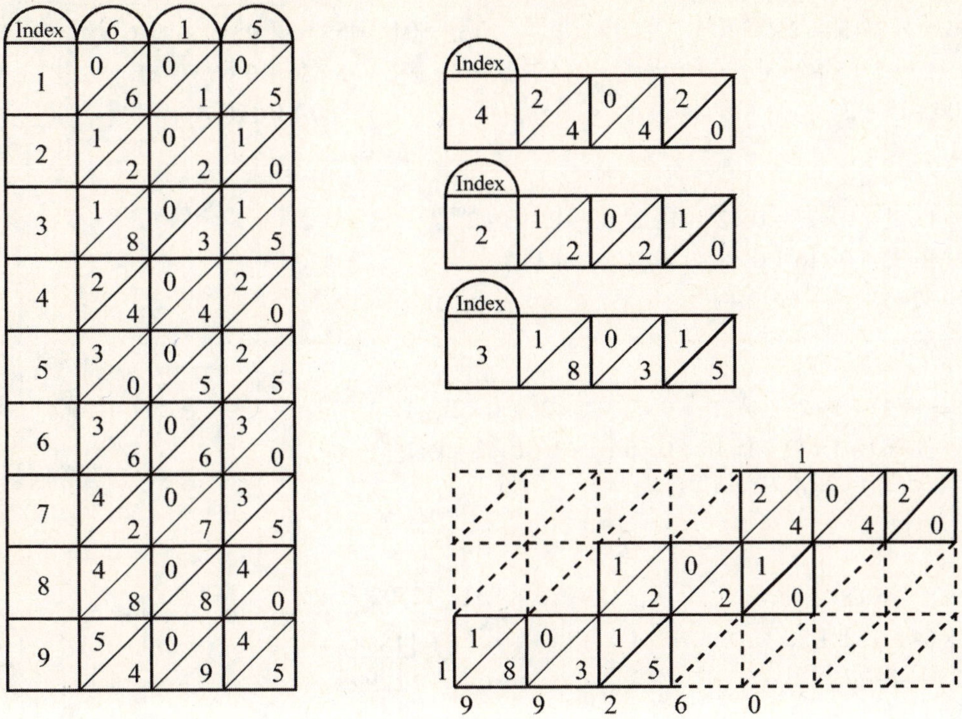

73. Numerals can be written more efficiently and there is no need for different symbols for different powers of the base.

75. In the galley method, all partial products are computed separately before being added to get the final product. In our method, several smaller partial products are combined into a single partial product before adding.

77. Answers may vary.

In writing , it could be difficult to tell whether there is one space or two spaces between the two groups of symbols. If there is one space, the number is $1\times 60+5=60+5=65$. If there are two spaces, the number is $1\times 60^2+5=3{,}600+5=3{,}605$.

79. Answers may vary.

The ancient Chinese system requires a separate symbol for each power of ten, e.g. one billion, ten billion, and so on. The Hindu–Arabic system uses different powers of 10, 10^9, 10^{10}, and so on. The positioning of symbols in the numeral indicates these powers of ten.

81. No solution provided.

83. 

85. 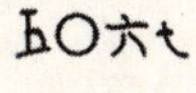

87. 

Section 5.3 Calculating in Other Bases

1. 4

3. five

5. It is based on ten and has place value.

7. 23_5 and 30_5

9. 1010_2 and 1100_2

11. EE_{16} and $F0_{16}$

13. $$\begin{aligned}432_5 &= 4\times5^2+3\times5^1+2\times5^0\\ &= 4\times25+3\times5+2\times1\\ &= 100+15+2\\ &= 117\end{aligned}$$

15. $$\begin{aligned}504_6 &= 5\times6^2+0\times6^1+4\times6^0\\ &= 5\times36+0\times6+4\times1\\ &= 180+0+4\\ &= 184\end{aligned}$$

17. $$\begin{aligned}100111_2 &= 1\times2^5+0\times2^4+0\times2^3+1\times2^2+1\times2^1+1\times2^0\\ &= 1\times32+0\times16+0\times8+1\times4+1\times2+1\times1\\ &= 32+0+0+4+2+1\\ &= 39\end{aligned}$$

19. $$\begin{aligned}1110101_2 &= 1\times2^6+1\times2^5+1\times2^4+0\times2^3+1\times2^2+0\times2^1+1\times2^0\\ &= 1\times64+1\times32+1\times16+0\times8+1\times4+0\times2+1\times1\\ &= 64+32+16+0+4+0+1\\ &= 117\end{aligned}$$

21. $$\begin{aligned}267_8 &= 2\times8^2+6\times8^1+7\times8^0\\ &= 2\times64+6\times8+7\times1\\ &= 128+48+7\\ &= 183\end{aligned}$$

23. $$\begin{aligned}704_8 &= 7\times8^2+0\times8^1+4\times8^0\\ &= 7\times64+0\times8+4\times1\\ &= 448+0+4\\ &= 452\end{aligned}$$

25. $$\begin{aligned}2F4_{16} &= 2\times16^2+15\times16^1+4\times16^0\\ &= 2\times256+15\times16+4\times1\\ &= 512+240+4\\ &= 756\end{aligned}$$

27. $$\begin{aligned}D08_{16} &= 13\times16^2+0\times16^1+8\times16^0\\ &= 13\times256+0\times16+8\times1\\ &= 3328+0+8\\ &= 3{,}336\end{aligned}$$

29. 2314_5

5	334	4
5	66	1
5	13	3
5	2	2
	0	

31. 12302_6

6	1838	2
6	306	0
6	51	3
6	8	2
6	1	1
	0	

33. 1100111_2

2	103	1
2	51	1
2	25	1
2	12	0
2	6	0
2	3	1
2	1	1
	0	

35. 1011110_2

2	94	0
2	47	1
2	23	1
2	11	1
2	5	1
2	2	0
2	1	1
	0	

37. 6513_8

8 | 3403	3
8 | 425	1
8 | 53	5
8 | 6	6
0	

39. $AE8_{16}$

16 | 2792	8
16 | 174	E (14)
16 | 10	A (10)
0	

41. DEA_{16}

16 | 3562	A (10)
16 | 222	E (14)
16 | 13	D (13)
0	

43.
$$\begin{array}{r} {}^{1} \\ 3412_5 \\ +\ \ 231_5 \\ \hline 4143_5 \end{array}$$

45.
$$\begin{array}{r} {}^{1\ 11} \\ 4013_5 \\ +\ 1242_5 \\ \hline 10310_5 \end{array}$$

47.
$$\begin{array}{r} {}^{1} \\ 2A18_{16} \\ +\ \ 43B_{16} \\ \hline 2E53_{16} \end{array}$$

49.
$$\begin{array}{r} {}^{11111} \\ 11011_2 \\ +\ 10101_2 \\ \hline 110000_2 \end{array}$$

51.
$$\begin{array}{r} {}^{3}_{1} \\ 2\not{4}12_5 \\ -\ \ 321_5 \\ \hline 2041_5 \end{array}$$

53.
$$\begin{array}{r} {}^{7}_{1} \\ A\not{8}3_{16} \\ -\ \ 43B_{16} \\ \hline 648_{16} \end{array}$$

55.
$$\begin{array}{r} {}^{0}_{1} \\ 11\not{1}011_2 \\ -\ 10101_2 \\ \hline 100110_2 \end{array}$$

57.
$$\begin{array}{r} 41_5 \\ \times\ 23_5 \\ \hline {}_{1}223_5 \\ 132 \\ \hline 2043_5 \end{array}$$

59.
$$\begin{array}{r} {}^{1}_{1} \\ 302_5 \\ \times\ 43_5 \\ \hline {}^{1}1411_5 \\ 2213 \\ \hline 24041_5 \end{array}$$

61. $114_5\ R\ 11_5$

$$\begin{array}{r} 114 \\ 24_5\,\overline{)\,3412_5} \\ \underline{24} \\ 101 \\ \underline{24} \\ 222 \\ \underline{211} \\ 11 \end{array}$$

63. $44_5\ R\ 24_5$

$$\begin{array}{r} 44 \\ 42_5\,\overline{)\,4132_5} \\ \underline{323} \\ 402 \\ \underline{323} \\ 24 \end{array}$$

65. 1355_8 $2ED_{16}$

1 011 101 101   10 1110 1101
1 3 5 5   2 E D

67. 1751_8 $3E9_{16}$

1 111 101 001   11 1110 1001
1 7 5 1   3 E 9

69. 10100110_2

2 4 6
010 100 110

71. 101000111110_2

A 3 E
1010 0011 1110

73. 754_{16}

3 5 2 4
011 101 010 100
0111 0101 0100
7 5 4

75. 1654_8

3 *A* *C*
0011 1010 1100
001 110 101 100
1 6 5 4

For Exercises 77 – 79 refer to the following table.

65	66	67	68	69	70	71	72	73	74	75	76	77
A	B	C	D	E	F	G	H	I	J	K	L	M

78	79	80	81	82	83	84	85	86	87	88	89	90
N	O	P	Q	R	S	T	U	V	W	X	Y	Z

77. CANDY

1000011	1000001	1001110	1000100	1011001
$2^6+2^1+2^0=$	$2^6+2^0=$	$2^6+2^3+2^2+2^1=$	$2^6+2^2=$	$2^6+2^4+2^3+2^0=$
$64+2+1=67$	$64+1=65$	$64+8+4+2=78$	$64+4=68$	$64+16+8+1=89$

79. LOVE

1001100	1001111	1010110	1000101
$2^6+2^3+2^2=$	$2^6+2^3+2^2+2^1+2^0=$	$2^6+2^4+2^2+2^1=$	$2^6+2^2+2^0=$
$64+8+4=76$	$64+8+4+2+1=79$	$64+16+4+2=86$	$64+4+1=69$

81. 60; 5

83. Base 8: 64 facts; The eight symbols 0, 1, … 7 are combined with themselves.
Base 16: 256 facts; The sixteen symbols 0, 1, …9, *A*, *B*, … *F* are combined with themselves.

85. The number 10 always represents the base, which is 16 in this case and a single symbol was needed to represent 10.

87. – 89. No solution provided.

91. 4123_5

$$\begin{aligned}201221_3 &= 2\times3^5+0\times3^4+1\times3^3+2\times3^2+2\times3^1+1\times3^0\\ &= 2\times243+0\times81+1\times27+2\times9+2\times3+1\times1\\ &= 486+27+18+6+1\\ &= 538\end{aligned}$$

Divisor	Quotient	Remainder
5	538	3
5	107	2
5	21	1
5	4	4
	0	

93. 3774_9

$$\begin{aligned} B05_{16} &= 11\times16^2 + 0\times16^1 + 5\times16^0 \\ &= 11\times256 + 0\times16 + 5\times1 \\ &= 3072 + 5 \\ &= 2821 \end{aligned}$$

$$\begin{array}{r|l l} 9 & 2821 & 4 \\ 9 & 313 & 7 \\ 9 & 34 & 7 \\ 9 & 3 & 3 \\ & 0 & \end{array}$$

95. If you choose A and B to represent 10 and 11, respectively, in base twelve then the first twenty-five counting numbers would be

$$1, 2, 3, 4, 5, 6, 7, 8, 9, A, B, 10_{12}, 11_{12}, 12_{12}, 13_{12},$$
$$14_{12}, 15_{12}, 16_{12}, 17_{12}, 18_{12}, 19_{12}, 1A_{12}, 1B_{12}, 20_{12}, 21_{12}.$$

Another choice that is commonly used is T and E to represent 10 and 11, respectively, in base twelve.

97. $\oplus\mp\neq\$ = 1023_6 = 1\times6^3 + 0\times6^2 + 2\times6^1 + 1\times6^0$
$= 1\times216 + 0\times36 + 2\times6 + 3\times1 = 216 + 0 + 12 + 3 = 231$

99.
$$\begin{array}{r} {\scriptstyle\oplus\ \ \oplus} \\ \oplus\neq\mp\% \\ +\quad \%\,\$\,@ \\ \hline \neq\oplus@\$ \end{array}$$

Section 5.4 Looking Deeper: Modular Systems

1. In doing modulo 12 computations, all multiples of 12 are the same as 0 on a 12 hour clock.

3. Answers may vary. One example is –3, which means moving counterclockwise 3 positions. This is the same as moving clockwise 9 positions so –3 and 9 are the same on a 12-hour clock.

5. 7 since $43 = 3\times12 + 7$

7. 4 since $39 = 5\cdot7 + 4$

9. True since $59 - 35 = 24$ is evenly divisible by 12.

11. True since $43 - 11 = 32$ is evenly divisible by 8.

13. False since $78 - 46 = 32$ is not evenly divisible by 10.

15. 5 since $8 + 9 \equiv 5 \pmod{12}$

17. 2 since $3 + 4 \equiv 2 \pmod{5}$

19. 7 since $4 - 9 \equiv 7 \pmod{12}$

21. 4 since $3 - 4 \equiv 4 \pmod{5}$

23. 0 since $4\times9 \equiv 0 \pmod{12}$

25. 3 since $2\times5 \equiv 3 \pmod{7}$

27. 10

$3+0 \equiv 1 \pmod{12}$; false

$3+1 \equiv 1 \pmod{12}$; false

$3+2 \equiv 1 \pmod{12}$; false

$3+3 \equiv 1 \pmod{12}$; false

$3+4 \equiv 1 \pmod{12}$; false

$3+5 \equiv 1 \pmod{12}$; false

$3+6 \equiv 1 \pmod{12}$; false

$3+7 \equiv 1 \pmod{12}$; false

$3+8 \equiv 1 \pmod{12}$; false

$3+9 \equiv 1 \pmod{12}$; false

$3+10 \equiv 1 \pmod{12}$; **true**

$3+11 \equiv 1 \pmod{12}$; false

29. 6

$6+0 \equiv 4 \pmod{8}$; false

$6+1 \equiv 4 \pmod{8}$; false

$6+2 \equiv 4 \pmod{8}$; false

$6+3 \equiv 4 \pmod{8}$; false

$6+4 \equiv 4 \pmod{8}$; false

$6+5 \equiv 4 \pmod{8}$; false

$6+6 \equiv 4 \pmod{8}$; **true**

$6+7 \equiv 4 \pmod{8}$; false

31. 6

$3-0 \equiv 4 \pmod{7}$; false

$3-1 \equiv 4 \pmod{7}$; false

$3-2 \equiv 4 \pmod{7}$; false

$3-3 \equiv 4 \pmod{7}$; false

$3-4 \equiv 4 \pmod{7}$; false

$3-5 \equiv 4 \pmod{7}$; false

$3-6 \equiv 4 \pmod{7}$; **true**

33. 3

$2-0 \equiv 8 \pmod{9}$; false

$2-1 \equiv 8 \pmod{9}$; false

$2-2 \equiv 8 \pmod{9}$; false

$2-3 \equiv 8 \pmod{9}$; **true**

$2-4 \equiv 8 \pmod{9}$; false

$2-5 \equiv 8 \pmod{9}$; false

$2-6 \equiv 8 \pmod{9}$; false

$2-7 \equiv 8 \pmod{9}$; false

$2-8 \equiv 8 \pmod{9}$; false

35. 4

$3\times 0 \equiv 5 \pmod{7}$; false

$3\times 1 \equiv 5 \pmod{7}$; false

$3\times 2 \equiv 5 \pmod{7}$; false

$3\times 3 \equiv 5 \pmod{7}$; false

$3\times 4 \equiv 5 \pmod{7}$; **true**

$3\times 5 \equiv 5 \pmod{7}$; false

$3\times 6 \equiv 5 \pmod{7}$; false

37. 3, 8

$4\times 0 \equiv 2 \pmod{10}$; false

$4\times 1 \equiv 2 \pmod{10}$; false

$4\times 2 \equiv 2 \pmod{10}$; false

$4\times 3 \equiv 2 \pmod{10}$; **true**

$4\times 4 \equiv 2 \pmod{10}$; false

$4\times 5 \equiv 2 \pmod{10}$; false

$4\times 6 \equiv 2 \pmod{10}$; false

$4\times 7 \equiv 2 \pmod{10}$; false

$4\times 8 \equiv 2 \pmod{10}$; **true**

$4\times 9 \equiv 2 \pmod{10}$; false

39. For $x \equiv 3 \pmod{7}$, x is in the congruence class containing 3 modulo 7 which is

3, 10, 17, **24**, 31, 38, ...

For $x \equiv 4 \pmod{5}$, x is in the congruence class containing 4 modulo 5 which is

4, 9, 14, 19, **24**, 29, ...

The smallest number that is found in both of these classes is 24.

41. For $x \equiv 2 \pmod{10}$, x is in the congruence class containing 2 modulo 10 which is

2, 12, 22, 32, **42**, 52, 62, 72...

For $x \equiv 2 \pmod{8}$, x is in the congruence class containing 2 modulo 8 which is

2, 10, 18, 26, 34, **42**, 50, 58 ...

For $x \equiv 0 \pmod{6}$, x is in the congruence class containing 0 modulo 6 which is

0, 6, 12, 18, 24, 30, 36, **42**, ...

The smallest number that is found in these three classes is 42.

43. 3 since $80 = 11 \cdot 7 + 3$

45. This problem is the same as finding the smallest positive integer that solves both $x \equiv 4 \pmod{6}$ and $x \equiv 3 \pmod{5}$.

For $x \equiv 4 \pmod{6}$, x is in the congruence class containing 4 modulo 6 which is

4, 10, 16, 22, **28**, 34, ...

For $x \equiv 3 \pmod{5}$, x is in the congruence class containing 3 modulo 5 which is

3, 8, 13, 18, 23, **28**, ...

The smallest number that is found in both of these classes is 28.

47. We are looking for years that are congruent to 2 modulo 12. Since this book was written in 2010 and Chris appears to be less than 30, we will start in the year $2010 - 29 = 1981$.

$1981 \equiv 2 \pmod{12}$; false
$1982 \equiv 2 \pmod{12}$; **true**
$1983 \equiv 2 \pmod{12}$; false
$1984 \equiv 2 \pmod{12}$; false
$1985 \equiv 2 \pmod{12}$; false
$1986 \equiv 2 \pmod{12}$; false
$1987 \equiv 2 \pmod{12}$; false
$1988 \equiv 2 \pmod{12}$; false
$1989 \equiv 2 \pmod{12}$; false
$1990 \equiv 2 \pmod{12}$; false
$1991 \equiv 2 \pmod{12}$; false
$1992 \equiv 2 \pmod{12}$; false

Chris was born in 1982.

Note: If we use 2011 for the starting year, the answer will still be 1982. If a starting year after 2011 is used, the answer will be 1982 + 12 = 1994.

49. 5

10×0	0
9×7	63
8×4	32
7×3	21
6×5	30
$5\times d$	$5d$
4×4	16
3×4	12
2×5	10
Total	$184+5d$

Since 0 is the check digit, we need to find the smallest integer that satisfies $11-0=0\equiv(184+5d)(\text{mod}\,11)$.

$0\equiv(184+5\times 0)(\text{mod}\,11)$; false
$0\equiv(184+5\times 1)(\text{mod}\,11)$; false
$0\equiv(184+5\times 2)(\text{mod}\,11)$; false
$0\equiv(184+5\times 3)(\text{mod}\,11)$; false
$0\equiv(184+5\times 4)(\text{mod}\,11)$; false
$0\equiv(184+5\times 5)(\text{mod}\,11)$; **true**
$0\equiv(184+5\times 6)(\text{mod}\,11)$; false
$0\equiv(184+5\times 7)(\text{mod}\,11)$; false
$0\equiv(184+5\times 8)(\text{mod}\,11)$; false
$0\equiv(184+5\times 9)(\text{mod}\,11)$; false
$0\equiv(184+5\times 10)(\text{mod}\,11)$; false

51. 6

10×0	0
9×0	0
8×6	48
7×1	7
6×3	18
5×4	20
4×9	36
$3\times d$	$3d$
2×0	0
Total	$129+3d$

Since 7 is the check digit, we need to find the smallest integer that satisfies $11-7=4\equiv(129+3d)(\text{mod}\,11)$.

$4\equiv(129+3\times 0)(\text{mod}\,11)$; false
$4\equiv(129+3\times 1)(\text{mod}\,11)$; false
$4\equiv(129+3\times 2)(\text{mod}\,11)$; false
$4\equiv(129+3\times 3)(\text{mod}\,11)$; false
$4\equiv(129+3\times 4)(\text{mod}\,11)$; false
$4\equiv(129+3\times 5)(\text{mod}\,11)$; false
$4\equiv(129+3\times 6)(\text{mod}\,11)$; **true**
$4\equiv(129+3\times 7)(\text{mod}\,11)$; false
$4\equiv(129+3\times 8)(\text{mod}\,11)$; false
$4\equiv(129+3\times 9)(\text{mod}\,11)$; false
$4\equiv(129+3\times 10)(\text{mod}\,11)$; false

53. Answers may vary. One possible answer is to perform the operations as usual and then count off the result on an *m*-hour clock.

55. Solve both congruences separately as in Exercise 54 and then find which solutions are common to both congruences.

57 – 59. No solution provided.

61. $12^5\ (\text{mod}\ 35)=248832\ (\text{mod}\ 35)=17$

63. $22^5\ (\text{mod}\ 35)=5153632\ (\text{mod}\ 35)=22$

65. Explanations may vary.
In this modulo 8 system, find 6 times what number (or numbers) is (are) congruent to 4.
If $4/6 \equiv x(\text{mod}\,8)$ then $4 \equiv 6x\,(\text{mod}\,8)$

$4 \equiv (6\times0)\,(\text{mod } 8)$; false
$4 \equiv (6\times1)\,(\text{mod } 8)$; false
$4 \equiv (6\times2)\,(\text{mod } 8)$; **true**
$4 \equiv (6\times3)\,(\text{mod } 8)$; false
$4 \equiv (6\times4)\,(\text{mod } 8)$; false
$4 \equiv (6\times5)\,(\text{mod } 8)$; false
$4 \equiv (6\times6)\,(\text{mod } 8)$; **true**
$4 \equiv (6\times7)\,(\text{mod } 8)$; false Answer: 4, 6

a) If $5/3\,(\text{mod}\,8) \equiv x$ then $5 \equiv 3x\,(\text{mod}\,8)$

$5 \equiv (3\times0)\,(\text{mod } 8)$; false
$5 \equiv (3\times1)\,(\text{mod } 8)$; false
$5 \equiv (3\times2)\,(\text{mod } 8)$; false
$5 \equiv (3\times3)\,(\text{mod } 8)$; false
$5 \equiv (3\times4)\,(\text{mod } 8)$; false
$5 \equiv (3\times5)\,(\text{mod } 8)$; false
$5 \equiv (3\times6)\,(\text{mod } 8)$; false
$5 \equiv (3\times7)\,(\text{mod } 8)$; **true** Answer: 7

b) If $2/6\,(\text{mod}\,8) \equiv x$ then $2 \equiv 6x\,(\text{mod}\,8)$

$2 \equiv (6\times0)\,(\text{mod } 8)$; false
$2 \equiv (6\times1)\,(\text{mod } 8)$; false
$2 \equiv (6\times2)\,(\text{mod } 8)$; false
$2 \equiv (6\times3)\,(\text{mod } 8)$; **true**
$2 \equiv (6\times4)\,(\text{mod } 8)$; false
$2 \equiv (6\times5)\,(\text{mod } 8)$; false
$2 \equiv (6\times6)\,(\text{mod } 8)$; false
$2 \equiv (6\times7)\,(\text{mod } 8)$; **true** Answer: 3, 7

Chapter Review Exercises

1. $1{,}000{,}000 + 200{,}000 + 30{,}000 + 2{,}000 + 200 + 10 = 1{,}232{,}210$

3. $37 = 1 + 4 + 32$

Powers of 2	Times 53
1	53
2	53 + 53 = 106
4	106 + 106 = 212
8	212 + 212 = 424
16	424 + 424 = 848
32	848 + 848 = 1,696

1,696 + 212 +53 = 1,961

5. $7\times1{,}000 + 5\times100 + 9\times10 + 3 = 7{,}000 + 500 + 90 + 3 = 7{,}593$

7. Answers may vary.
The Roman system had the subtraction principle, which allowed them to write certain numbers more efficiently. They also had a multiplication principle.

9. $11{,}292 = 3\times 60^2 + 8\times 60 + 12$  or

$11{,}292 = 3\times 60^2 + (10-2)\times 60 + 12$ 

11. 4,738

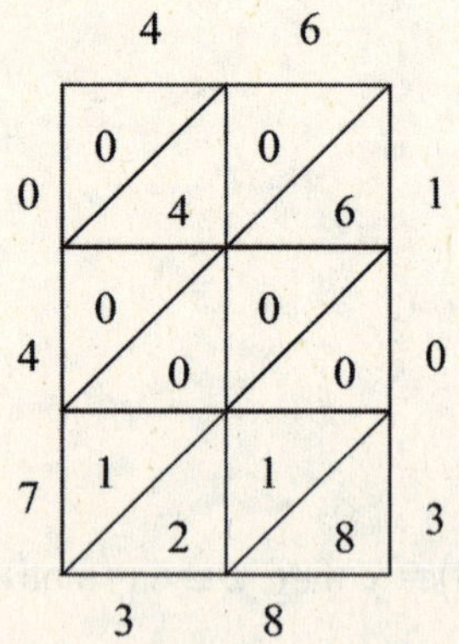

13. a) $1\times 20\times 18 + 15 = 360 + 15 = 375$

b) $3\times 20\times 18\times 20 + 18\times 20\times 18 = 21{,}600 + 6{,}480 = 28{,}080$

15. 6513_8

8 | 3403   3
8 | 425   1
8 | 53   5
8 | 6   6
0

17. 134_5 R 20_5

$$\begin{array}{r} 134 \\ 23_5\,\overline{)4312_5} \\ \underline{23} \\ 201 \\ \underline{124} \\ 222 \\ \underline{202} \\ 20 \end{array}$$

19. 1431_5

$$\begin{aligned} 463_7 &= 4\times 7^2 + 6\times 7^1 + 3\times 7^0 \\ &= 4\times 49 + 6\times 7 + 3\times 1 \\ &= 196 + 42 + 3 \\ &= 241 \end{aligned}$$

5 | 241   1
5 | 48   3
5 | 9   4
5 | 1   1
0

21. a) True since $72 - 54 = 18$ is evenly divisible by 6.

b) False since $75 - 29 = 46$ is not evenly divisible by 11.

23. a) 3, 7, 11

$3\times0\equiv9\pmod{12}$; false
$3\times1\equiv9\pmod{12}$; false
$3\times2\equiv9\pmod{12}$; false
$3\times3\equiv9\pmod{12}$; **true**
$3\times4\equiv9\pmod{12}$; false
$3\times5\equiv9\pmod{12}$; false
$3\times6\equiv9\pmod{12}$; false
$3\times7\equiv9\pmod{12}$; **true**
$3\times8\equiv9\pmod{12}$; false
$3\times9\equiv9\pmod{12}$; false
$3\times10\equiv9\pmod{12}$; false
$3\times11\equiv9\pmod{12}$; **true**

b) $\varnothing$

$4\times0\equiv3\pmod{8}$; false
$4\times1\equiv3\pmod{8}$; false
$4\times2\equiv3\pmod{8}$; false
$4\times3\equiv3\pmod{8}$; false
$4\times4\equiv3\pmod{8}$; false
$4\times5\equiv3\pmod{8}$; false
$4\times6\equiv3\pmod{8}$; false
$4\times7\equiv3\pmod{8}$; false

Chapter Test

1. MMMDCLXXXV

3. $$\begin{aligned}264_7 &= 2\times7^2+6\times7^1+4\times7^0\\ &=2\times49+6\times7+4\times1\\ &=98+42+4\\ &=144\end{aligned}$$ and $$\begin{aligned}A3E_{16} &= 10\times16^2+3\times16^1+14\times16^0\\ &=10\times256+3\times16+14\times1\\ &=2560+48+14\\ &=2{,}622\end{aligned}$$

5. a) $1\times20\times18+19=360+19=379$

b) $3\times20\times18+12\times20=1{,}080+240=1{,}320$

7.

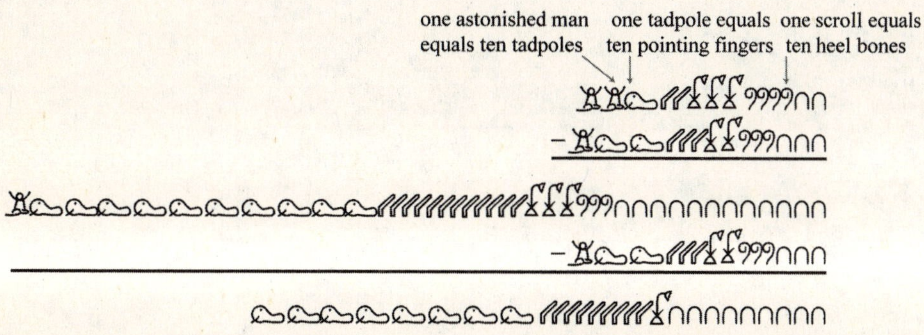

9. 3 since $57=9\times6+3$

11.
$$\begin{array}{r} 101101_2\\ +110101_2\\ \hline 1100010_2\end{array}$$

13. 15,946

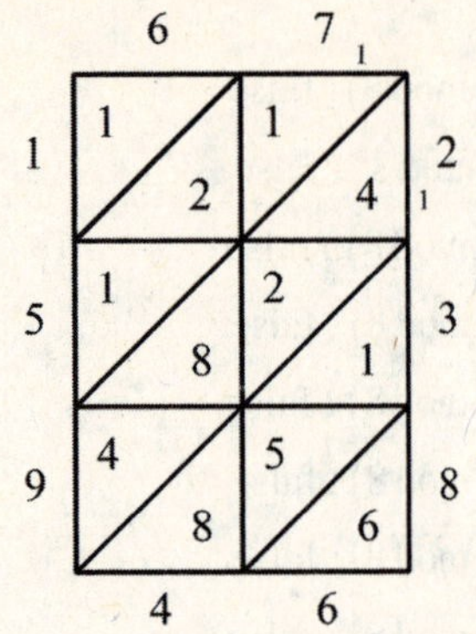

15. a) 8 since $9+12\equiv 8 \pmod{13}$

b) 8 since $6-10\equiv 8 \pmod{12}$

c) 2 since $5\times 4\equiv 2 \pmod{6}$

17. Answers may vary. Possible answers include: More symbols were required to write their numerals.

19. 110_5 $R2_5$

$$\begin{array}{r} 110_5 \\ 24_5\overline{)3142_5} \\ \underline{24} \\ 24 \\ \underline{24} \\ \underline{2} \\ 2 \end{array}$$

21. 6536_8 $\quad D5E_{16}$

110 101 011 110
6 5 3 6

1101 0101 1110
D 5 *E*

Chapter 6

Number Theory and the Real Number System: Understanding the Numbers All Around Us

Section 6.1 Number Theory

1. The square root of 83 is between 9 and 10. Since 7 is the largest prime less than 10, there will be no prime factors.

3. When each branch ended in a prime number.

5. He made many contributions to geography.

7. true

9. false

11. false

13. true

15. 53, 59, 61, 67, 71, 73, 79, 83, 89, 97

17. Since $81 < 95 < 100$, $9 < \sqrt{95} < 10$. Thus $a = 9$.

19. Since $144 < 153 < 169$, $12 < \sqrt{153} < 13$. Thus $a = 12$.

21. 231 is composite. Possible factorizations into the product of two natural numbers include 1×231, 3×77, 7×33, and 11×21.

For Exercise 15 refer to the following.

~~1~~	2	3	~~4~~	5	~~6~~	7	~~8~~	~~9~~	~~10~~
11	~~12~~	13	~~14~~	~~15~~	~~16~~	17	~~18~~	19	~~20~~
~~21~~	~~22~~	23	~~24~~	~~25~~	~~26~~	~~27~~	~~28~~	29	~~30~~
31	~~32~~	~~33~~	~~34~~	~~35~~	~~36~~	37	~~38~~	~~39~~	~~40~~
41	~~42~~	43	~~44~~	~~45~~	~~46~~	47	~~48~~	~~49~~	~~50~~
~~51~~	~~52~~	53	~~54~~	~~55~~	~~56~~	~~57~~	~~58~~	59	~~60~~
61	~~62~~	~~63~~	~~64~~	~~65~~	~~66~~	67	~~68~~	~~69~~	~~70~~
71	~~72~~	73	~~74~~	~~75~~	~~76~~	~~77~~	~~78~~	79	~~80~~
~~81~~	~~82~~	83	~~84~~	~~85~~	~~86~~	~~87~~	~~88~~	89	~~90~~
~~91~~	~~92~~	~~93~~	~~94~~	~~95~~	~~96~~	97	~~98~~	~~99~~	~~100~~
101	~~102~~	103	~~104~~	~~105~~	~~106~~	107	~~108~~	109	~~110~~
~~111~~	~~112~~	113	~~114~~	~~115~~	~~116~~	~~117~~	~~118~~	~~119~~	~~120~~

23. 113 is prime.

25. 197 is prime.

27. 119 is composite. Possible factorizations into the product of two natural numbers include 1×119 and 7×17.

29. 141,270 is divisible by **2** because the last digit, 0, is divisible by 2.
141,270 is divisible by **3** because the sum of the digits, 15, is divisible by 3.
141,270 is not divisible by 4 because the number formed by the last two digits, 70, is not divisible by 4.
141,270 is divisible by **5** because the last digit, 0, is a 0 or a 5.
141,270 is divisible by **6** because the number is divisible by both 2 and 3.
141,270 is not divisible by 8 because the number formed by the last three digits, 270, is not divisible by 8.
141,270 is not divisible by 9 because the sum of the digits, 15, is not divisible by 9.
141,270 is divisible by **10** because the last digit, 0, is 0.

31. 47,385 is not divisible by 2 because the last digit, 5, is not divisible by 2.
47,385 is divisible by **3** because the sum of the digits, 27, is divisible by 3.
47,385 is not divisible by 4 because the number formed by the last two digits, 85, is not divisible by 4.
47,385 is divisible by **5** because the last digit, 5, is a 0 or a 5.
47,385 is not divisible by 6 because the number is not divisible by both 2 and 3.
47,385 is not divisible by 8 because the number formed by the last three digits, 385, is not divisible by 8.
47,385 is divisible by **9** because the sum of the digits, 27, is divisible by 9.
47,385 is not divisible by 10 because the last digit, 5, is not 0.

33. Since any number divisible by both 2 and 3 is also divisible by 6, the smallest possible number would be $2\times3\times5=30$.

35. Since 10 is the product of 2 and 5, any number divisible by 4 (2×2) and 5 would be divisible by 10. The smallest possible number would be $4\times5=20$.

37. Answers may vary. Possible values for *a* are 4, 12, 20, 28, 36, etc.

39. Answers may vary. Possible values for *a* are 20, 60, 100, 140, 180, etc.

41. $980=2^2\times5\times7^2$

43. $9900=2^2\times3^2\times5^2\times11$

45. $621=3^3\times23$

47. $319=11\times29$

49. $20=2^2\times5$ and $24=2^3\times3$
$GCD(20,24)=2^2=4$
$LCM(20,24)=2^3\times3\times5=120$

51. $56=2^3\times7$ and $70=2\times5\times7$
$GCD(56,70)=2\times7=14$
$LCM(56,70)=2^3\times5\times7=280$

53. $216=2^3\times3^3$ and $288=2^5\times3^2$
$GCD(216,288)=2^3\times3^2=72$
$LCM(216,288)=2^5\times3^3=864$

55. $147=3\times7^2$ and $567=3^4\times7$
$GCD(147,567)=3\times7=21$
$LCM(147,567)=3^4\times7^2=3969$

57. 28

$$\begin{array}{r} 2 \\ 280\,\overline{)\,588} \\ \underline{560} \\ \textcircled{28} \end{array} \qquad \begin{array}{r} 10 \\ 28\,\overline{)\,280} \\ \underline{280} \\ 0 \end{array}$$

59. 45

$$\begin{array}{r} 3 \\ 495\,\overline{)\,1575} \\ \underline{1485} \\ 90 \end{array} \qquad \begin{array}{r} 5 \\ 90\,\overline{)\,495} \\ \underline{450} \\ \textcircled{45} \end{array} \qquad \begin{array}{r} 2 \\ 45\,\overline{)\,90} \\ \underline{24} \\ 0 \end{array}$$

61. $GCD(12,27)=3$

$$\begin{array}{r} 2 \\ 12\,\overline{)\,27} \\ \underline{24} \\ \textcircled{3} \end{array} \qquad \begin{array}{r} 4 \\ 3\,\overline{)\,12} \\ \underline{12} \\ 0 \end{array}$$

$$LCM(12,27)=\frac{12\cdot27}{3}=4\cdot27=108$$

63. $GCD(90,120)=30$

$$\begin{array}{r} 1 \\ 90\,\overline{)\,120} \\ \underline{90} \\ \textcircled{30} \end{array} \qquad \begin{array}{r} 3 \\ 30\,\overline{)\,90} \\ \underline{90} \\ 0 \end{array}$$

$$LCM(90,120)=\frac{90\cdot120}{30}=3\cdot120=360$$

65. $LCM(13,17) = 221$, so both types of cicadas will emerge at the same time in $2010 + 221 = 2231$.

67. In this exercise we need to find $LCM(20,35)$ which is 140.; 140 minutes

69. In this exercise we need to find $GCD(30,36)$ which is 6.; 6 packages

71. In this exercise we need to find $GCD(21,33)$ which is 3.; 3 feet by 3 feet

73. In this exercise we need to find $LCM(48,54)$ which is 432.; 432 hours

75. 17

77. The largest powers of any primes that occur in either number are multiplied.

79. Answers may vary.
Four will always divide a multiple of 100, so all we have to do to check if four divides a number is to determine if four divides the number formed by the last two digits. So four automatically divides 36,800. Therefore, we only have to check if four divides 24, which it does.

81. In each case, the two numbers have common factors greater than one.

83. not prime

85. prime

87. i) $\frac{9{,}808{,}358}{46 \cdot 90} \approx 2369.17$ pages.

ii) $2369.17 \times 11.5 = 27245.45$ inches; $\frac{27245.45}{12} = 2270.45$ feet.

iii) $\frac{2270.45}{5280} = 0.43$ miles.

89. If you refer to the table for Exercise 15, you can easily determine that the next three twin primes are: 59 and 61; 71 and 73; 101 and 103.

91. No; Counterexamples may vary.
Both 4 and 6 divide 12, but $4 \times 6 = 24$, which does not divide 12.

93. Fifteen will divide a number if both three and five divide the number.

Section 6.2 The Integers

1. They are balanced on opposite sides of zero.

3. If for the past three days, you spent five dollars, then three days ago you have $15 more.

5. They dealt primarily with geometry which only involved positive quantities.

7. $(+8)+(-5)=+3$

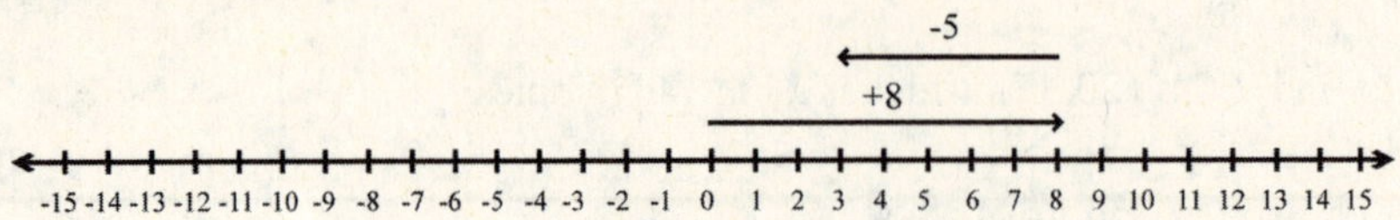

9. $(+4)+(+6)=+10$

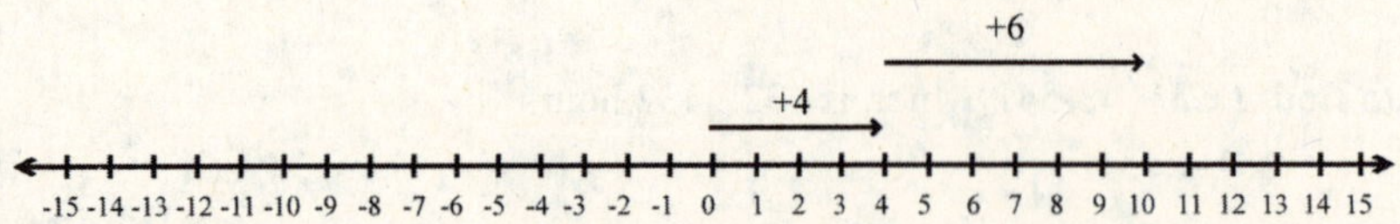

11. It is not practical to draw a number line when calculating with such large numbers; however, you can still imagine movements along the number line. If you visualize moving 128 units to the left and then 137 units to the right, you can see that the net effect is that you moved $137-128=9$ units to the right, so $(-128)+(+137)=+9$.

13. It is not practical to draw a number line when calculating with such large numbers; however, you can still imagine movements along the number line. If you visualize moving 57 units to the right and then 38 units to the left, you can see that the net effect is that you moved $57-38=19$ units to the right, so $(+57)+(-38)=+19$.

15. $(+18)-(-5)=(+18)+(+5)=+23$

17. $(+6)-(+19)=(+6)+(-19)=-13$

19. $(-28)-(+37)=(-28)+(-37)=-65$

21. $(+32)-(-18)=(+32)+(+18)=+50$

23. $(-5)(-7)=+35$

25. $(-7)(+8)=-56$

27. $(+8)(+6)=+48$

29. $(+19)(-2)=-38$

31. $(-9)(+7)=-63$

33. $(-8)(-9)=+72$

35. $+16=(-2)\cdot c$; $c=-8$

37. $-14=(-2)\cdot c$; $c=+7$

39. -3

41. $+15$

43. $+5$

45. -6

47. $+6$

49. -4

51. $(-2)(-3)+(+4)(-8)=$
$(+6)+(-32)=-26$

53. $(-4)(-2)+(+4)(-5)=$
$(+8)+(-20)=-12$

55. $(-6-(-14))\cdot\left(\dfrac{-12+(-6)}{7-4}\right)=(+8)\cdot\left(\dfrac{-18}{3}\right)=(+8)\cdot(-6)=-48$

57. $\left(\frac{-1-(-5)}{4-6}\right)\cdot\left(\frac{-4+(-2)}{-5+3}\right)=\left(\frac{4}{-2}\right)\cdot\left(\frac{-6}{-2}\right)=(-2)\cdot(3)=-6$

59. true

61. false; Counterexamples may vary; $(-5)+(+8)=+3$

63. false; counterexamples may vary; $\frac{-6}{-3}=+2$

65. true

67. $29,028-(-28,232)=57,260$ feet

69. $(-28,232)-(-35,433)=7,201$ feet

71. $136-(-129)=265$ degrees

73. $59-(-7)=66$ floors

75. $1368-(-551)=1,919$ years

77. $65+4(-7)=65+(-28)=37$ degrees

79. The notion of adding signed numbers was used to define subtraction.

81. Yes; when multiplying or dividing signed numbers, if both numbers have the same sign, the result is positive; if the numbers have different signs, the result is negative.

83. There are an infinite number of integers that are not whole numbers. You can choose from –1, –2, –3, –4,…

85. 0

87. not possible; The whole numbers are 0, 1, 2, 3, … All natural numbers appear in this list.

89. If $\frac{8}{0}=x$, then $8=0\cdot x$, which is not possible.

91. For the next four days you will spend \$6, so in four days, your net change in finances will be $-\$24$. Thus, $(+4)(-6)=-24$.

93. For the past three days you received \$7, so three days ago, you had \$21 less. Thus $(-3)(+7)=-21$.

95. No solution provided.

97. Since $a-8=-6$, we have $a=2$.; Since $c+1=-6$, we have $c=-7$.; Since $e+0=-6$, we have $e=-6$.; Since $b+e+3=-6$ and $e=-6$, we have $b=-3$.; Since $b+d+5$ and $b=-3$, we have $d=-8$.

	−9	5	4	e	$e+0$
	a	−4	b	−1	$a+b+(-5)$
	−2	0	1	−5	−6
	3	c	d	6	$c+d+9$
$b+e+3$	$a-8$	$c+1$	$b+d+5$	$e+0$	−6

Since $a+b+(-5)=(2)+(-3)+(-5)=-6$, $c+d+9=(-7)+(-8)+(9)=-6$, and $b+e+3=(-3)+(-6)+3=-6$, the values satisfy all rows, columns, and diagonals.

99. The sum of the entries, 333, is nine times the center number.

101. The reasoning is similar to Exercise 100, except now there is no center number. So put a point, c, in the middle of the 4-by-4 square and notice that any two numbers that are symmetric with respect to c have the same total. There are eight such pairs in a 4-by-4 box; therefore add two numbers in the box that are symmetric with respect to the center, c, and multiply this total by eight to get the total of all sixteen numbers.

Section 6.3 The Rational Numbers

1. Cross multiply and see if the products are the same.

3. denominators

5. It gives and answer in fractional form.

7. equal; $\frac{2}{3}=\frac{8}{12}$ since $2\cdot12=3\cdot8$ $(24=24)$

9. not equal; $\frac{12}{14}\neq\frac{14}{16}$ since $12\cdot16\neq14\cdot14$ $(192\neq196)$

11. not equal; $\frac{22}{14}\neq\frac{30}{21}$ since $22\cdot21\neq14\cdot30$ $(462\neq420)$

13. equal; $\frac{5}{14}=\frac{15}{42}$ since $5\cdot42=14\cdot15$ $(210=210)$

15. $\frac{15}{35}=\frac{\cancel{5}\cdot3}{\cancel{5}\cdot7}=\frac{3}{7}$

17. $-\frac{24}{72}=-\frac{\cancel{24}\cdot1}{\cancel{24}\cdot3}=-\frac{1}{3}$

19. $\frac{225}{350}=\frac{\cancel{25}\cdot9}{\cancel{25}\cdot14}=\frac{9}{14}$

21. $\frac{143}{154}=\frac{\cancel{11}\cdot13}{\cancel{11}\cdot14}=\frac{13}{14}$

23. $\frac{2}{3}+\frac{1}{2}=\frac{2\cdot2+3\cdot1}{3\cdot2}=\frac{4+3}{6}=\frac{7}{6}$

25. $\frac{1}{6}-\frac{1}{2}=\frac{1}{6}-\frac{1\cdot3}{2\cdot3}=\frac{1}{6}-\frac{3}{6}=-\frac{2}{6}=-\frac{1}{3}$

27. $\frac{7}{16}-\frac{1}{3}=\frac{7\cdot3-16\cdot1}{16\cdot3}=\frac{21-16}{48}=\frac{5}{48}$

29. $\frac{5}{24}+\frac{7}{18}=\frac{5\cdot3}{24\cdot3}+\frac{7\cdot4}{18\cdot4}=\frac{15}{72}+\frac{28}{72}=\frac{43}{72}$

31. $\frac{3}{4}+\frac{5}{6}+\frac{7}{8}=\frac{3\cdot3}{4\cdot3}+\frac{5\cdot2}{6\cdot2}+\frac{7}{8}=\frac{9}{12}+\frac{10}{12}+\frac{7}{8}=\frac{19}{12}+\frac{7}{8}=\frac{19\cdot2}{12\cdot2}+\frac{7\cdot3}{8\cdot3}=\frac{38}{24}+\frac{21}{24}=\frac{59}{24}$

33. $\frac{1}{8}-\frac{2}{3}+\frac{1}{2}=\frac{1\cdot3}{8\cdot3}-\frac{2\cdot8}{3\cdot8}+\frac{1}{2}=\frac{3}{24}-\frac{16}{24}+\frac{1}{2}=\frac{-13}{24}+\frac{1}{2}=\frac{-13}{24}+\frac{1\cdot12}{2\cdot12}=\frac{-13}{24}+\frac{12}{24}=\frac{-1}{24}=-\frac{1}{24}$

35. $\frac{\overset{1}{\cancel{2}}}{3}\cdot\frac{1}{\underset{1}{\cancel{2}}}=\frac{1}{3}$

37. $\frac{1}{6}\div\frac{1}{2}=\frac{1}{\underset{3}{\cancel{6}}}\cdot\frac{\overset{1}{\cancel{2}}}{1}=\frac{1}{3}$

39. $\frac{7}{8}\div\left(-\frac{5}{24}\right)=\frac{7}{\underset{1}{\cancel{8}}}\cdot\left(-\frac{\overset{3}{\cancel{24}}}{5}\right)=-\frac{21}{5}$

41. $\frac{7}{\underset{6}{\cancel{18}}}\cdot\left(-\frac{\overset{1}{\cancel{3}}}{4}\right)=-\frac{7}{24}$

43. $\left(\frac{7}{\underset{6}{\cancel{18}}}\cdot\left(-\frac{\overset{1}{\cancel{3}}}{4}\right)\right)\div\left(\frac{7}{9}\right)=-\frac{7}{24}\div\frac{7}{9}=-\frac{\overset{1}{\cancel{7}}}{\underset{8}{\cancel{24}}}\cdot\frac{\overset{3}{\cancel{9}}}{\underset{1}{\cancel{7}}}=-\frac{3}{8}$

45. $\left(\frac{11}{30}\div\left(-\frac{1}{6}\right)\right)\cdot\left(\frac{15}{4}\right)=\left(\frac{11}{\underset{5}{\cancel{30}}}\cdot\left(-\frac{\overset{1}{\cancel{6}}}{1}\right)\right)\cdot\left(\frac{15}{4}\right)=-\frac{11}{\underset{1}{\cancel{5}}}\,\frac{\overset{3}{\cancel{15}}}{4}=-\frac{33}{4}$

47. $\frac{5}{3}\cdot\left(\frac{8}{15}+\frac{2}{3}\right)=\frac{5}{3}\cdot\left(\frac{8}{15}+\frac{2\cdot5}{3\cdot5}\right)=\frac{5}{3}\cdot\left(\frac{8}{15}+\frac{10}{15}\right)=\frac{\overset{1}{\cancel{5}}}{\underset{1}{\cancel{3}}}\cdot\left(\frac{\overset{\overset{2}{\cancel{6}}}{\cancel{18}}}{\underset{\underset{1}{\cancel{3}}}{\cancel{15}}}\right)=\frac{2}{1}=2$

49. $\frac{22}{27}\cdot\left(\frac{3}{11}+\frac{2}{3}\right)=\frac{22}{27}\cdot\left(\frac{3\cdot3+11\cdot2}{11\cdot3}\right)=\frac{22}{27}\cdot\left(\frac{9+22}{33}\right)=\frac{\overset{2}{\cancel{22}}}{27}\cdot\left(\frac{31}{\underset{3}{\cancel{33}}}\right)=\frac{62}{81}$

51. $\frac{7}{30}\div\left(\frac{1}{6}-\frac{3}{14}\right)=\frac{7}{30}\div\left(\frac{1\cdot7}{6\cdot7}-\frac{3\cdot3}{14\cdot3}\right)=\frac{7}{30}\div\left(\frac{7}{42}-\frac{9}{42}\right)=\frac{7}{30}\div\left(-\frac{2}{42}\right)=\frac{7}{\underset{5}{\cancel{30}}}\cdot\left(-\frac{\overset{7}{\cancel{42}}}{2}\right)=-\frac{49}{10}$

53. $\frac{79}{5}=15\frac{4}{5}$

55. $\frac{29}{3}=9\frac{2}{3}$

57. $\begin{array}{r} 6 \\ 4\overline{)27} \\ \underline{24} \\ 3 \end{array}$; $6\frac{3}{4}$

59. $\begin{array}{r} 8 \\ 15\overline{)121} \\ \underline{120} \\ 1 \end{array}$; $8\frac{1}{15}$

61. $\frac{2\cdot4+3}{4}=\frac{8+3}{4}=\frac{11}{4}$

63. $\frac{9\cdot6+1}{6}=\frac{54+1}{6}=\frac{55}{6}$

65. 0.75

$$\begin{array}{r} 0.75 \\ 4\overline{)3.00} \\ \underline{28} \\ 20 \\ \underline{20} \\ 0 \end{array}$$

67. 0.1875

$$\begin{array}{r} 0.1875 \\ 16\overline{)3.0000} \\ \underline{16} \\ 140 \\ \underline{128} \\ 120 \\ \underline{112} \\ 80 \\ \underline{80} \\ 0 \end{array}$$

69. $0.\overline{81}$

$$\begin{array}{r} 0.81 \\ 11\overline{)9.00} \\ \underline{88} \\ 20 \\ \underline{11} \\ 90 \end{array}$$

71. $0.\overline{307692}$

$$\begin{array}{r} 0.307692 \\ 13\overline{)4.000000} \\ \underline{39} \\ 10 \\ \underline{0} \\ 100 \\ \underline{91} \\ 90 \\ \underline{78} \\ 120 \\ \underline{117} \\ 30 \\ \underline{26} \\ 40 \end{array}$$

73. $0.64 = \dfrac{64}{100} = \dfrac{16 \cdot 4}{25 \cdot 4} = \dfrac{16}{25}$

75. $0.836 = \dfrac{836}{1000} = \dfrac{209 \cdot 4}{250 \cdot 4} = \dfrac{209}{250}$

77. $12.2 = \dfrac{122}{10} = \dfrac{61 \cdot 2}{5 \cdot 2} = \dfrac{61}{5}$

79.
$$\begin{array}{rl} 10 \cdot x = & 4.444444... \\ - \quad x = & -0.444444... \\ \hline \end{array}$$
$$9 \cdot x = 4 \Rightarrow x = \frac{4}{9}$$

81.
$$\begin{array}{rl} 1000 \cdot x = & 189.189189.... \\ - \quad x = & -\ 0.189189... \\ \hline \end{array}$$
$$999 \cdot x = 189 \Rightarrow x = \frac{189}{999} = \frac{7 \cdot 27}{37 \cdot 27} = \frac{7}{37}$$

83. $100 \cdot x = 31.81818...$

$- \quad x = -0.31818...$

$$99 \cdot x = 31.5 \Rightarrow x = \frac{31.5}{99} = \frac{31.5 \cdot 10}{99 \cdot 10} = \frac{315}{990} = \frac{7 \cdot 45}{22 \cdot 45} = \frac{7}{22}$$

$1000 \cdot x = 318.1818...$

or $\quad - 10 \cdot x = - 3.1818...$

$$990 \cdot x = 315 \Rightarrow x = \frac{315}{990} = \frac{7 \cdot 45}{22 \cdot 45} = \frac{7}{22}$$

85. Andre spends $\frac{1}{3} + \frac{1}{4} + \frac{1}{6} = \frac{1 \cdot 4 + 3 \cdot 1}{3 \cdot 4} + \frac{1}{6} = \frac{4+3}{12} + \frac{1}{6} = \frac{7}{12} + \frac{1}{6} =$ $\frac{7}{12} + \frac{1 \cdot 2}{6 \cdot 2} = \frac{7}{12} + \frac{2}{12} = \frac{9}{12} = \frac{3 \cdot 3}{4 \cdot 3} = \frac{3}{4}$ of his paycheck on rent, food, and utilities. That leaves $1 - \frac{3}{4} = \frac{4}{4} - \frac{3}{4} = \frac{1}{4}$ of his paycheck for other expenses.

87. $\frac{1}{3} + \frac{1}{4} = \frac{1 \cdot 4 + 3 \cdot 1}{3 \cdot 4} = \frac{4+3}{12} = \frac{7}{12}$ of the purse goes to the winner and the person in second place. That leaves $1 - \frac{7}{12} = \frac{12}{12} - \frac{7}{12} = \frac{5}{12}$ for the last four players. Each would get $\frac{1}{4} \cdot \frac{5}{12} = \frac{5}{48}$ of the purse.

89. $\frac{9}{25} + \frac{11}{50} = \frac{9 \cdot 2}{25 \cdot 2} + \frac{11}{50} = \frac{18}{50} + \frac{11}{50} = \frac{18+11}{50} = \frac{29}{50}$

91. The height of the room is $17 \cdot 12 = 204$ inches. We need to find $204 \div 8\frac{1}{2}$ and determine if it represents a natural number. You could perform long division or simplify $\frac{204}{8\frac{1}{2}}$. Using the second approach we have $\frac{204}{8\frac{1}{2}} = \frac{204}{\frac{17}{2}} = \frac{\overset{12}{\cancel{204}}}{1} \cdot \frac{2}{\underset{1}{\cancel{17}}} = \frac{24}{1} = 24$. Thus we will have 24 rows and the tiles do not need to be cut.

93. In order to increase the recipe that serves 4 to a recipe that serves 10, you must increase the ingredients by 2.5 or $2\frac{1}{2}$ times. You should use $2\frac{1}{2} \cdot 1\frac{1}{2} = \frac{5}{2} \cdot \frac{3}{2} = \frac{15}{4} = 3\frac{3}{4}$ tablespoons of lemon juice.

95. $5\frac{1}{4}=5.25$ and $7\frac{1}{8}=7.125$

The unit cost of the $5\frac{1}{4}$-ounce tube would be about 49.5 cents per ounce (rounded).

$$\begin{array}{r} 0.4952 \\ 5.25\overline{)2.600000} \\ \underline{2100} \\ 5000 \\ \underline{4725} \\ 2750 \\ \underline{2625} \\ 1250 \\ \underline{1050} \\ 200 \end{array}$$

The unit cost of the $7\frac{1}{8}$-ounce tube would be about 50.5 cents per ounce (rounded).

$$\begin{array}{r} 0.5052 \\ 7.125\overline{)3.6000000} \\ \underline{35625} \\ 3750 \\ \underline{0} \\ 37500 \\ \underline{35625} \\ 18750 \\ \underline{14250} \\ 4500 \end{array}$$

The smaller tube is a better buy.

97. $1-\left(\frac{3}{5}+\frac{7}{40}\right)=1-\left(\frac{3\cdot 8}{5\cdot 8}+\frac{7}{40}\right)=1-\left(\frac{24}{40}+\frac{7}{40}\right)=\frac{40}{40}-\frac{31}{40}=\frac{9}{40}$

99. He can get $12\div 3\frac{3}{8}=12\div\frac{27}{8}=\frac{\overset{4}{\cancel{12}}}{1}\cdot\frac{8}{\underset{9}{\cancel{27}}}=\frac{32}{9}=3\frac{5}{9}$ strips from a twelve-foot strip and $15\div 3\frac{3}{8}=15\div\frac{27}{8}=\frac{\overset{5}{\cancel{15}}}{1}\cdot\frac{8}{\underset{9}{\cancel{27}}}=\frac{40}{9}=4\frac{4}{9}$ strips from a fifteen-foot strip. Since $\frac{5}{9}>\frac{4}{9}$, there is more waste from the twelve-foot strips.

101. Cancel common factors from the numerators and denominators.

103. Divide the denominator, d, into the numerator to get a quotient q and a remainder r. Then, write the mixed number as $q\frac{r}{d}$.

105. Responses may vary. In finding the decimal expansion of $\frac{a}{b}$ (assume for simplicity that a and b are positive), there are at most $b-1$ remainders when performing the division steps. So when dividing the denominator into the numerator, the remainders have to repeat, which will cause the digits in the quotient to also repeat.

107. $\dfrac{\frac{3}{2}}{\frac{1}{4}}=\dfrac{\frac{3}{2}\cdot\frac{4}{1}}{\frac{1}{4}\cdot\frac{4}{1}}=\dfrac{\frac{3}{2}\cdot\frac{4}{1}}{1}=\frac{3}{\underset{1}{\cancel{2}}}\cdot\frac{\overset{2}{\cancel{4}}}{1}=\frac{6}{1}=6$

109. $\dfrac{\frac{5}{12}}{\frac{3}{8}}=\dfrac{\frac{5}{12}\cdot\frac{8}{3}}{\frac{3}{8}\cdot\frac{8}{3}}=\dfrac{\frac{5}{12}\cdot\frac{8}{3}}{1}=\frac{5}{\underset{3}{\cancel{12}}}\cdot\frac{\overset{2}{\cancel{8}}}{3}=\frac{10}{9}$

111. No solution provided.

113. Each piece would be $10\frac{7}{8} \div 4 = \frac{87}{8} \div \frac{4}{1} = \frac{87}{8} \cdot \frac{1}{4} = \frac{87}{32} = 2\frac{23}{32}$ feet

115. There are many approaches to this exercise. One approach is to first determine the distance from the edge of the mirror to the edge of the wall with the mirror centered on the wall. We can do this by finding the difference between the length of the wall and the mirror and then dividing this difference by 2. We can then add to this measurement one-third of the length of the mirror. This will yield the desired measurement.

Distance from edge of mirror to edge of wall:

$$(72 - 40\tfrac{1}{2}) \div 2 = (\tfrac{72}{1} - \tfrac{81}{2}) \div 2 = (\tfrac{72 \cdot 2}{1 \cdot 2} - \tfrac{81}{2}) \div 2 = (\tfrac{144}{2} - \tfrac{81}{2}) \div 2 = (\tfrac{63}{2}) \div 2 = \tfrac{63}{2} \cdot \tfrac{1}{2} = \tfrac{63}{4} = 15\tfrac{3}{4} \text{ inches}$$

One third of the length the mirror: $\frac{1}{3} \cdot \left(40\frac{1}{2}\right) = \frac{1}{\cancelto{1}{3}} \cdot \frac{\cancelto{27}{81}}{2} = \frac{27}{2} = 13\frac{1}{2}$ inches

Distance from the edge of the wall to a nail would therefore be:

$$15\frac{3}{4} + 13\frac{1}{2} = 15 + \frac{3}{4} + 13 + \frac{1}{2} = (15+13) + \left(\frac{3}{4} + \frac{1}{2}\right) = 28 + \left(\frac{3}{4} + \frac{1 \cdot 2}{2 \cdot 2}\right) =$$

$$= 28 + \left(\frac{3}{4} + \frac{2}{4}\right) = 28 + \frac{5}{4} = 28 + 1\frac{1}{4} = 29\frac{1}{4} \text{ inches}$$

Section 6.4 The Real Number System

1. Irrational numbers have nonrepeating expansions; rational numbers have repeating or terminating expansions.

3. $\sqrt{ab} = \sqrt{a}\sqrt{b}$, $\sqrt{\frac{a}{b}} = \frac{\sqrt{a}}{\sqrt{b}}$

5. Using approximations can introduce errors, which can eventually lead to an incorrect answer.

7. rational

9. irrational

11. rational

13. irrational

15. Examples may vary; $\sqrt{9} = 3$

17. $\sqrt{18} = \sqrt{9 \cdot 2} = \sqrt{9}\sqrt{2} = 3\sqrt{2}$

19. $\sqrt{75} = \sqrt{25 \cdot 3} = \sqrt{25}\sqrt{3} = 5\sqrt{3}$

21. $\sqrt{48} = \sqrt{16 \cdot 3} = \sqrt{16}\sqrt{3} = 4\sqrt{3}$

23. $\sqrt{189} = \sqrt{9 \cdot 21} = \sqrt{9}\sqrt{21} = 3\sqrt{21}$

25. $3\sqrt{5} + 8\sqrt{5} = 11\sqrt{5}$

27. not possible

29. $\sqrt{20} + 6\sqrt{5} = \sqrt{4 \cdot 5} + 6\sqrt{5} = \sqrt{4}\sqrt{5} + 6\sqrt{5} = 2\sqrt{5} + 6\sqrt{5} = 8\sqrt{5}$

31. not possible

$\sqrt{50} + 2\sqrt{75} = \sqrt{25 \cdot 2} + 2\sqrt{25 \cdot 3} = \sqrt{25}\sqrt{2} + 2\sqrt{25}\sqrt{3} = 5\sqrt{2} + 2 \cdot 5\sqrt{3} = 5\sqrt{2} + 10\sqrt{3}$

33. $\sqrt{18}\sqrt{2} = \sqrt{18 \cdot 2} = \sqrt{9 \cdot 2 \cdot 2} = \sqrt{9}\sqrt{2 \cdot 2} = 3\sqrt{4} = 3 \cdot 2 = 6$

35. $\sqrt{12}\sqrt{15}=\sqrt{12\cdot 15}=\sqrt{4\cdot 3\cdot 3\cdot 5}=\sqrt{4}\sqrt{3\cdot 3\cdot 5}=2\sqrt{9\cdot 5}=2\sqrt{9}\sqrt{5}=2\cdot 3\sqrt{5}=6\sqrt{5}$

37. $\sqrt{28}\sqrt{21}=\sqrt{28\cdot 21}=\sqrt{4\cdot 7\cdot 7\cdot 3}=\sqrt{4}\sqrt{7\cdot 7\cdot 3}=2\sqrt{49\cdot 3}=2\sqrt{49}\sqrt{3}=2\cdot 7\sqrt{3}=14\sqrt{3}$

39. $\dfrac{\sqrt{24}}{\sqrt{6}}=\sqrt{\dfrac{24}{6}}=\sqrt{4}=2$

41. $\dfrac{\sqrt{32}}{\sqrt{18}}=\sqrt{\dfrac{32}{18}}=\sqrt{\dfrac{16}{9}}=\dfrac{\sqrt{16}}{\sqrt{9}}=\dfrac{4}{3}$

43. $\dfrac{\sqrt{96}}{\sqrt{72}}=\dfrac{\sqrt{16\cdot 6}}{\sqrt{36\cdot 2}}=\dfrac{4\sqrt{6}}{6\sqrt{2}}=\dfrac{4}{6}\cdot\dfrac{\sqrt{6}}{\sqrt{2}}=\dfrac{2}{3}\sqrt{\dfrac{6}{2}}=\dfrac{2}{3}\sqrt{3}=\dfrac{2\sqrt{3}}{3}$

45. $\dfrac{3}{\sqrt{5}}=\dfrac{3\cdot\sqrt{5}}{\sqrt{5}\cdot\sqrt{5}}=\dfrac{3\sqrt{5}}{\sqrt{5\cdot 5}}=\dfrac{3\sqrt{5}}{\sqrt{25}}=\dfrac{3\sqrt{5}}{5}$

47. $\dfrac{12}{\sqrt{6}}=\dfrac{12\cdot\sqrt{6}}{\sqrt{6}\cdot\sqrt{6}}=\dfrac{12\sqrt{6}}{\sqrt{6\cdot 6}}=\dfrac{12\sqrt{6}}{\sqrt{36}}=\dfrac{12\sqrt{6}}{6}=2\sqrt{6}$

49. $\dfrac{10}{\sqrt{22}}=\dfrac{10\sqrt{22}}{\sqrt{22}\cdot\sqrt{22}}=\dfrac{10\sqrt{22}}{\sqrt{22\cdot 22}}=\dfrac{10\sqrt{22}}{\sqrt{484}}=\dfrac{10\sqrt{22}}{22}=\dfrac{5\sqrt{22}}{11}$

51. $\dfrac{3\sqrt{10}}{\sqrt{72}}=\dfrac{3\cdot\sqrt{10}\sqrt{2}}{\sqrt{36\cdot 2}\cdot\sqrt{2}}=\dfrac{3\sqrt{20}}{\sqrt{144}}=\dfrac{3\sqrt{4\cdot 5}}{12}=\dfrac{3\sqrt{4}\cdot\sqrt{5}}{12}=\dfrac{3\cdot 2\sqrt{5}}{12}=\dfrac{6\sqrt{5}}{12}=\dfrac{\sqrt{5}}{2}$

An easier approach for this exercise however is to simplify as much as possible before rationalizing the denominator.

$\dfrac{3\sqrt{10}}{\sqrt{72}}=3\sqrt{\dfrac{10}{72}}=3\sqrt{\dfrac{5}{36}}=3\dfrac{\sqrt{5}}{\sqrt{36}}=3\dfrac{\sqrt{5}}{6}=\dfrac{\sqrt{5}}{2}$

53. Answers may vary.

0.435; 0.43512112111211112…

55. Answers may vary.

$0.\overline{4578}=0.4578\underline{4}5784578...$

$0.45\overline{78}=0.4578\underline{7}8787878...$

0.45785; 0.45785121121112111112…

57. Answers may vary.

$\dfrac{4}{7}=0.\overline{571428}$

$\dfrac{5}{7}=0.\overline{714285}$

0.6; 0.612112111211112…

59. Answers may vary.

$0.121\underline{1}2111211112...$

$0.\overline{12}=0.121\underline{2}121212...$

0.12113; 0.1211312112111211112…

61. a, d, b, c

a) $0.345\underline{345000}000$

b) $0.345\underline{345345}345...$

c) $0.345\underline{454545}454...$

d) $0.345\underline{345343}434...$

63. a, c, d, b

a) $0.\underline{444}4444...$

b) $0.\underline{555}5555...$

c) $0.\underline{454}5454...$

d) $0.\underline{554}5454...$

65. distributive

67. commutative (addition)

69. associative (addition)

71. identity element for addition

73. commutative (addition)

75. $D=\sqrt{2(6+42)}=\sqrt{2\cdot 48}=\sqrt{2\cdot 3\cdot 16}=\sqrt{6\cdot 16}=\sqrt{16}\cdot\sqrt{6}=4\sqrt{6}\approx 9.8$ miles.

77. $v=2\sqrt{5\cdot 120}=2\sqrt{600}=2\sqrt{100\cdot 6}=2\sqrt{100}\cdot\sqrt{6}=2\cdot 10\cdot\sqrt{6}=20\sqrt{6}\approx 49.0$ mph.

79. $T=2\pi\sqrt{\frac{20}{32}}=2\pi\sqrt{\frac{5}{8}}=2\pi\sqrt{\frac{5\cdot 2}{8\cdot 2}}=2\pi\sqrt{\frac{10}{16}}=2\pi\frac{\sqrt{10}}{4}=\pi\frac{\sqrt{10}}{2}\approx 4.97$ seconds

81. $F=\sqrt{(120)^2+(160)^2}=\sqrt{40000}=200$ pounds

83. $t=\frac{\sqrt{1600}}{4}=\frac{40}{4}=10$ seconds

85. perfect squares

87. The Always Principle

89. false; $\sqrt{2}$ is irrational; 1.414212362 is a rational, decimal approximation of $\sqrt{2}$.

91. false; Counterexample: $\sqrt{25}=5$, which is rational.

93. true; $\sqrt{3}\sqrt{5}=\sqrt{15}$, and 15 is not a perfect square.

95. true; If a is a rational number and b is an irrational number and $a+b=c$, then if c is rational then we could say that $b=c-a$. But $c-a$ is the difference of two rational numbers and therefore rational. This would contradict the fact that b is an irrational number.

97. – 99. No solution provided.

101. a) we can cancel common factors from the numerator and denominator
b) square both sides of a)
c) multiply through equation by b^2
d) 2 divides the left side of c), so 2 divides the right side of c)
e) if a were odd then a^2 would be odd, but it is not
f) a is even
g) substitute $2k$ for a in c)
h) simplify $(2k)^2=(2k)(2k)=4k^2$
i) divide both sides of h) by 2
j) 2 divides b^2, since 2 divides $2k^2$
k) if b were odd, then b^2 would be odd
l) We assumed that $\frac{a}{b}$ was reduced, so a and b cannot both be even.

103. 2

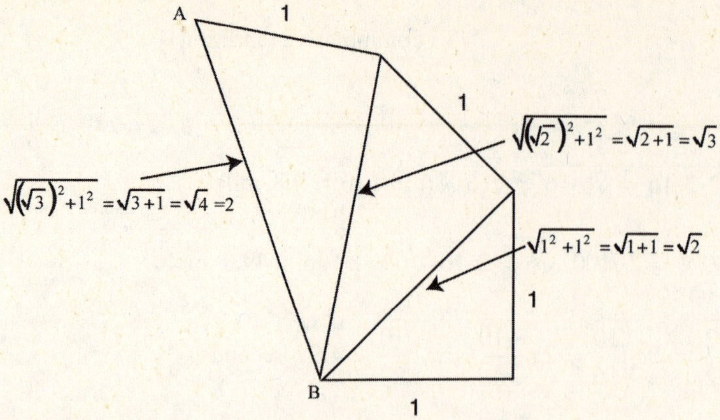

105. Answers may vary.
These numbers are possible lengths of the three sides of a right triangle. If a right triangle has legs of lengths 3 and 4 and hypotenuse of length 5, then $3^2+4^2=5^2$.

Section 6.5 Exponents and Scientific Notation

1. $x^2 \cdot x^3 = xx \cdot xxx = xxxxx = x^5$; $x^m \cdot x^n = x^{m+n}$

3. $\dfrac{x^8}{x^3} = \dfrac{\cancel{xxx}xxxxx}{\cancel{xxx}} = xxxxx = x^5$; $\dfrac{x^m}{x^n} = x^{m-n}$

5. Growth of money; growth of debt; spread of diseases; growth of interest in a new product.

7. 32

9. $-2^4 = -(2^4) = -(16) = -16$

11. $-3^2 = -(3^2) = -(9) = -9$

13. 9

15. 1

17. 0

19. $3^{2+4} = 3^6 = 729$

21. $(7^2)^3 = 7^{2\cdot 3} = 7^6 = 117{,}649$

23. $5^4 \cdot 5^{-6} = 5^{4+(-6)} = 5^{-2} = \dfrac{1}{5^2} = \dfrac{1}{25}$

25. $(3^2)^{-3} = 3^{2\cdot(-3)} = 3^{-6} = \dfrac{1}{3^6} = \dfrac{1}{729}$

27. $(-3)^{-2}(-3)^3 = (-3)^{-2+3} = (-3)^1 = -3$

29. $\dfrac{5^9}{5^7} = 5^{9-7} = 5^2 = 25$

31. $\dfrac{6^{-2}}{6^{-4}} = 6^{-2-(-4)} = 6^2 = 36$

33. $\dfrac{-3^6}{-3^9} = \dfrac{3^6}{3^9} = 3^{6-9} = 3^{-3} = \dfrac{1}{3^3} = \dfrac{1}{27}$

35. $(3^{-4})^0 = 3^{-4\cdot 0} = 3^0 = 1$

37. $\dfrac{2^9}{4^3} = \dfrac{2^9}{(2^2)^3} = \dfrac{2^9}{2^6} = 2^{9-6} = 2^3 = 8$

39. 4.356×10^6

41. 7.83×10^2

43. 2.4×10^{-3}

45. 8.0×10^{-3}

47. 32,500

49. 0.00178

51. 63

53. 0.00000045

55. 1,000,000

57. 2.381×10^{7}

59. 8.4×10^{2}

61. $(3\times10^{6})(2\times10^{5})=(3\cdot2)\times(10^{6}\cdot10^{5})=6.0\times10^{6+5}=6.0\times10^{11}$

63. $(1.2\times10^{-3})(3\times10^{5})=(1.2\cdot3)\times(10^{-3}\cdot10^{5})=3.6\times10^{-3+5}=3.6\times10^{2}$

65. $(8\times10^{-2})\div(2\times10^{3})=\dfrac{8\times10^{-2}}{2\times10^{3}}=\dfrac{8}{2}\times\dfrac{10^{-2}}{10^{3}}=4\times10^{-2-3}=4.0\times10^{-5}$

67. $$\frac{(5.44\times10^{8})(2.1\times10^{-3})}{(3.4\times10^{6})}=\frac{5.44\cdot2.1}{3.4}\times\frac{10^{8}\cdot10^{-3}}{10^{6}}=\frac{11.424}{3.4}\times\frac{10^{8+(-3)}}{10^{6}}=3.36\times\frac{10^{5}}{10^{6}}$$
$$=3.36\times10^{5-6}=3.36\times10^{-1}$$

69. $$\frac{(9.6368\times10^{3})(4.15\times10^{-6})}{(1.52\times10^{4})}=\frac{9.6368\cdot4.15}{1.52}\times\frac{10^{3}\cdot10^{-6}}{10^{4}}=\frac{39.99272}{1.52}\times\frac{10^{3+(-6)}}{10^{4}}=26.311\times\frac{10^{-3}}{10^{4}}$$
$$=26.311\times10^{-3-4}=26.311\times10^{-7}=2.6311\times10^{-6}$$

71. $$67{,}300{,}000\times1{,}200=(6.73\times10^{7})(1.2\times10^{3})=(6.73\cdot1.2)\times(10^{7}\cdot10^{3})$$
$$=8.076\times10^{7+3}=8.076\times10^{10}$$

73. $$6{,}800{,}000\times2{,}300{,}000=(6.8\times10^{6})(2.3\times10^{6})=(6.8\cdot2.3)\times(10^{6}\cdot10^{6})$$
$$=15.64\times10^{6+6}=15.64\times10^{12}=1.564\times10^{13}$$

75. $$0.00016\times0.0025=(1.6\times10^{-4})(2.5\times10^{-3})=(1.6\cdot2.5)\times(10^{-4}\cdot10^{-3})$$
$$=4.0\times10^{-4+(-3)}=4.0\times10^{-7}$$

77. $\dfrac{930.2\times10^{9}}{249\times10^{6}}=\dfrac{9.302\times10^{11}}{2.49\times10^{8}}=\dfrac{9.302}{2.49}\times\dfrac{10^{11}}{10^{8}}\approx3.736\times10^{11-8}=3.736\times10^{3}=\$3{,}736$ per person.

79. $$18{,}736\times10^{9}-4{,}970\times10^{9}=(18{,}736-4{,}970)\times10^{9}=13{,}766\times10^{9}=1.3766\times10^{4}\times10^{9}$$
$$=1.3766\times10^{4+9}=1.3766\times10^{13}=\$13{,}766\text{ billion}$$

81. (a) $454\cdot130=59020=5.902\times10^{4}$

(b) $\dfrac{5.902\times10^{4}}{1\times10^{-3}}=\dfrac{5.902}{1}\times\dfrac{10^{4}}{10^{-3}}=5.902\times10^{4-(-3)}=5.902\times10^{7}=59.02\times10^{6}$

(c) The 130 pound person weighs as much as 59 million mosquitoes.

83. $\dfrac{439.3\times10^9}{301\times10^6}=\dfrac{4.393\times10^{11}}{3.01\times10^8}=\dfrac{4.393}{3.01}\times\dfrac{10^{11}}{10^8}\approx1.459\times10^{11-8}=1.459\times10^3=\$1,459$ per person

85. Assuming there are 365 days in a year, we have $60\times60\times24\times365$ seconds in one year.

$$\frac{1.0\times10^9}{60\times60\times24\times365}=\frac{1.0\times10^9}{31,536,000}=\frac{1.0\times10^9}{3.1536\times10^7}=\frac{1.0}{3.1536}\times\frac{10^9}{10^7}\approx0.317\times10^{9-7}$$
$$=0.317\times10^2=3.17\times10^1=31.7 \text{ years}$$

87. $\dfrac{304\times10^6}{3.5\times10^6}=\dfrac{3.04\times10^8}{3.5\times10^6}=\dfrac{3.04}{3.5}\times\dfrac{10^8}{10^6}\approx0.869\times10^{8-6}=0.869\times10^2$
$=8.69\times10^1=86.9$ people per square mile.

89. $\dfrac{93,000,000}{25,000}=\dfrac{9.3\times10^7}{2.5\times10^4}=\dfrac{9.3}{2.5}\times\dfrac{10^7}{10^4}=3.72\times10^{7-4}=3.72\times10^3=3,720$ hours

91. $\dfrac{3.6\times10^9}{93\times10^6}=\dfrac{3.6\times10^9}{9.3\times10^7}=\dfrac{3.6}{9.3}\times\dfrac{10^9}{10^7}\approx0.387\times10^{9-7}=0.387\times10^2=3.87\times10^1=38.7$ times larger

93. $$\frac{8.424\times10^{11}}{(2.6\times10^6)(10,000)}=\frac{8.424\times10^{11}}{(2.6\times10^6)(1\times10^4)}=\frac{8.424\times10^{11}}{2.6\times10^{6+4}}=\frac{8.424}{2.6}\times\frac{10^{11}}{10^{10}}$$
$=3.24\times10^{11-10}=3.24\times10^1$ =32.4 seconds

95. Examples may vary. Refer to the solution to Exercise 1.

97. Examples may vary. Refer to the solution to Exercise 3.

99. In evaluating -2^4, we raise 2 to the fourth power to get 16 and then negate the result to get –16. In evaluating $(-2)^4$, we first negate 2 and then raise that negative number to the fourth power to get 16.

101. No solution provided.

103. No; Examples may vary.

$$(2^3)^4\neq2^{(3^4)}$$
$$8^4\neq2^{81}$$
$$4,096\neq2,417,851,639,229,258,349,412,352$$

105. No solution provided.

Section 6.6 Looking Deeper: Sequences

1. We considered the terms of the sequence to all be of the form 3 plus some multiple of 4 to see a pattern.

3. Answers will vary.

5. This is an arithmetic sequence. There is a common difference of 3 between consecutive terms. The next two terms are 17 and 20.

7. This is a geometric sequence. There is a common ratio of 3 between consecutive terms. The next two terms are 648 and 1,944.

9. This is a geometric sequence. There is a common ratio of $\frac{1}{2}$ between consecutive terms. The next two terms are $\frac{1}{16}$ and $\frac{1}{32}$.

11. This is an arithmetic sequence. There is a common difference of -5 between consecutive terms. The next two terms are -10 and -15.

13. This is a geometric sequence. There is a common ratio of 2 between consecutive terms. The next two terms are 32 and 64.

15. This is an arithmetic sequence. There is a common difference of 0.5 between consecutive terms. The next two terms are 3.5 and 4.0.

17. a) Since $a_1 = 5$, $d = 3$, and $n = 11$, we have $a_{11} = 5 + (11-1)\cdot 3 = 5 + 10\cdot 3 = 5 + 30 = 35$.

b) Since $a_1 = 5$, $a_{11} = 35$, and $n = 11$, the sum of the first 11 terms is $\frac{11(5+35)}{2} = \frac{11\cdot 40}{2} = 11\cdot 20 = 220$.

19. a) Since $a_1 = 2$, $d = 6$, and $n = 15$, we have $a_{15} = 2 + (15-1)\cdot 6 = 2 + 14\cdot 6 = 2 + 84 = 86$.

b) Since $a_1 = 2$, $a_{15} = 86$, and $n = 15$, the sum of the first 15 terms is $\frac{15(2+86)}{2} = \frac{15\cdot 88}{2} = 15\cdot 44 = 660$.

21. a) Since $a_1 = 1$, $d = 0.5$, and $n = 20$, we have $a_{20} = 1 + (20-1)\cdot 0.5 = 1 + 19\cdot 0.5 = 1 + 9.5 = 10.5$.

b) Since $a_1 = 1$, $a_{20} = 10.5$, and $n = 20$, the sum of the first 20 terms is $\frac{20(1+10.5)}{2} = \frac{20\cdot 11.5}{2} = 10\cdot 11.5 = 115$.

23. We need to find the sum of the first 21 terms of this sequence and subtract the sum of the first 13 terms. Since $a_1 = 5$, $d = 3$, and $n = 21$, we have $a_{21} = 5 + (21-1)\cdot 3 = 5 + 20\cdot 3 = 5 + 60 = 65$. Thus, the sum of the first 21 terms is $\frac{21(5+65)}{2} = \frac{21\cdot 70}{2} = 21\cdot 35 = 735$. Since $a_1 = 5$, $d = 3$, and $n = 13$, we have $a_{13} = 5 + (13-1)\cdot 3 = 5 + 12\cdot 3 = 5 + 36 = 41$. Thus the sum of the first 13 terms is $\frac{13(5+41)}{2} = \frac{13\cdot 46}{2} = 13\cdot 23 = 299$. The desired sum is therefore $735 - 299 = 436$.

25. a) Since $a_1 = 1$, $r = 3$, and $n = 11$, we have $a_{11} = 1\cdot 3^{11-1} = 1\cdot 3^{10} = 1\cdot 59{,}049 = 59{,}049$.

b) Since $a_1 = 1$, $r = 3$, and $n = 11$, the sum of the first 11 terms is $\frac{1\cdot(3^{11}-1)}{3-1} = \frac{1\cdot(177{,}147-1)}{2} = \frac{1\cdot 177{,}146}{2} = \frac{177{,}146}{2} = 88{,}573$.

27. a) Since $a_1 = 1$, $r = \frac{1}{2}$, and $n = 7$, we have $a_7 = 1\cdot\left(\frac{1}{2}\right)^{7-1} = 1\cdot\left(\frac{1}{2}\right)^6 = 1\cdot\frac{1}{64} = \frac{1}{64}$.

b) Since $a_1 = 1$, $r = \frac{1}{2}$, and $n = 7$, the sum of the first 7 terms is $\dfrac{1\cdot\left(\left(\frac{1}{2}\right)^7 - 1\right)}{\frac{1}{2} - 1} = \dfrac{1\cdot\left(\frac{1}{128} - 1\right)}{-\frac{1}{2}} =$

$$\frac{1\cdot\left(-\frac{127}{128}\right)}{-\frac{1}{2}} = \frac{-\frac{127}{128}}{-\frac{1}{2}} = \left(-\frac{127}{128}\right)\cdot\left(-\frac{2}{1}\right) = \frac{127}{64}.$$

29. a) Since $a_1 = 2$, $r = 0.1$, and $n = 6$, we have $a_6 = 2\cdot 0.1^{6-1} = 2\cdot 0.1^5 = 2\cdot 0.00001 = 0.00002$.

b) Since $a_1 = 2$, $r = 0.1$, and $n = 6$, the sum of the first 6 terms is $\dfrac{2\cdot\left(0.1^6 - 1\right)}{0.1 - 1} = \dfrac{2\cdot(0.000001 - 1)}{-0.9} = \dfrac{2\cdot(-0.999999)}{-0.9} = 2(1.11111) = 2.22222.$

31. 144

$F_{13} = F_{11} + F_{12}$

$233 = F_{12} + 89$

$144 = F_{12}$

33. 377

$F_{15} = F_{13} + F_{14}$

$610 = 233 + F_{14}$

$377 = F_{14}$

35. $a_1 = 2{,}875$, $d = 35$, and $n = 12$, thus $a_{12} = 2{,}875 + (12-1)\cdot 35 = 2{,}875 + 11\cdot 35 = 2{,}875 + 385 = 3{,}260$.

Therefore, $\dfrac{12(2{,}875 + 3{,}260)}{2} = \dfrac{12\cdot 6{,}135}{2} = 6\cdot 6{,}135 = \$36{,}810$ is the amount earned in one year.

37. In this exercise, you need to find the seventh term of this geometric sequence. The reason for the seventh term is because you are initially making a deposit. The amount in the account in six years is the amount of money you have starting the seventh year. The common ratio would be 1.035 because you retain the amount plus receive a percentage of the amount from one year to the next. Thus the amount you will receive will be $1{,}200(1.035)^{7-1} = 1{,}200(1.035)^6 \approx \$1{,}475.11$.

39. In this geometric sequence, the height after the fifth bounce would be the sixth term where $a_1 = 8$ and $r = \frac{7}{8}$. $a_6 = 8\cdot\left(\frac{7}{8}\right)^{6-1} = 8\cdot\left(\frac{7}{8}\right)^5 = 8\cdot\dfrac{16{,}807}{32{,}768} = \dfrac{16{,}807}{4{,}096} \approx 4.10$ feet

41. You are finding the tenth term in a geometric sequence where $a_1 = 185$, $r = 1.0091$, and $n = 13$. The common ratio is 1.0091 because you retain the population from one year to the next and then add a percentage of the population. Thus the population in 2020 would be $185\cdot 1.0091^{13-1} = 185\cdot 1.0091^{12} \approx$ 206.24 million.

43. – 45. Answers will vary.

47. No solution provided.

49. Answers may vary.

Consider the sum $a_k + a_{k+1} + \ldots + a_{n-1} + a_n$ and the same sum written in reverse order as $a_n + a_{n-1} + \ldots + a_{k+1} + a_k$. If we place these sums under one another and add, we get

$$\begin{array}{ccccccc} a_k & + & a_{k+1} & +\ldots+ & a_{n-1} & + & a_n \\ a_n & + & a_{n-1} & +\ldots+ & a_{k+1} & + & a_k \\ \hline \end{array}$$
$$(a_k + a_n) + (a_{k+1} + a_{n-1}) + \ldots + (a_{n-1} + a_{k+1}) + (a_n + a_k).$$

49. (continued)

It can be shown that each term represents the same quantity, namely $a_k + a_n$. For example $a_{k+1} + a_{n-1} = (a_k + d) + (a_n - d) = a_k + a_n$, where d is the common difference. This term is being added $n - k + 1$ times, so $(n - k + 1)(a_k + a_n)$ is twice the desired sum, so the formula would be $\frac{(n-k+1)(a_k+a_n)}{2}$.

An alternate, lengthier approach would be to find the sum of the first n terms of the original sequence and subtract from that the sum of the first $k-1$ terms.

$$\frac{n(a_1+a_n)}{2} - \frac{(k-1)(a_1+a_{k-1})}{2}$$

Since $a_n = a_1 + (n-1)\cdot d = a_1 + nd - d$ and

$a_{k-1} = a_1 + ((k-1)-1)\cdot d = a_1 + (k-2)\cdot d = a_1 + kd - 2d$, we have

$$\frac{n(a_1+a_n)}{2} - \frac{(k-1)(a_1+a_{k-1})}{2} = \frac{n(a_1+a_1+nd-d)}{2} - \frac{(k-1)(a_1+a_1+kd-2d)}{2}$$

$$= \frac{n(2a_1+nd-d)}{2} - \frac{(k-1)(2a_1+kd-2d)}{2}$$

$$= \frac{2na_1+n^2d-nd}{2} - \frac{2ka_1+k^2d-2kd-2a_1-kd+2d}{2}$$

$$= \frac{na_1+na_1+n^2d-nd-2ka_1-k^2d+2kd+2a_1+kd-2d}{2}$$

$$= \frac{n(a_1+nd-d)+na_1-ka_1+nkd-nkd-ka_1-k^2d+kd+kd+a_1+(a_1+kd-d)-d+nd-nd}{2}$$

$$= \frac{n(a_1+nd-d)-ka_1-ka_1+n(a_1+kd-d)-nkd-k^2d+kd+kd+(a_1+nd-d)+(a_1+kd-d)}{2}$$

$$= \frac{n(a_1+(n-1)d)-ka_1-ka_1+n(a_1+(k-1)d)-nkd-k^2d+kd+kd+(a_1+(n-1)d)+(a_1+(k-1)d)}{2}$$

$$= \frac{na_n-ka_1-ka_1+na_k-nkd-k^2d+kd+kd+a_n+a_k}{2}$$

$$= \frac{na_n+a_n-k(a_1+nd-d)-k(a_1+kd-d)+na_k+a_k}{2}$$

$$= \frac{na_n+a_n-k(a_1+(n-1)d)-k(a_1+(k-1)d)+na_k+a_k}{2}$$

$$= \frac{na_n+a_n-ka_n-ka_k+na_k+a_k}{2} = \frac{na_n-ka_n+a_n+na_k-ka_k+a_k}{2} = \frac{(n-k+1)a_n+(n-k+1)a_k}{2}$$

$$= \frac{(n-k+1)(a_n+a_k)}{2}$$

51. $F_1 + F_2 + ... + F_n = F_{n+2} - 1$

n	F_n	$F_1 + F_2 + ... + F_n$
1	1	$1 = 2 - 1 = F_3 - 1$
2	1	$2 = 3 - 1 = F_4 - 1$
3	2	$4 = 5 - 1 = F_5 - 1$
4	3	$7 = 8 - 1 = F_6 - 1$
5	5	$12 = 13 - 1 = F_7 - 1$
6	8	$20 = 21 - 1 = F_8 - 1$
7	13	$33 = 34 - 1 = F_9 - 1$
8	21	$54 = 55 - 1 = F_{10} - 1$
9	34	$88 = 89 - 1 = F_{11} - 1$
10	55	$143 = 144 - 1 = F_{12} - 1$
11	89	$232 = 233 - 1 = F_{13} - 1$
12	144	$376 = 377 - 1 = F_{14} - 1$
13	233	$609 = 610 - 1 = F_{15} - 1$
14	377	
15	610	

53. $F_5 = \dfrac{\left(\dfrac{1+\sqrt{5}}{2}\right)^5 + \left(\dfrac{1-\sqrt{5}}{2}\right)^5}{\sqrt{5}} = \dfrac{\dfrac{80\sqrt{5}+176}{32} + \dfrac{176-80\sqrt{5}}{32}}{\sqrt{5}} = \dfrac{\dfrac{352}{32}}{\sqrt{5}} = \dfrac{11}{\sqrt{5}} \approx 4.9193$, rounded to the nearest whole number would be 5.

55. $F_9 = \dfrac{\left(\dfrac{1+\sqrt{5}}{2}\right)^9 + \left(\dfrac{1-\sqrt{5}}{2}\right)^9}{\sqrt{5}} = \dfrac{\dfrac{8,704\sqrt{5}+19,456}{512} + \dfrac{19,456-8,704\sqrt{5}}{512}}{\sqrt{5}} = \dfrac{\dfrac{38,912}{512}}{\sqrt{5}} = \dfrac{76}{\sqrt{5}} \approx 33.9882$, rounded to the nearest whole number would be 34.

Chapter Review Exercises

1. Refer to the table for Exercise 15 in the solutions for Section 6.1,
71, 73, 79, 83, 89

3. 191 is prime. You need to check for divisibility by 2, 3, 5, 7, 11, and 13.
441 is composite. $441 = 3^2 \times 7^2$

5. $1,584 = 2^4 \times 3^2 \times 11$ and $1,320 = 2^3 \times 3 \times 5 \times 11$; $GCD(396,330) = 2^3 \times 3 \times 11 = 264$
$LCM(1584,1320) = 2^4 \times 3^2 \times 5 \times 11 = 7,920$

7. 2,772 is divisible by **9** because the sum of the digits, 18, is divisible by 9.
2,772 is divisible by **4** because the number formed by the last two digits, 72, is divisible by 4.
2,772 is not divisible by 8 because the number formed by the last three digits, 270, is not divisible by 8.

9. a) 9
b) 11
c) 72
d) −16

11. If $\dfrac{-24}{-8}=c$, then $-24=(-8)\cdot c$. Thus $c=+3$.

13. If $\dfrac{5}{0}=c$, then $5=0\cdot c$, which is not possible.

15. \$67; For the past four days you lose \$6, which can be thought of as $(-4)(-6)$. Therefore, four days ago you had \$24 more. Thus, $(-4)(-6)=24$. .

17. a) $\dfrac{4}{9}\cdot\left(\dfrac{3}{4}-\dfrac{1}{3}\right)=\dfrac{4}{9}\cdot\left(\dfrac{3\cdot3-4\cdot1}{4\cdot3}\right)=\dfrac{4}{9}\cdot\left(\dfrac{9-4}{12}\right)=\dfrac{\overset{1}{\cancel{4}}}{9}\cdot\left(\dfrac{5}{\underset{3}{\cancel{12}}}\right)=\dfrac{5}{27}$

b) $\dfrac{3}{7}\div\left(\dfrac{2}{3}+\dfrac{3}{14}\right)=\dfrac{3}{7}\div\left(\dfrac{2\cdot14+3\cdot3}{3\cdot14}\right)=\dfrac{3}{7}\div\left(\dfrac{28+9}{42}\right)=\dfrac{3}{7}\div\left(\dfrac{37}{42}\right)=\dfrac{3}{\underset{1}{\cancel{7}}}\,\dfrac{\overset{6}{\cancel{42}}}{37}=\dfrac{18}{37}$

19. $\left(3\dfrac{1}{2}\right)\left(4\dfrac{1}{7}\right)=\dfrac{7}{2}\cdot\dfrac{29}{7}=\dfrac{\overset{1}{\cancel{7}}}{2}\cdot\dfrac{29}{\underset{1}{\cancel{7}}}=\dfrac{29}{2}=14\dfrac{1}{2}$

21. $1\dfrac{1}{2}\div12=\left(\dfrac{3}{2}\right)\cdot\left(\dfrac{1}{12}\right)=\left(\dfrac{\overset{1}{\cancel{3}}}{2}\right)\cdot\left(\dfrac{1}{\underset{4}{\cancel{12}}}\right)=\left(\dfrac{1}{2}\right)\cdot\left(\dfrac{1}{4}\right)=\dfrac{1}{8}$ pounds

23. No solution provided.

25. a) $\sqrt{108}=\sqrt{36\cdot3}=\sqrt{36}\sqrt{3}=6\sqrt{3}$

b) $\dfrac{\sqrt{15}}{\sqrt{5}}=\sqrt{\dfrac{15}{5}}=\sqrt{3}$

27. Examples may vary.

$$8\div(4\div2)\neq(8\div4)\div2$$
$$8\div2\neq2\div2$$
$$4\neq1$$

29. In evaluating -2^4, we raise 2 to the fourth power to get 16 and then negate the result to get–16. In evaluating $(-2)^4$, we first negate 2 and then raise that negative number to the fourth power to get positive 16.

31. $1.325\times10^6=1{,}325{,}000$ and $8.63\times10^{-5}=0.0000863$

33. $\dfrac{6.6\times10^9}{109\times10^6}=\dfrac{6.6\times10^9}{1.09\times10^8}=\dfrac{6.6}{1.09}\times\dfrac{10^9}{10^8}\approx6.055\times10^{9-8}=6.055\times10^1=60.55$ times

35. a) Since $a_1=4$, $d=3$, and $n=30$, we have $a_{30}=4+(30-1)\cdot3=4+29\cdot3=4+87=91$.

b) Since $a_1=4$, $a_{11}=91$, and $n=30$, the sum of the first 30 terms is $\dfrac{30(4+91)}{2}=\dfrac{30\cdot95}{2}=$

$15\cdot95=1{,}425$.

37. $F_{10}=55;\ F_{15}=610$

$F_{13}=F_{11}+F_{12}$
$233=F_{11}+144$
$89=F_{11}$
$F_{12}=F_{10}+F_{11}$
$144=F_{10}+89$
$55=F_{10}$

$F_{14}=F_{12}+F_{13}=144+233=377$
$F_{15}=F_{13}+F_{14}=233+377=610$

Chapter Test

1. Refer to the table for Exercise 15 in the solutions for Section 6.1.
101,103,107,109,113

3. 241 is prime. You need to check for divisibility by 2, 3, 5, 7, 11, and 13.
539 is composite. $539=7^2\times11$

5. $1716=2^2\times3\times11\times13$ and $936=2^3\times3^2\times13$; $GCD(1716,936)=2^2\times3\times13=156$
$LCM(1716,936)=2^3\times3^2\times11\times13=10,296$

7. a) –6
b) +3
c) –54
d) +8

9. If you spend \$5 for the next seven days, you will have \$35 less. Thus $(+7)(-5)=-35$.

11. $50-(-19)=69$ degrees.

13. $F_{14}=377;\ F_{20}=6,765$

$F_{17}=F_{15}+F_{16}$
$1597=F_{15}+987$
$610=F_{15}$
$F_{16}=F_{14}+F_{15}$
$987=F_{14}+610$
$377=F_{14}$

$F_{18}=F_{16}+F_{17}987+1597=2584$
$F_{19}=F_{17}+F_{18}=1597+2584=4181$
$F_{20}=F_{18}+F_{19}=2584+4181=6765$

15. a) $\frac{3}{5}\cdot\left(\frac{7}{9}-\frac{3}{4}\right)=\frac{3}{5}\cdot\left(\left(\frac{7\cdot4-3\cdot9}{9\cdot4}\right)\right)=\frac{3}{5}\cdot\left(\frac{28-27}{36}\right)=\frac{\overset{1}{\cancel{3}}}{5}\cdot\left(\frac{1}{\underset{12}{\cancel{36}}}\right)=\frac{1}{60}$

b) $\frac{4}{5}\div\left(\frac{2}{5}-\frac{1}{4}\right)=\frac{4}{5}\div\left(\frac{2\cdot4-1\cdot5}{5\cdot4}\right)=\frac{4}{5}\div\left(\frac{8-5}{20}\right)=\frac{4}{5}\div\left(\frac{3}{20}\right)=\frac{4}{\underset{1}{\cancel{5}}}\ \frac{\overset{4}{\cancel{20}}}{3}=\frac{16}{3}$

17. $4\frac{1}{2}\div3\frac{3}{8}=\frac{9}{2}\div\frac{27}{8}=\frac{\overset{1}{\cancel{9}}}{\underset{1}{\cancel{2}}}\cdot\frac{\overset{4}{\cancel{8}}}{\underset{3}{\cancel{27}}}=\frac{4}{3}$

19. a) Since $a_1 = 11$, $d = 9$, and $n = 20$, we have $a_{20} = 11 + (20-1)\cdot 9 = 11 + 19\cdot 9 = 11 + 171 = 182$.

b) Since $a_1 = 11$, $a_{20} = 182$, and $n = 20$, the sum of the first 20 terms is $\dfrac{20(11+182)}{2} = \dfrac{20\cdot 193}{2} =$

$10\cdot 193 = 1{,}930$.

21. No solution provided.

23. a) $\sqrt{180} = \sqrt{36\cdot 5} = \sqrt{36}\sqrt{5} = 6\sqrt{5}$

b) $\dfrac{3}{\sqrt{15}} = \dfrac{3\cdot\sqrt{15}}{\sqrt{15}\cdot\sqrt{15}} = \dfrac{\overset{1}{\cancel{3}}\cdot\sqrt{15}}{\underset{5}{\cancel{15}}} = \dfrac{\sqrt{15}}{5}$

25. a) $2^7\cdot 2^{-4} = 2^{7+(-4)} = 2^3 = 8$

b) $\left(3^2\right)^{-2} = 3^{2(-2)} = 3^{-4} = \dfrac{1}{3^4} = \dfrac{1}{81}$

c) $\dfrac{5^7}{5^4} = 5^{7-4} = 5^3 = 125$

d) $\dfrac{3^{-2}}{3^{-5}} = 3^{-2-(-5)} = 3^3 = 27$

27. In evaluating -3^2, 3 is raised to the second power to get 9 before being negated to get –9. In evaluating $(-3)^2$, 3 is negated first and then that negative number is raised to the second power to get 9.

29. $\dfrac{1.3\times 10^9}{2.9\times 10^8} = \dfrac{1.3\times 10^9}{2.9\times 10^8} = \dfrac{1.3}{2.9}\times\dfrac{10^9}{10^8} \approx 0.44827\times 10^{9-8} = 4.48\times 10^1$ = 44.8 times larger

Chapter 7

Algebraic Models: How Do We Approximate Reality?

Section 7.1 Linear Equations

1. Add, subtract, multiply, and divide both sides by the same quantity.

3. -315 ; 12.50 and 15

5. A geometry in which we represent geometric objects such as points, lines, and circles by numbers and equations.

7.
$$\begin{aligned} 3x+4&=5x-6\\ -2x+4&=-6\\ -2x&=-10\\ x&=5 \end{aligned}$$

9.
$$\begin{aligned} 4-2y&=8+3y\\ 4&=8+5y\\ -4&=5y\\ \frac{-4}{5}&=y\\ y&=-\frac{4}{5} \end{aligned}$$

11.
$$\begin{aligned} \frac{1}{2}x+4&=\frac{3}{4}x-6\\ 4\cdot\left[\frac{1}{2}x+4\right]&=\left[\frac{3}{4}x-6\right]\cdot 4\\ 2x+16&=3x-24\\ -x+16&=-24\\ -x&=-40\\ x&=40 \end{aligned}$$

13.
$$\begin{aligned} \frac{1}{3}y+4&=\frac{1}{4}y+3\\ 12\cdot\left[\frac{1}{3}y+4\right]&=\left[\frac{1}{4}y+3\right]\cdot 12\\ 4y+48&=3y+36\\ y+48&=36\\ y&=-12 \end{aligned}$$

15.
$$\begin{aligned} 0.2x+6&=3x-0.4\\ 10\cdot[0.2x+6]&=[3x-0.4]\cdot 10\\ 2x+60&=30x-4\\ -28x+60&=-4\\ -28x&=-64\\ x&=\frac{-64}{-28}=\frac{16}{7} \end{aligned}$$

17.
$$\begin{aligned} 0.3y+2&=0.5y-3\\ 10\cdot[0.3y+2]&=[0.5y-3]\cdot 10\\ 3y+20&=5y-30\\ 20&=2y-30\\ 50&=2y\\ 25&=y\\ y&=25 \end{aligned}$$

19.
$$\begin{aligned} P&=2l+2w\\ P-2l&=2w\\ \frac{P-2l}{2}&=w\\ w&=\frac{P-2l}{2} \end{aligned}$$

21.
$$\begin{aligned} z&=\frac{x-\mu}{\sigma}\\ \sigma\cdot z&=\left[\frac{x-\mu}{\sigma}\right]\cdot\sigma\\ \sigma z&=x-\mu\\ \sigma z-x&=-\mu\\ -(\sigma z-x)&=\mu\\ x-\sigma z&=\mu\\ \mu&=x-\sigma z \end{aligned}$$

23.
$$A = P(1+rt)$$
$$A = P + \text{Pr}t$$
$$A - P = Prt$$
$$\frac{A-P}{Pt} = \frac{Prt}{Pt} = r$$
$$r = \frac{A-P}{Pt}$$

25. $2x+3y=6$
$$2x = -3y+6$$
$$x = \frac{-3y+6}{2}$$

27. $V = lwh$
$$\frac{V}{wh} = \frac{lwh}{wh} = l$$
$$l = \frac{V}{wh}$$

29.
$$S = 2\pi rh + 2\pi r^2$$
$$S - 2\pi r^2 = 2\pi rh$$
$$\frac{S-2\pi r^2}{2\pi r} = \frac{2\pi rh}{2\pi r} = h$$
$$h = \frac{S-2\pi r^2}{2\pi r}$$

31. x-intercept (when $y=0$)
$$3x + 2\cdot 0 = 12$$
$$3x + 0 = 12$$
$$3x = 12$$
$$x = 4 \Rightarrow (4,0)$$

y-intercept (when $x=0$)
$$3\cdot 0 + 2y = 12$$
$$0 + 2y = 12$$
$$2y = 12$$
$$y = 6 \Rightarrow (0,6)$$

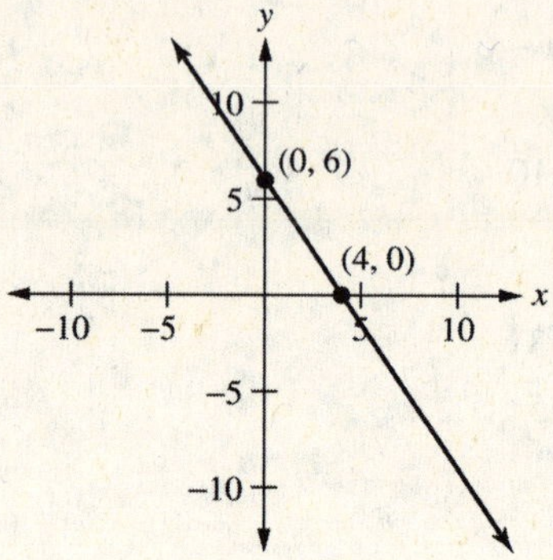

33. x-intercept (when $y=0$)
$$4x - 3\cdot 0 = 16$$
$$4x - 0 = 16$$
$$4x = 16$$
$$x = 4 \Rightarrow (4,0)$$

y-intercept (when $x=0$)
$$4\cdot 0 - 3y = 16$$
$$0 - 3y = 16$$
$$-3y = 16$$
$$y = -\frac{16}{3} \Rightarrow \left(0, -\tfrac{16}{3}\right)$$

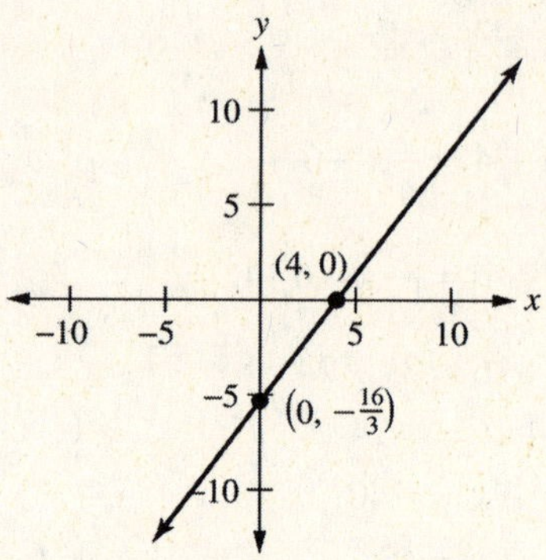

35. x-intercept (when $y = 0$)

$$\frac{1}{3}x + \frac{1}{2}\cdot 0 = 3$$
$$\frac{1}{3}x + 0 = 3$$
$$\frac{1}{3}x = 3$$
$$x = 9 \Rightarrow (9, 0)$$

y-intercept (when $x = 0$)

$$\frac{1}{3}\cdot 0 + \frac{1}{2}y = 3$$
$$0 + \frac{1}{2}y = 3$$
$$\frac{1}{2}y = 3$$
$$y = 6 \Rightarrow (0, 6)$$

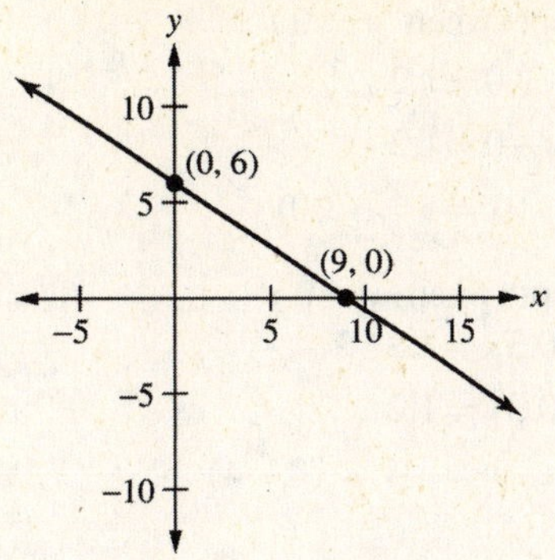

37. x-intercept (when $y = 0$)

$$\frac{1}{6}x - 2\cdot 0 = \frac{3}{4} \Rightarrow \frac{1}{6}x - 0 = \frac{3}{4}$$
$$\frac{1}{6}x = \frac{3}{4} \Rightarrow 12\cdot\left[\frac{1}{6}x\right] = \left[\frac{3}{4}\right]\cdot 12$$
$$2x = 9 \Rightarrow x = \frac{9}{2} \Rightarrow \left(\tfrac{9}{2}, 0\right)$$

y-intercept (when $x = 0$)

$$\frac{1}{6}\cdot 0 - 2y = \frac{3}{4} \Rightarrow \frac{1}{6}x - 0 = \frac{3}{4}$$
$$0 - 2y = \frac{3}{4}$$
$$-2y = \frac{3}{4} \Rightarrow y = -\frac{3}{8} \Rightarrow \left(0, -\tfrac{3}{8}\right)$$

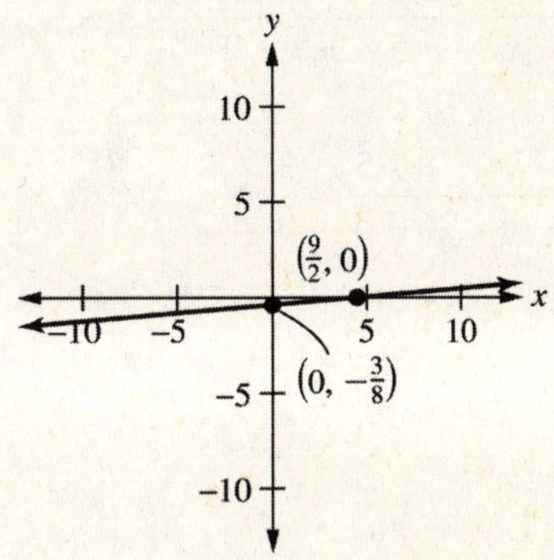

39. x-intercept (when $y = 0$)

$$0.2x = 4\cdot 0 + 1.6$$
$$0.2x = 0 + 1.6$$
$$0.2x = 1.6$$
$$x = 8 \Rightarrow (8, 0)$$

y-intercept (when $x = 0$)

$$0.2\cdot 0 = 4y + 1.6$$
$$0 = 4y + 1.6$$
$$-1.6 = 4y$$
$$-0.4 = y \Rightarrow (0, -0.4)$$

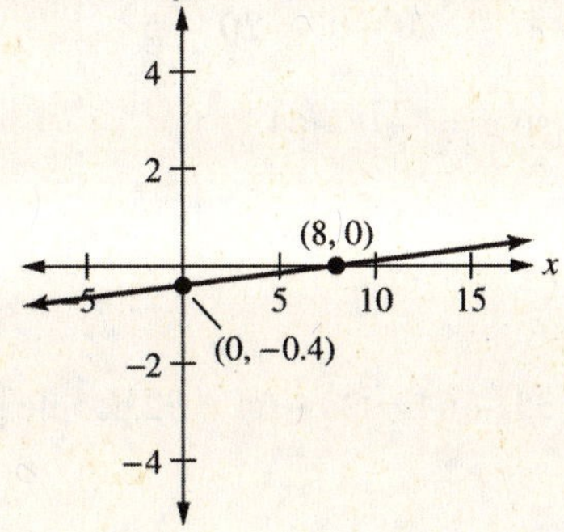

41. x-intercept (when $y=0$)

$$0.4x-0.3\cdot 0=1.2$$
$$0.4x-0=1.2$$
$$x=3 \Rightarrow (3,0)$$

y-intercept (when $x=0$)

$$0.4\cdot 0-0.3y=1.2$$
$$0.-0.3y=1.2$$
$$y=-4 \Rightarrow (0,-4)$$

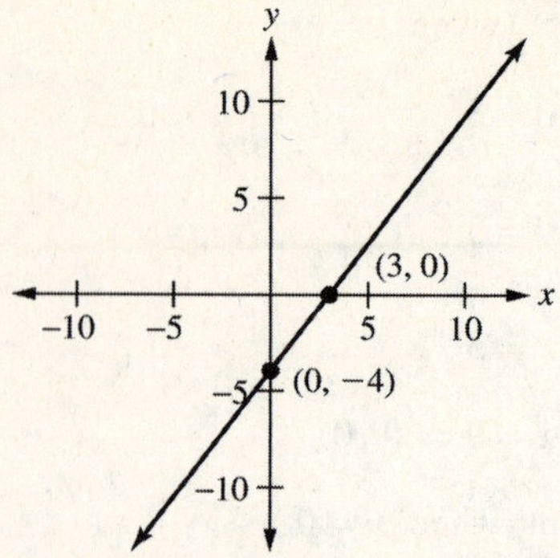

43. $m=\dfrac{y_2-y_1}{x_2-x_1}=\dfrac{8-5}{6-2}=\dfrac{3}{4}$

45. $m=\dfrac{y_2-y_1}{x_2-x_1}=\dfrac{2-6}{8-3}=\dfrac{-4}{5}=-\dfrac{4}{5}$

47. $m=\dfrac{y_2-y_1}{x_2-x_1}=\dfrac{1-(-4)}{5-3}=\dfrac{5}{2}$

49. $m=\dfrac{y_2-y_1}{x_2-x_1}=\dfrac{8-5}{6-6}=\dfrac{3}{0}$;

slope does not exist

51. a and d

53. f

55. The rise is zero.

57. y-intercept is $(0,-3)$; slope is 4

59. y-intercept is $(0,-3)$; slope is –5

61. Let h be the number of hours Jillian used the club last year.

$$95+2.50h=515$$
$$2.50h=420$$
$$h=168$$

168 hours

63. Let p be the number of pages.

$$16+1.30(p-10)=44.60$$
$$16+1.30p-13=44.60$$
$$3+1.30p=44.60$$
$$1.30p=41.60$$
$$p=32 \text{ pages}$$

65. Let d be the population density and t be years since 1980.

$$m=\frac{d_2-d_1}{t_2-t_1}=\frac{80-64}{20-0}=\frac{16}{20}=\frac{4}{5};\ b=64$$
$$\text{so } d=\left(\frac{4}{5}\right)t+64$$

67. For 2015, $t=35$;

$$d=\left(\frac{4}{5}\right)(35)+64=28+64=92 \text{ people per square mile.}$$

69. Let c be college expenses and t be years since 2007.

$$m=(0.041)(11900)=488;\ b=11900$$
$$\text{so } c=488t+11900$$

In 2015: $c=488(8)+11900=3904+11900=\$15,804$

71. Let n be the net pay.
Let g be the gross pay.

$n = g - 10 - 0.0675g - 0.14g - 0.021g - 17$

$n = 0.7715g - 27$

$n = 0.7715 \cdot 400 - 27$

$n = 308.6 - 27$

$n = 281.60$

\$281.60

73. Let d be the number of hours worked in the diner.
Let p be the number of hours worked in the pet store.

$5.60d + 7.35p = 133$

75. Let t be the number of shares with Time-Warner.
Let c be the number of shares with CDW.

$35t + 55c = 14,500$

77. Let e be the number of end tables to be constructed.
Let c be the number of coffee tables to be constructed.

$9e + 15c = 342$

79. Let x be the number of tokens.

$9 + 0.70x = 0.85x$

$9 = 0.15x$

$60 = x$

after 60 tokens

81. the y-coordinate

83. the x- and y-intercepts

85. Let x be the number of computer systems Chuck sells.
$y = 45x + 225$ is the amount Chuck would receive from Buy More.
$y = 20x + 400$ is the amount Chuck would receive from Circuit Town.
To find when both deals are the same, we solve $45x + 225 = 20x + 400$.

$$45x + 225 = 20x + 400$$
$$25x + 225 = 400$$
$$25x = 175$$
$$x = 7$$

If Chuck sells 7 computer systems, then the two deals are the same. Because he would receive more per computer system sold, if Chuck sells more than seven systems, then Buy More is better for him.

87. – 89. No solution provided.

91. amount of interest on the CD: $5000(0.038) = 190$
amount of tax paid to the government: $190(0.14) = 26.60$
amount realized on the CD: $190 - 26.60 = 163.40$
Let r be the interest rate needed to be earned on the bonds.

$163.40 = 5000 \cdot r$

$0.03268 = r$

The bonds would need to earn 3.268%.

93. a) $d = \frac{y}{105}$

b) $y = 105d$

95. $e = 0.68d \Rightarrow \dfrac{e}{0.68} = d$

$p = 2380d \Rightarrow \dfrac{p}{2380} = d$

$\dfrac{e}{0.68} = \dfrac{p}{2380} \Rightarrow p = \dfrac{2380}{0.68}e \Rightarrow$

$p = 3{,}500e$

97. $m = 0.08 = \dfrac{18}{x}$

$x = \dfrac{18}{0.08} = 225 \text{ inches}$

$225 - 12 = 213 \text{ inches}$

$\dfrac{213}{12} = 17.75 \text{ feet}$

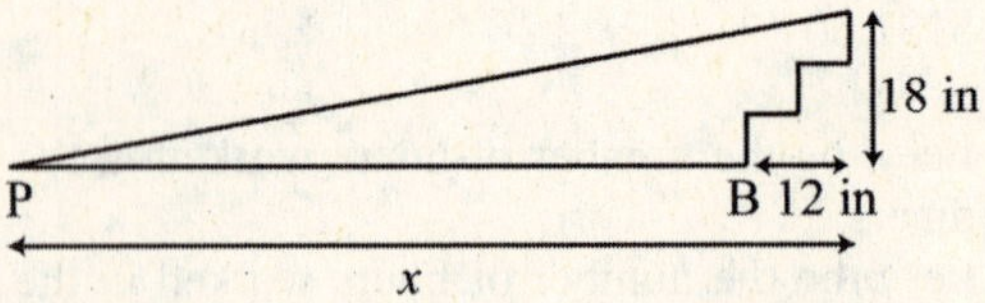

Section 7.2 Modeling with Linear Equations

1. Write the equation of the line in standard form; state the x- and y-intercepts; specify the slope and y-intercept.

3. Two points were again used to find the equation of the line.

5. $y = mx + b$

$y = 3x + b$

$1 = 3 \cdot 2 + b \Rightarrow 1 = 6 + b \Rightarrow -5 = b$

$y = 3x - 5$

7. $y = mx + b$

$y = 4x + b$

$6 = 4 \cdot (-2) + b \Rightarrow 6 = -8 + b \Rightarrow 14 = b$

$y = 4x + 14$

9. $y = mx + b$

$y = -2x + b$

$3 = -2 \cdot 4 + b \Rightarrow 3 = -8 + b \Rightarrow 11 = b$

$y = -2x + 11$

11. $y = mx + b$

$y = -5x + b$

$-1 = -5 \cdot (-6) + b \Rightarrow -1 = 30 + b \Rightarrow$

$b = -31$

$y = -5x - 31$

13. $m = \dfrac{y_2 - y_1}{x_2 - x_1} = \dfrac{9-3}{5-2} = \dfrac{6}{3} = 2$

$y = mx + b$

$y = 2x + b$

$3 = 2 \cdot 2 + b \Rightarrow 3 = 4 + b \Rightarrow -1 = b$

$y = 2x - 1$

15. $m = \dfrac{y_2 - y_1}{x_2 - x_1} = \dfrac{10-12}{9-17} = \dfrac{-2}{-8} = \dfrac{1}{4}$

$y = mx + b$

$y = \dfrac{1}{4}x + b$

$10 = \dfrac{1}{4} \cdot 9 + b \Rightarrow 10 = \dfrac{9}{4} + b \Rightarrow \dfrac{31}{4} = b$

$y = \dfrac{1}{4}x + \dfrac{31}{4}$

17. $m=\frac{y_2-y_1}{x_2-x_1}=\frac{2-(-4)}{-8-11}=\frac{6}{-19}=-\frac{6}{19}$

$y=mx+b$

$y=-\frac{6}{19}x+b$

$2=-\frac{6}{19}\cdot(-8)+b \Rightarrow 2=\frac{48}{19}+b\Rightarrow$

$b=-\frac{10}{19}$

$y=-\frac{6}{19}x-\frac{10}{19}$

19. $m=\frac{y_2-y_1}{x_2-x_1}=\frac{-1-(-8)}{-4-(-6)}=\frac{7}{2}$

$y=mx+b$

$y=\frac{7}{2}x+b$

$-1=\frac{7}{2}\cdot(-4)+b$

$-1=-14+b$

$13=b$

$y=\frac{7}{2}x+13$

21. a) Let x be the number of years after 2007.
Let y be the life expectancy in years.
$y=0.3x+80.97$

b) $y=0.3\cdot 13+80.97=3.9+80.97=84.87$ years

23. a) Let x be the number of years after 2005.
Let y be the number of male students enrolled in college (in millions).
$y=0.133x+7.4$

b) $y=0.133\cdot 15+7.4=1.995+7.400=9.395$
9,395,000 estimated male students enrolled

c) $10=0.133\cdot t+7.4\Rightarrow 2.6=0.133\cdot t\Rightarrow t=\frac{2.6}{0.133}\Rightarrow t\approx 20\Rightarrow 2005+20=2025$

25. a) Let x be the number of years after 2005.
Let y be the price of tuition.
$y=760x+11,400$

b) $y=760\cdot 10+11,400=7,600+11,400=19,000$
\$19,000 estimated price of tuition

27. a) $m=\frac{v_2-v_1}{t_2-t_1}=\frac{466-495}{2-0}$

$=\frac{-29}{2}=-14.5$

b) $v=mt+b$

$v=-14.5t+b$

$495=-14.5\cdot 0+b$

$495=0+b$

$495=b$

$v=-14.5t+495$

c) $v=-14.5\cdot 13+495=-188.5+495$

$=306.5$

29. Let x be the number of years after 1998.

Let y be the number of database administrators (in thousands).

The given information corresponds to the two ordered pairs (0,87) and (10,155).

$m=\frac{y_2-y_1}{x_2-x_1}=\frac{155-87}{10-0}=\frac{68}{10}=6.8$

$y=mx+b$

$y=6.8x+b$

$87=6.8\cdot 0+b$

$87=0+b$

$87=b$

$y=6.8x+87$

$y=6.8\cdot 17+87=115.6+87=202.6$

202,600 estimated database administrators in 2015

31. Let x be the number of years after 1900.
Let y be the number of hours of sleep.
The given information corresponds to the two ordered pairs (0,8.5) and (102,6.9).

$$m=\frac{y_2-y_1}{x_2-x_1}=\frac{6.9-8.5}{104-0}=\frac{-1.6}{104}=-\frac{16}{1040}=-\frac{1}{65}\approx-0.0154$$

$$y=mx+b$$
$$y=-0.0154x+b$$
$$8.5=-0.0154\cdot 0+b$$
$$8.5=0+b$$
$$8.5=b$$
$$y=-0.0154x+8.5$$
$$0=-0.0154\cdot x+8.5$$
$$0.0154x=8.5$$
$$x\approx 552$$

In the year $1900+552=2452$, it is estimated that Americans won't be sleeping at all on a weeknight.

33. a) $m=\dfrac{l_2-l_1}{t_2-t_1}=\dfrac{151.3-146}{3-0}=\dfrac{5.3}{3}\approx 1.77$

b) $l=mt+b$

$$l=1.77t+b$$
$$146=1.3\cdot 0+b$$
$$146=0+b$$
$$146=b$$
$$l=1.77t+146$$

c) The U.S. civilian labor force will reach 200 million in the year 2034.

$$200=1.77t+146$$
$$54=1.77t$$
$$t=\frac{54}{1.77}\approx 30.51$$
$$2003+30.51=2033.51$$

35. the slope of the line

37. – 39. No solution provided.

41.

	2003 (year 0)	2004 (year 1)	2005 (year 2)	2006 (year 3)
Actual data	146	147	149.2	151.3
Values predicted by your model	$l=1.77\cdot 0+146$ $=0+146$ $=146$	$l=1.77\cdot 1+146$ $=1.77+146$ $=147.77$	$l=1.77\cdot 2+146$ $=3.54+146$ $=149.54$	$l=1.77\cdot 3+146$ $=5.31+146$ $=151.3$
Values predicted by line of best fit	$l=1.81\cdot 0+145.66$ $=0+145.66$ $=145.66$	$l=1.81\cdot 0+145.66$ $=1.81+145.66$ $=147.47$	$l=1.81\cdot 2+145.66$ $=3.62+145.66$ $=149.28$	$l=1.81\cdot 3+145.66$ $=5.43+145.66$ $=151.09$

43. No solution provided.

Section 7.3 Modeling with Quadratic Equations

1. To solve a quadratic equation.

3. Even if we use accurate data, if we are using the wrong model, then out predictions are not valid.

5. By using the quadratic formula, $x=\dfrac{-b\pm\sqrt{b^2-4ac}}{2a}$

$$x=\frac{-(-10)\pm\sqrt{(-10)^2-4(1)(16)}}{2(1)}=\frac{10\pm\sqrt{100-64}}{2}=\frac{10\pm\sqrt{36}}{2}=\frac{10\pm 6}{2}$$

$$x=\frac{10+6}{2}=\frac{16}{2}=8 \text{ or } x=\frac{10-6}{2}=\frac{4}{2}=2$$

7. $x=\dfrac{-(-5)\pm\sqrt{(-5)^2-4(2)(3)}}{2(2)}=\dfrac{5\pm\sqrt{25-24}}{4}=\dfrac{5\pm\sqrt{1}}{4}=\dfrac{5\pm 1}{4}$

$$x=\frac{5+1}{4}=\frac{6}{4}=\frac{3}{2} \text{ or } x=\frac{5-1}{4}=\frac{4}{4}=1$$

9. $x=\dfrac{-(7)\pm\sqrt{(7)^2-4(3)(-6)}}{2(3)}=\dfrac{-7\pm\sqrt{49-(-72)}}{6}=\dfrac{-7\pm\sqrt{121}}{6}=\dfrac{-7\pm 11}{6}$

$$x=\frac{-7+11}{6}=\frac{4}{6}=\frac{2}{3} \text{ or } x=\frac{-7-11}{6}=\frac{-18}{6}=-3$$

11. $x=\dfrac{-(-17)\pm\sqrt{(-17)^2-4(5)(-12)}}{2(5)}=\dfrac{17\pm\sqrt{289-(-240)}}{10}=\dfrac{17\pm\sqrt{529}}{10}=\dfrac{17\pm 23}{10}$

$$x=\frac{17+23}{10}=\frac{40}{10}=4 \text{ or } x=\frac{17-23}{10}=\frac{-6}{10}=-\frac{3}{5}$$

13. Opening: down since $a<0$

Vertex: $x=\dfrac{-b}{2a}=\dfrac{-6}{2(-1)}=\dfrac{-6}{-2}=3$. Substituting this value for *x*, we obtain

$y=-3^2+6\cdot 3-8=-9+18-8=1$ as the *y*-coordinate of the vertex.

The vertex is therefore $(3,1)$.

x-intercepts: Set $y=0$. We get $0=-x^2+6x-8$ and by using the quadratic formula, we obtain

$$x=\frac{-(6)\pm\sqrt{(6)^2-4(-1)(-8)}}{2(-1)}=\frac{-6\pm\sqrt{36-32}}{-2}=\frac{-6\pm\sqrt{4}}{-2}=\frac{-6\pm 2}{-2}$$

$$x=\frac{-6+2}{-2}=\frac{-4}{-2}=2 \text{ or } x=\frac{-6-2}{-2}=\frac{-8}{-2}=4$$

The *x*-intercepts of the graph are therefore $(2,0)$ and $(4,0)$.

y-intercept: Set $x=0$. Thus $y=-0^2+6\cdot 0-8=-8$. The *y*-intercept is therefore $(0,-8)$.

13. (continued)

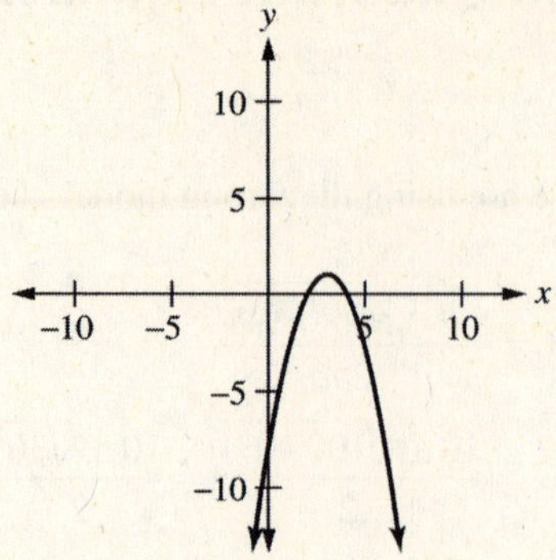

15. Opening: down since $a<0$

Vertex: $x=\frac{-b}{2a}=\frac{-8}{2(-4)}=\frac{-8}{-8}=1$. Substituting this value for x, we obtain

$y=-4\cdot1^2+8\cdot1+5=-4+8+5=9$ as the y-coordinate of the vertex.

The vertex is therefore $(1,9)$.

x-intercepts: Set $y=0$. We get $0=-4x^2+8x+5$ and by using the quadratic formula, we obtain

$$x=\frac{-(8)\pm\sqrt{(8)^2-4(-4)(5)}}{2(-4)}=\frac{-8\pm\sqrt{64+80}}{-8}=\frac{-8\pm\sqrt{144}}{-8}=\frac{-8\pm12}{-8}$$

$$x=\frac{-8+12}{-8}=\frac{4}{-8}=-\frac{1}{2} \text{ or } x=\frac{-8-12}{-8}=\frac{-20}{-8}=\frac{5}{2}$$

The x-intercepts of the graph are therefore $\left(-\frac{1}{2},0\right)$ and $\left(\frac{5}{2},0\right)$.

y-intercept: Set $x=0$. Thus $y=-4\cdot0^2+8\cdot0+5=5$. The y-intercept is therefore $(0,5)$.

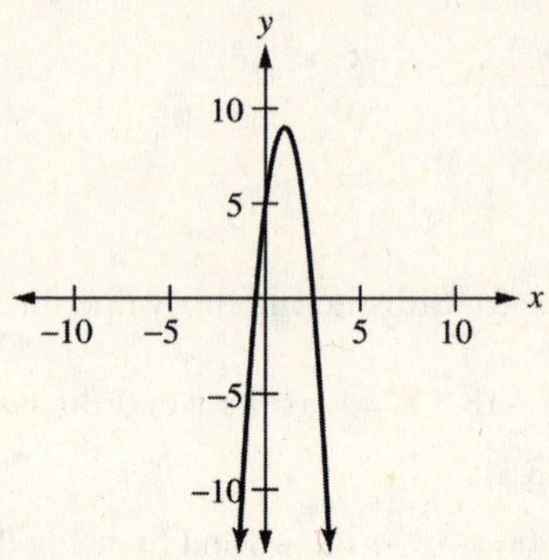

17. Opening: up since $a>0$

Vertex: $x=\frac{-b}{2a}=\frac{-(-4)}{2(4)}=\frac{4}{8}=\frac{1}{2}$. Substituting this value for x, we obtain

$y=4\cdot\left(\frac{1}{2}\right)^2-4\cdot\frac{1}{2}-2=1-2-2=-3$ as the y-coordinate of the vertex.

The vertex is therefore $\left(\frac{1}{2},-3\right)$.

17. (continued)

x-intercepts: Set $y=0$. We get $0=4x^2-4x-2$ and by using the quadratic formula, we obtain

$$x=\frac{-(-4)\pm\sqrt{(-4)^2-4(4)(-2)}}{2(4)}=\frac{4\pm\sqrt{16+32}}{8}=\frac{4\pm\sqrt{48}}{8}=\frac{4\pm4\sqrt{3}}{8}=\frac{1\pm\sqrt{3}}{2}$$

$$x=\frac{1+\sqrt{3}}{2}\approx1.4 \text{ or } x=\frac{1-\sqrt{3}}{2}=\approx-0.4$$

The x-intercepts of the graph are therefore $\left(\frac{1+\sqrt{3}}{2},0\right)$ and $\left(\frac{1-\sqrt{3}}{2},0\right)$.

y-intercept: Set $x=0$. Thus $y=4\cdot0^2-4\cdot0-2=-2$. The y-intercept is therefore $(0,-2)$.

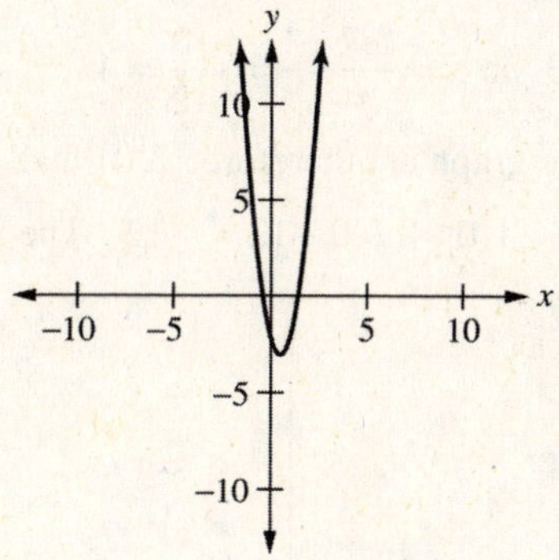

19. Opening: up since $a>0$

Vertex: $x=\frac{-b}{2a}=\frac{-(7)}{2(3)}=\frac{-7}{6}=-\frac{7}{6}$. Substituting this value for x, we obtain

$$y=3\cdot\left(-\frac{7}{6}\right)^2+7\cdot\left(-\frac{7}{6}\right)-6=\frac{49}{12}+\left(-\frac{49}{6}\right)-6=-\frac{121}{12}$$ as the y-coordinate of the vertex.

The vertex is therefore $\left(-\frac{7}{6},-\frac{121}{12}\right)$.

x-intercepts: Set $y=0$. We get $0=3x^2+7x-6$ and by using the quadratic formula, we obtain

$$x=\frac{-(7)\pm\sqrt{(7)^2-4(3)(-6)}}{2(3)}=\frac{-7\pm\sqrt{49+72}}{6}=\frac{-7\pm\sqrt{121}}{6}=\frac{-7\pm11}{6}=$$

$$x=\frac{-7+11}{6}=\frac{4}{6}=\frac{2}{3} \text{ or } x=\frac{-7-11}{6}=-3$$

The x-intercepts of the graph are therefore $\left(\frac{2}{3},0\right)$ and $(-3,0)$.

y-intercept: Set $x=0$. Thus $y=3\cdot0^2+7\cdot0-6=-6$. The y-intercept is therefore $(0,-6)$.

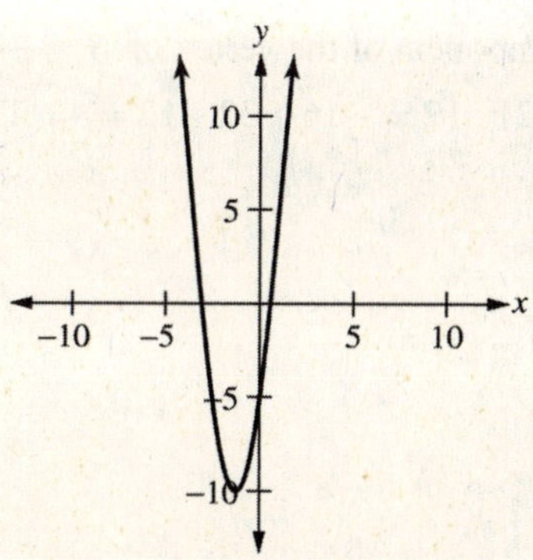

21. Opening: down since $a < 0$

Vertex: $x = \frac{-b}{2a} = \frac{-(7)}{2(-1)} = \frac{-7}{-2} = \frac{7}{2}$. Substituting this value for x, we obtain

$y = -\left(\frac{7}{2}\right)^2 + 7 \cdot \left(\frac{7}{2}\right) - 12 = -\frac{49}{4} + \frac{49}{2} - 12 = \frac{1}{4}$ as the y-coordinate of the vertex.

The vertex is therefore $\left(\frac{7}{2}, \frac{1}{4}\right)$.

x-intercepts: Set $y = 0$. We get $0 = -x^2 + 7x - 12$ and by using the quadratic formula, we obtain

$$x = \frac{-(7) \pm \sqrt{(7)^2 - 4(-1)(-12)}}{2(-1)} = \frac{-7 \pm \sqrt{49-48}}{-2} = \frac{-7 \pm \sqrt{1}}{-2} = \frac{-7 \pm 1}{-2}$$

$$x = \frac{-7+1}{-2} = \frac{-6}{-2} = 3 \text{ or } x = \frac{-7-1}{-2} = \frac{-8}{-2} = 4$$

The x-intercepts of the graph are therefore $(3,0)$ and $(4,0)$.

y-intercept: Set $x = 0$. Thus $y = -1 \cdot 0^2 + 7 \cdot 0 - 12 = -12$. The y-intercept is therefore $(0, -12)$.

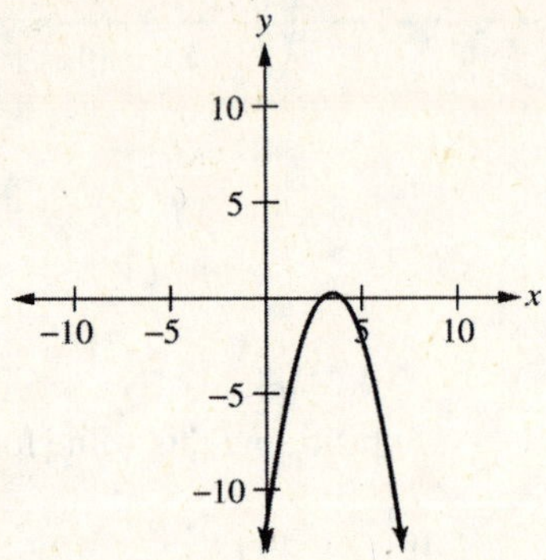

23. a) Week 2: $-0.125 \cdot 2^2 + 1.75 \cdot 2 + 2 = -0.5 + 3.5 + 2 = 5$;
Week 4: $-0.125 \cdot 4^2 + 1.75 \cdot 4 + 2 = -2 + 7 + 2 = 7$
Week 6: $-0.125 \cdot 6^2 + 1.75 \cdot 6 + 2 = -4.5 + 10.5 + 2 = 8$

b) We need to find the vertex of $A = -0.125x^2 + 1.75x + 2$. $x = \frac{-b}{2a} = \frac{-(1.75)}{2(-0.125)} = \frac{-1.75}{-0.25} = 7$

Substituting this value for x, we obtain
$A = -0.125 \cdot (7)^2 + 1.75 \cdot (7) + 2 = -6.125 + 12.25 + 2 = 8.125$.

25. a) We need to find the first component of the vertex of $S = -4n^2 + 16n - 12$
$n = \frac{-b}{2a} = \frac{-(16)}{2(-4)} = \frac{-16}{-8} = 2$ implies in week 2.

b) We need to find the second component of the vertex of $S = -4n^2 + 16n - 12$. Substituting $n = 2$, we obtain $S = -4 \cdot (2)^2 + 16 \cdot (2) - 12 = -16 + 32 - 12 = 4$. This implies \$4 million.

25. (continued)

c) This corresponds to solving the equation $0 = -4n^2 + 16n - 12$ and by using the quadratic formula, we obtain

$$n = \frac{-(16) \pm \sqrt{(16)^2 - 4(-4)(-12)}}{2(-4)} = \frac{-16 \pm \sqrt{256 - 192}}{-8} = \frac{-16 \pm \sqrt{64}}{-8} = \frac{-16 \pm 8}{-8}$$

$$n = \frac{-16+8}{-8} = \frac{-8}{-8} = 1 \text{ or } n = \frac{-16-8}{-8} = \frac{-24}{-8} = 3$$

We can expect the sales to drop to zero in week 3.

27. a) Vertex: $t = \frac{-b}{2a} = \frac{-0}{2(-16)} = \frac{0}{-32} = 0$. Substituting this value for *t*, we obtain

$H = 160 - 16 \cdot 0^2 = 160 - 0 = 160$ as the second coordinate of the vertex, $(0,160)$

t-intercepts: We need to solve $0 = 160 - 16t^2$ and by using the quadratic formula, we obtain

$$t = \frac{-(0) \pm \sqrt{(0)^2 - 4(-16)(160)}}{2(-16)} = \frac{0 \pm \sqrt{10240}}{-32} = \frac{\pm 32\sqrt{10}}{-32} = \pm\sqrt{10} \approx \pm 3.16$$

The *t*-intercepts of the graph are therefore $(-\sqrt{10},0)$ and $(\sqrt{10},0)$.

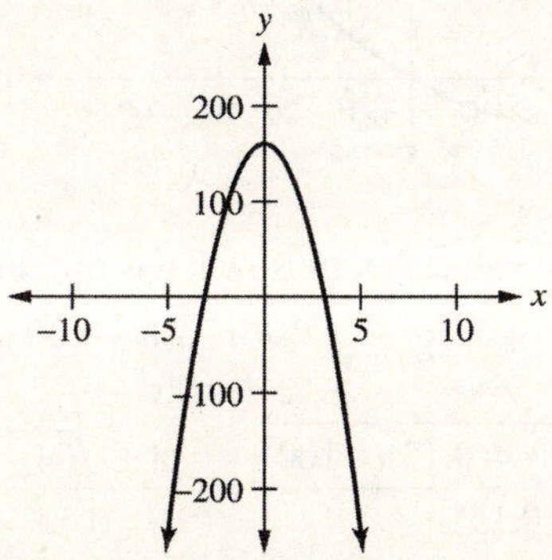

b) *t* (time) cannot be negative.

c) Approximately after 3.16 seconds

29. a) $1.52(0)^2 - 3.2(0) + 196 = 196$; $(0,196)$ is a solution to the equation

$1.52(10)^2 - 3.2(10) + 196 = 316$; $(10,316)$ is a solution to the equation

$1.52(20)^2 - 3.2(20) + 196 = 740$; $(20,740)$ is a solution to the equation

b) $1.52(30)^2 - 3.2(30) + 196 = 1,468$

The equation predicts there will be 1,468 prisoners in 2000; which is greater than the actual number is 1,391.

31. $b^2 - 4ac$; If it is negative, there are no real roots for the equation; if it is zero, there is one real root; if it is positive, there are two distinct real roots.

33. The crate begins falling a point above the ground and picks up speed as it falls. A linear equation would not be appropriate, because the rate of change of the distance of the crate above the ground is not constant.

35. a) Vertex: $t=\frac{-b}{2a}=\frac{-8}{2(0.15)}=\frac{-8}{0.3}=-\frac{80}{3}\approx-26.7$. Substituting this value for *t*, we obtain

$d=0.15\left(-\frac{80}{3}\right)^2+8\cdot\left(-\frac{80}{3}\right)=\frac{320}{3}+\frac{-640}{3}=-\frac{320}{3}\approx-106.7$ as the second coordinate of the vertex, $\left(-\frac{80}{3},-\frac{320}{3}\right)$.

t-intercepts: We need to solve $0=0.15t^2+8t$ and by using the quadratic formula, we obtain

$$t=\frac{-(8)\pm\sqrt{(8)^2-4(0.15)(0)}}{2(0.15)}=\frac{-8\pm\sqrt{64-0}}{0.30}=\frac{-8\pm\sqrt{64}}{0.30}=\frac{-8\pm8}{0.30}$$

$$t=\frac{-8-8}{0.30}\approx-53.3 \text{ or } t=\frac{-8+8}{0.30}=0.$$

The *t*-intercepts of the graph are therefore approximately $(-53.3,0)$ and $(0,0)$

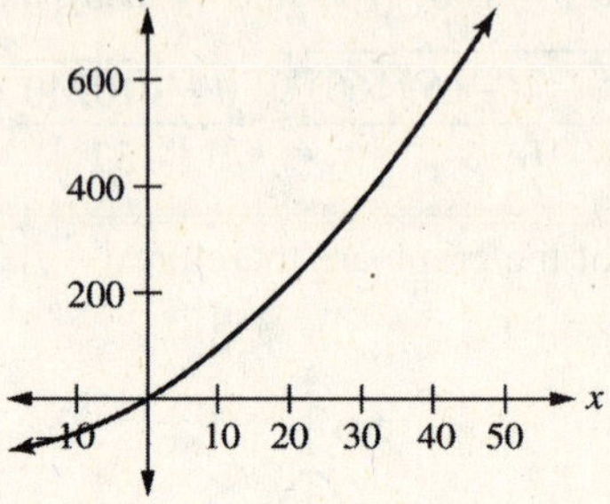

b) *t* (time) cannot be negative.

c) Considering only non-negative values, Dom Bowden is running more slowly when t is close to 0.

d) We need to solve $100=0.15t^2+8t$ or $0=0.15t^2+8t-100$ and by using the quadratic formula, we obtain

$$x=\frac{-(8)\pm\sqrt{(8)^2-4(0.15)(-100)}}{2(0.15)}=\frac{-8\pm\sqrt{64+60}}{0.30}=\frac{-8\pm\sqrt{124}}{0.30}$$

$$t=\frac{-8-\sqrt{124}}{0.30}\approx-63.79 \text{ or } t=\frac{-8+\sqrt{124}}{0.30}\approx10.45$$

Since we disregard the negative solution, it would take approximately 10.45 seconds for Dom Bowden to finish the race.

37. – 39. No solution provided.

41. a) In 2004: $y=2.5625(4)^2+10.975(4)+220.45=305.35$ which is close to 301.

In 2004: $y=2.5625(6)^2+10.975(6)+220.45=378.55$ which is close to 380.

41. (continued)

b) We need to solve $2.5625x^2 + 10.975x + 220.45 = 1000$ or $2.5625x^2 + 10.975x - 779.55 = 0$ and by using the quadratic formula, we obtain

$$x = \frac{-(10.975) \pm \sqrt{(10.975)^2 - 4(2.5625)(-779.55)}}{2(2.5625)}$$

$$x = \frac{-10.975 \pm \sqrt{120.450625 + 7990.3875}}{5.125} = \frac{-10.975 \pm \sqrt{8100.838125}}{5.125}$$

$$x = \frac{-10.975 - \sqrt{8100.838125}}{5.125} \approx -19.7; x = \frac{-10.975 + \sqrt{8100.838125}}{5.125} \approx 15.4$$

Since we disregard the negative solution, it would take approximately 15.4 years for Medicare expenditures to reach $1 trillion., which would occur between 2000 + 15 = 2015 and 2000 + 16 = 2016.

Section 7.4 Exponential Equations and Growth

1. A is the amount after compounding, P is the principle, r is the interest rate, and n is the number of years the money is compounded.

3. The initial population and the growth rate.

5. As the years of compounding increases, neither type of model was very accurate.

7. $A = 1{,}000(1+0.05)^1 = 1{,}000 \cdot 1.05 = \$1{,}050$

9. $A = 4{,}000(1+0.025)^1 = 4{,}000 \cdot 1.025 = \$4{,}100$

11. $A = 5{,}000(1+0.05)^5 = 5{,}000 \cdot 1.05^5 \approx \$6{,}381.40$

13. $A = 4{,}000(1+0.08)^2 = 4{,}000 \cdot 1.08^2 = \$4{,}665.60$

15. 3.24%

$$22000 = 15000(1+r)^{12}$$

$$\frac{22000}{15000} = (1+r)^{12}$$

$$(1+r)^{12} = \frac{22}{15}$$

$$1+r = \sqrt[12]{\frac{22}{15}}$$

$$r = \sqrt[12]{\frac{22}{15}} - 1 \approx 0.0324$$

17. 3.33%

$$13000 = 10000(1+r)^8$$

$$\frac{13000}{10000} = (1+r)^8$$

$$(1+r)^8 = 1.3$$

$$1+r = \sqrt[8]{1.3}$$

$$r = \sqrt[8]{1.3} - 1 \approx 0.0333$$

19. $P(1+r)^n = 1{,}129(1+0.016)^{13} = 1{,}129 \cdot 1.016^{13} \approx 1{,}387.8$ (million)

21. $P(1+r)^n = 14(1+0.0173)^{13} = 14 \cdot 1.0173^{13} \approx 17.5$ (million)

23. 0.47%

$$1{,}322 = 1{,}285(1+r)^6$$
$$(1+r)^6 = \frac{1{,}322}{1{,}285}$$
$$1+r = \sqrt[6]{\frac{1{,}322}{1{,}285}}$$
$$r = \sqrt[6]{\frac{1{,}322}{1{,}285}} - 1 \approx 0.0047$$

25. 0.97%

$$124 = 117(1+r)^6$$
$$(1+r)^6 = \frac{124}{117}$$
$$1+r = \sqrt[6]{\frac{124}{117}}$$
$$r = \sqrt[6]{\frac{124}{117}} - 1 \approx 0.0097$$

27.
$$5^x = 20$$
$$\log 5^x = \log 20$$
$$x \log 5 = \log 20$$
$$x = \frac{\log 20}{\log 5} \approx \frac{1.30103}{0.69897} \approx 1.86$$

29.
$$3^x = 10$$
$$\log 3^x = \log 10$$
$$x \log 3 = \log 10$$
$$x = \frac{\log 10}{\log 3} \approx \frac{1}{0.47712} \approx 2.10$$

31.
$$10^x = 3.2$$
$$\log 10^x = \log 3.2$$
$$x \log 10 = \log 3.2$$
$$x = \frac{\log 3.2}{\log 10} \approx \frac{0.50515}{1} \approx 0.51$$

33.
$$(3.4)^x = 6.85$$
$$\log (3.4)^x = \log 6.85$$
$$x \log 3.4 = \log 6.85$$
$$x = \frac{\log 6.85}{\log 3.4} \approx \frac{0.83569}{0.53148} \approx 1.57$$

35.
$$P_{4+1} = [1 + (0.03) \cdot (1 - 0.36)] \cdot 0.36$$
$$P_5 = [1 + (0.03) \cdot 0.64] \cdot 0.36$$
$$P_5 = [1 + 0.0192] \cdot 0.36$$
$$P_5 = 1.0192 \cdot 0.36$$
$$P_5 \approx 0.3669$$

At the end of the fifth year, the population is at approximately 36.69% of its maximum capacity.

37.
$$P_{8+1} = [1 + (0.045) \cdot (1 - 0.72)] \cdot 0.72$$
$$P_9 = [1 + (0.045) \cdot 0.28] \cdot 0.72$$
$$P_9 = [1 + 0.0126] \cdot 0.72$$
$$P_9 = 1.0126 \cdot 0.72$$
$$P_9 \approx 0.7291$$

At the end of the ninth year, the population is at approximately 72.91% of its maximum capacity.

39.
$$P_{0+1} = [1 + (0.08) \cdot (1 - 0.30)] \cdot 0.30$$
$$P_1 = [1 + (0.08) \cdot 0.70] \cdot 0.30$$
$$P_1 = [1 + 0.056] \cdot 0.30$$
$$P_1 = 1.056 \cdot 0.30$$
$$P_1 = 0.3168$$

$$P_{1+1} = [1 + (0.08) \cdot (1 - 0.3168)] \cdot 0.3168$$
$$P_2 = [1 + (0.08) \cdot 0.6832] \cdot 0.3168$$
$$P_2 = [1 + 0.054656] \cdot 0.3168$$
$$P_2 = 1.054656 \cdot 0.3168$$
$$P_2 \approx 0.3341150208$$

$$P_{2+1} \approx [1 + (0.08) \cdot (1 - 0.3341150208)] \cdot 0.3341150208$$
$$P_3 \approx [1 + (0.08) \cdot 0.6658849792] \cdot 0.3341150208$$
$$P_3 \approx [1 + 0.053270798336] \cdot 0.3341150208$$
$$P_3 \approx 1.053270798336 \cdot 0.3341150208$$
$$P_3 \approx 0.3519 \approx 35.2\%$$

41. $P_{0+1} = [1+(0.10)\cdot(1-0.25)]\cdot 0.25$
$P_1 = [1+(0.10)\cdot 0.75]\cdot 0.25$
$P_1 = [1+0.075]\cdot 0.25$
$P_1 = 1.075\cdot 0.25$
$P_1 = 0.26875$

$P_{1+1} = [1+(0.10)\cdot(1-0.26875)]\cdot 0.26875$
$P_2 = [1+(0.10)\cdot 0.73125]\cdot 0.26875$
$P_2 = [1+0.073125]\cdot 0.26875$
$P_2 = 1.073125\cdot 0.26875$
$P_2 \approx 0.28840234375$

$P_{2+1} \approx [1+(0.10)\cdot(1-0.28840234375)]\cdot 0.28840234375$
$P_3 \approx [1+(0.10)\cdot 0.71159765625]\cdot 0.28840234375$
$P_3 \approx [1+0.071159765625]\cdot 0.28840234375$
$P_3 \approx 1.071159765625\cdot 0.28840234375$
$P_3 \approx 0.3089 \approx 30.9\%$

43. $P_{0+1} = [1+(0.045)\cdot(1-0.60)]\cdot 0.60$
$P_1 = [1+(0.045)\cdot 0.40]\cdot 0.60$
$P_1 = [1+0.018]\cdot 0.60$
$P_1 = 1.018\cdot 0.60$
$P_1 = 0.6108$

$P_{1+1} = [1+(0.045)\cdot(1-0.6108)]\cdot 0.6108$
$P_2 = [1+(0.045)\cdot 0.3892]\cdot 0.6108$
$P_2 = [1+0.017514]\cdot 0.6108$
$P_2 = 1.017514\cdot 0.6108$
$P_2 = 0.6214975512$

$P_{2+1} \approx [1+(0.045)\cdot(1-0.6214975512)]\cdot 0.6214975512$
$P_3 \approx [1+(0.045)\cdot 0.3785024488]\cdot 0.6214975512$
$P_3 \approx [1+0.017032610196]\cdot 0.6214975512$
$P_3 \approx 1.017032610196\cdot 0.6214975512$
$P_3 \approx 0.632 = 63.2\%$

45. $20,000 = 10,000(1+0.05)^n$
$2 = 1.05^n$
$\log 2 = \log 1.05^n$
$\log 2 = n\cdot\log 1.05$
$\frac{\log 2}{\log 1.05} = n$
$n \approx \frac{0.30103}{0.02119} \approx 14.21$ years

47. $602 = 301(1+0.009)^n$
$2 = 1.009^n$
$\log 2 = \log 1.009^n$
$\log 2 = n\cdot\log 1.009$
$\frac{\log 2}{\log 1.009} = n$
$n \approx \frac{0.30103}{0.00389} \approx 77.385$ years

By the year $2007+78=2085$, the population should double.

49. $1,030(1+(-0.005))^{13} = 1,030\cdot 0.995^{13} \approx 965$ (million)

51.
$$50 = 100(1 + (-0.0035))^n$$
$$\frac{1}{2} = 0.9965^n$$
$$\log\frac{1}{2} = \log 0.9965^n$$
$$\log\frac{1}{2} = n \cdot \log 0.9965$$
$$\frac{\log\frac{1}{2}}{\log 0.9965} = n$$
$$n \approx \frac{-0.30103}{-0.00152} \approx 197.70$$

after approximately 198 years

53. $500(1+(-0.15))^3 = 500 \cdot 0.85^3 \approx 307$ mg

55. After 5 hours: $500(1+(-0.15))^5 \approx 222$

After 6 hours: $500(1+(-0.15))^6 \approx 189$

After 7 hours: $500(1+(-0.15))^7 \approx 160$

After 8 hours: $500(1+(-0.15))^8 \approx 136$

Therefore after approximately 8 hours the patient should stop feeling numbness.

57. a) $c = 120(1.0299)^t$

Let c be the cost of the shoes.

Let t be the number of years since 2004.

$$c = a \cdot b^t$$
$$135 = 120 \cdot b^4$$
$$b^4 = \frac{135}{120}$$
$$b = \sqrt[4]{\frac{135}{120}} \approx 1.0299$$

b) $c = 120(1.0299)^{10} = \$161.21$

59. 80.55%

$$1712 = 4.65(1+r)^{10}$$
$$(1+r)^{10} = \frac{1712}{4.65}$$
$$1 + r = \sqrt[10]{\frac{1712}{4.65}}$$
$$r = \sqrt[10]{\frac{1712}{4.65}} - 1 \approx 0.8055$$

61. 4.5%

$$1302 = 1000(1+r)^6$$
$$(1+r)^6 = \frac{1302}{1000}$$
$$1 + r = \sqrt[6]{\frac{1302}{1000}}$$
$$r = \sqrt[6]{\frac{1302}{1000}} - 1 \approx 0.045$$

63. It reverses the operation of raising 10 to a power.

65. When the rate of growth of a quantity is proportional to the amount present.

67. No solution provided.

69. $P_0 = \dfrac{200}{800} = 0.25$

$$P_{0+1} = [1+(0.18)\cdot(1-0.25)]\cdot 0.25$$
$$P_1 = [1+(0.18)\cdot 0.75]\cdot 0.25$$
$$P_1 = [1+0.135]\cdot 0.25$$
$$P_1 = 1.135\cdot 0.25$$
$$P_1 = 0.28375$$
$$\Rightarrow 0.28375\cdot 800 \approx 227 \text{ fish}$$

$$P_{1+1} = [1+(0.18)\cdot(1-0.28375)]\cdot 0.28375$$
$$P_2 = [1+(0.18)\cdot 0.71625]\cdot 0.28375$$
$$P_2 = [1+0.128925]\cdot 0.28375$$
$$P_2 = 1.128925\cdot 0.28375$$
$$P_2 \approx 0.32033246875$$
$$\Rightarrow 0.32033246875\cdot 800 \approx 256 \text{ fish}$$

$$P_{2+1} \approx [1+(0.18)\cdot(1-0.32033246875)]\cdot 0.32033246875$$
$$P_3 \approx [1+(0.18)\cdot 0.67966753125]\cdot 0.32033246875$$
$$P_3 \approx [1+0.122340155625]\cdot 0.32033246875$$
$$P_3 \approx 1.122340155625\cdot 0.32033246875$$
$$P_3 \approx 0.359521992829$$
$$\Rightarrow 0.359521992829\cdot 800 \approx 288 \text{ fish}$$

$$P_{3+1} \approx [1+(0.18)\cdot(1-0.359521992829)]\cdot 0.359521992829$$
$$P_4 \approx [1+(0.18)\cdot 0.640478007171]\cdot 0.359521992829$$
$$P_4 \approx [1+0.115286041291]\cdot 0.359521992829$$
$$P_4 \approx 1.11528604129\cdot 0.359521992829$$
$$P_4 \approx 0.400969860139$$
$$\Rightarrow 0.400969860139\cdot 800 \approx 321 \text{ fish}$$

The population will exceed 300 by the end of the fourth year. Fishing can begin toward the end of the fourth year.

Section 7.5 Proportions and Variation

1. The exchange rate between Mexican pesos and American dollars was compared to the ratio of American dollars and Mexican pesos needed to purchase the mummified piranha.

3. Step 1: Find the constant of variation, *k*. Step 2: Use *k* to rewrite the variation equation and then substitute values to find the desired quantity.

5. Rewrite $24:x=18:3$ as $\frac{24}{x}=\frac{18}{3}$.

$$\frac{24}{x}=\frac{18}{3}$$
$$24\cdot 3=x\cdot 18$$
$$72=18x$$
$$4=x$$
$$x=4$$

7. $$\frac{50}{4}=\frac{x}{5}$$
$$50\cdot 5=4\cdot x$$
$$250=4x$$
$$\frac{250}{4}=x$$
$$x=62.5$$

9. Rewrite $x:12=3:2$ as $\frac{x}{12}=\frac{3}{2}$.

$$\frac{x}{12}=\frac{3}{2}$$
$$x\cdot 2=12\cdot 3$$
$$2x=36$$
$$x=18$$

11. $$\frac{30}{40}=\frac{27}{x}$$
$$30\cdot x=40\cdot 27$$
$$30x=1{,}080$$
$$x=36$$

13. $$y=kx \qquad y=5x$$
$$37.5=k\cdot 7.5 \qquad y=5\cdot 13$$
$$\tfrac{37.5}{7.5}=k \qquad y=65$$
$$k=5$$

15. $$r=\frac{k}{s} \qquad r=\frac{8}{s}$$
$$12=\frac{k}{2/3} \qquad r=\frac{8}{8}$$
$$12\cdot\tfrac{2}{3}=k \qquad r=1$$
$$k=8$$

17. $$a=kb^2 \qquad a=\tfrac{4}{9}b^2$$
$$16=k\cdot 6^2 \qquad a=\tfrac{4}{9}\cdot 15^2$$
$$16=k\cdot 36 \qquad a=\tfrac{4}{9}\cdot 225$$
$$\frac{16}{36}=k \qquad a=100$$
$$k=\tfrac{4}{9}$$

19. $$D=\frac{k}{C} \qquad D=\frac{\frac{3}{2}}{C}$$
$$3\cdot 2=4k \qquad D=\frac{\frac{3}{2}}{24}$$
$$6=4k \qquad D=\tfrac{3}{2}\cdot\tfrac{1}{24}=\tfrac{1}{16}$$
$$k=\tfrac{6}{4}=\tfrac{3}{2}$$

21. $$r=kxy \qquad r=1.25xy$$
$$12.5=k\cdot 2\cdot 5 \qquad r=1.25\cdot 8\cdot 2.5$$
$$12.5=k\cdot 10 \qquad r=25$$
$$\frac{12.5}{10}=k$$
$$k=1.25$$

23. $$y=kwx^2 \qquad y=\tfrac{7}{2}wx^2$$
$$504=k\cdot 4\cdot 6^2 \qquad 6{,}860=\tfrac{7}{2}\cdot 10\cdot x^2$$
$$504=k\cdot 4\cdot 36 \qquad 6{,}860=35\cdot x^2$$
$$504=k\cdot 144 \qquad x^2=\frac{6{,}860}{35}=196$$
$$\frac{504}{144}=k \qquad x=\sqrt{196}=14$$
$$k=\tfrac{7}{2}$$

25. $$y=k\frac{w}{x} \qquad y=\frac{20}{3}\cdot\frac{w}{x}$$
$$4=k\frac{6}{10} \qquad y=\frac{20}{3}\cdot\frac{3}{15}$$
$$40=6k \qquad y=\frac{4}{3}$$
$$\frac{40}{6}=k$$
$$k=\tfrac{20}{3}$$

27.

$$y = k\frac{x^2 w}{z} \qquad y = 6\cdot\frac{x^2 w}{z}$$
$$15 = k\frac{2.5^2\cdot 8}{20} \qquad y = 6\cdot\frac{8^2\cdot 7}{14}$$
$$300 = k(6.25\cdot 8) \qquad y = 6\cdot\frac{64\cdot 7}{14}$$
$$300 = 50k \qquad y = 6\cdot\frac{64}{2}$$
$$\frac{300}{50} = k$$
$$k = 6 \qquad y = 6\cdot 32 = 192$$

In the solutions to Exercises 29 – 39, your proportion could be set up differently.

29.

$$\frac{150}{6} = \frac{65}{x}$$
$$150\cdot x = 6\cdot 65$$
$$150x = 390$$
$$x = \frac{390}{150}$$
$$x = 2.6 \text{ mg}$$

31.

$$\frac{30}{48} = \frac{x}{56}$$
$$30\cdot 56 = 48\cdot x$$
$$1{,}680 = 48x$$
$$\frac{1{,}680}{48} = x$$
$$x = 35 \text{ miles}$$

33. The front lawn is $200\cdot 650 = 130{,}000$ square feet.

$$\frac{1.5}{60{,}000} = \frac{x}{130{,}000}$$
$$1.5\cdot 130{,}000 = 60{,}000\cdot x$$
$$195{,}000 = 60{,}000x$$
$$\frac{195{,}000}{60{,}000} = x$$
$$x = 3.25 \text{ hours}$$

35.

$$\frac{\text{number of tagged bald eagles in population}}{\text{number of bald eagles in population}} = \frac{\text{number of tagged bald eagles in sample}}{\text{number of bald eagles in sample}}$$
$$\frac{400}{n} = \frac{8}{240}$$
$$400\cdot 240 = n\cdot 8$$
$$96{,}000 = 8n$$
$$\frac{96{,}000}{8} = n$$
$$n = 12{,}000$$

37. $$\frac{\text{number of tagged grizzly bears in population}}{\text{number of grizzly bears in population}} = \frac{\text{number of tagged grizzly bears in sample}}{\text{number of grizzly bears in sample}}$$

$$\frac{55}{1{,}000} = \frac{n}{95}$$
$$55 \cdot 95 = 1{,}000 \cdot n$$
$$5{,}225 = 1{,}000n$$
$$\frac{5{,}225}{1{,}000} = n$$
$$n = 5.225$$

5 or 6 tagged grizzly bears could be expected.

39. Let t be the time of the trip.
Let s be the speed during the trip.

$$t = \frac{k}{s} \qquad t = \frac{120}{s}$$
$$2 = \frac{k}{60} \qquad t = \frac{120}{65} = \frac{24}{13}$$
$$2 \cdot 60 = k \qquad t \approx 1.846 \text{ hours}$$
$$k = 120$$

You would save $(2 - 1.846) \times 60 = 9.24$ minutes by traveling at 65 miles per hour.

41. Let w be the width of the beam.
Let d be the depth of the beam.
Let l be the length of the beam.
Let s be the weight the beam can support.

$$s = k\frac{w \cdot d^2}{l} \qquad s = \frac{7 \cdot w \cdot d^2}{l}$$
$$672 = k\frac{6 \cdot 8^2}{4} \qquad s = \frac{7 \cdot 4 \cdot 6^2}{8}$$
$$672 = k\frac{384}{4} \qquad s = \frac{7 \cdot 4 \cdot 36}{8}$$
$$672 = k \cdot 96$$
$$k = \frac{672}{96} = 7 \qquad s = \frac{1008}{8} = 126$$

The beam could support 126 pounds.

43. Let w be the width of the beam.
Let d be the depth of the beam.
Let l be the length of the beam.
Let s be the weight the beam can support.

$$s = k\frac{w \cdot d^2}{l} \qquad s = \frac{60 \cdot w \cdot d^2}{l}$$
$$1,280 = k\frac{4 \cdot 8^2}{12} \qquad 540 = \frac{60 \cdot 3 \cdot 6^2}{l}$$
$$1,280 = k\frac{256}{12} \qquad 540 = \frac{6,480}{l}$$
$$k = \frac{1,280}{1} \cdot \frac{12}{256} = 60 \qquad l = \frac{6,480}{540} = 12$$

The beam would be 12 feet long.

45. Let w be the amount of water Mario uses for irrigation in gallons.
Let r be the amount of rainfall in inches.

$$w = \frac{k}{r} \qquad w = \frac{90,000}{r}$$
$$30,000 = \frac{k}{3} \qquad w = \frac{90,000}{5}$$
$$30,000 \cdot 3 = k \qquad w = 18,000 \text{ gal}$$
$$k = 90,000$$

47. Let d be the distance in feet.
Let t be the time in seconds.

$$d = kt^2 \qquad d = 16t^2$$
$$144 = k \cdot 3^2 \qquad d = 16 \cdot 5^2$$
$$144 = k \cdot 9 \qquad d = 16 \cdot 25$$
$$\frac{144}{9} = k \qquad d = 400 \text{ feet}$$
$$k = 16$$

49. Let h be the amount of heating oil in gallons.
Let t be the temperature in degrees.

$$h = \frac{k}{t} \qquad h = \frac{21,168}{t}$$
$$504 = \frac{k}{42} \qquad h = \frac{21,168}{36}$$
$$504 \cdot 42 = k \qquad h = 588 \text{ gallons}$$
$$k = 21,168$$

51. Let p be the pressure in pounds per square inch.
Let v be the volume in cubic inches.

$$p = \frac{k}{v} \qquad p = \frac{480}{v}$$
$$4 = \frac{k}{120} \qquad p = \frac{480}{75}$$
$$4 \cdot 120 = k \qquad p = 6.4 \text{ psi}$$
$$k = 480$$

53. y increases.

55. inversely

57. In $m:n$, we are comparing m objects with n objects. In $n:(m+n)$, we are comparing the n objects with the total of all the objects.

59. No solution provided.

61. The ratio of PC's to all computers on campus is 7:9. Explanations and examples may vary.

63. The strength stays the same.

$$s=k\frac{(2w)\cdot d^2}{2l}=k\frac{w\cdot d^2}{l}$$

65. The beam is four times as strong.

$$s=k\frac{(2w)\cdot(2d)^2}{2l}=k\frac{4w\cdot d^2}{l}=4\left(k\frac{w\cdot d^2}{l}\right)$$

Section 7.6 Functions

1. f is the function that relates c, the cost of gasoline, to n, the number of miles driven.

3. p is the base price of the car. $f(p)$ is the total price of the car after taxes and fees.

5. "y equals f of x."

7. "g of x equals 8."

9.
$f(x)=3x+4$
$f(2)=3(2)+4$
$f(2)=6+4$
$f(2)=10$

11.
$h(a)=3a^2-2a+5$
$h(5)=3(5)^2-2(5)+5$
$h(5)=3(25)-10+5$
$h(5)=75-10+5$
$h(5)=70$

13.
$g(t)=\sqrt{t-4}$
$g(40)=\sqrt{(40)-4}$
$g(40)=\sqrt{36}$
$g(40)=6$

15.
$f(x)=\sqrt{x-9}$
$f(5)=\sqrt{(5)-9}$
$f(5)=\sqrt{-4}$
Does not exist.

17. $\{3,4,5,6,7,9\}$

19. 8

21. 9

23. Does not exist.

25. Does not exist.

27. –2

29. not a function

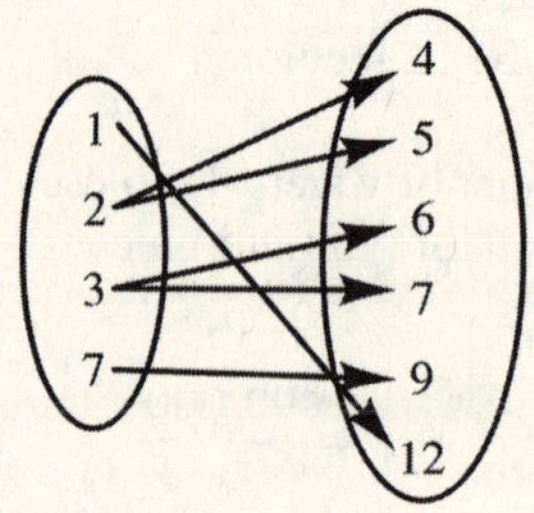

31. function

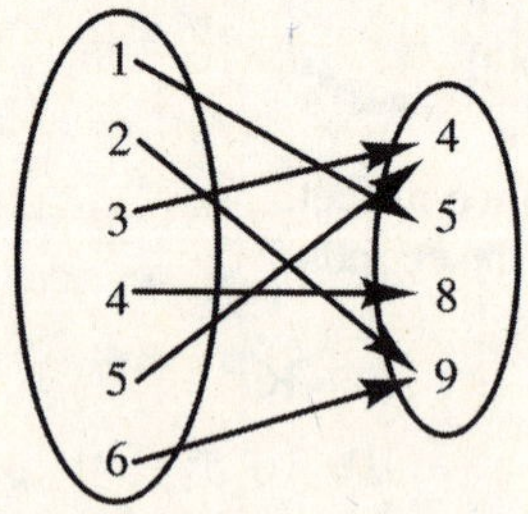

33. function

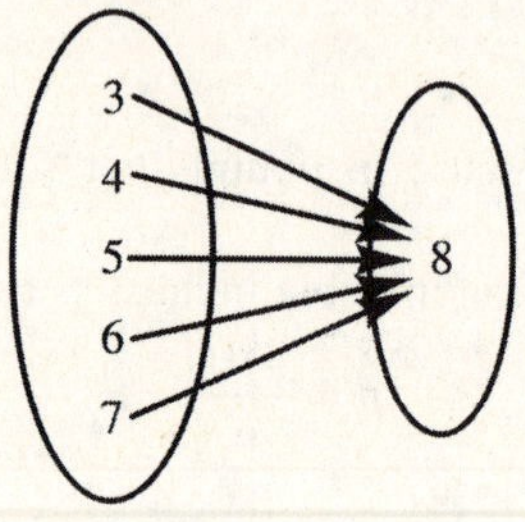

35. yes; $f(x)=5x+3$
$f(-4)=5(-4)+3=-20+3=-17$
$f(0)=5(0)+3=0+3=3$
$f(2)=5(2)+3=10+3=13$

37. no; $\{-2,-2\},\{-2,2\}$

39. yes; $f(x)=\sqrt{x+3}$

$f(-7)=\sqrt{(-7)+3}=\sqrt{-4}$; Does not exist.

$f(0)=\sqrt{(0)+3}=\sqrt{3}$
$f(6)=\sqrt{(6)+3}=\sqrt{9}=3$

41. no; $\{1,\sqrt{2}\},\{1,-\sqrt{2}\}$

43. Function; it passes vertical line test.

45. Not a function; it fails vertical line test.

47. $y=2x^2-5x+3 \Rightarrow f(x)=2x^2-5x+3$

49. $y=3x^2-5x \Rightarrow f(x)=3x^2-5x$

51. Yes

53. No, a movie may have many stars.

55. No, a person may have many people in their immediate family.

57. No, some TV programs may appear several nights per week.

59. $d(t)=104t+64$
$d(14)=104(14)+64$
$d(14)=1456+64$
$d(14)=1520$

61. $f(t)=1.25t+131.95$
$f(3)=1.25(3)+131.95$
$f(3)=3.75+131.95$
$f(3)=135.7$

63. $S(n)=-13n^2+169n$
$S(6.5)=-13(6.5)^2+169(6.5)$
$S(6.5)=-13(42.25)+1098.5$
$S(6.5)=-549.25+1098.5$
$S(6.5)=549.25$

65. $A(n)=97n+975$
$A(5)=97(5)+975$
$A(5)=485+975$
$A(5)=1460$

67. Domain element means the same thing as independent variable. Range element means the same thing as dependent variable.

69. One domain element has two or more distinct range elements associated with it.

71. – 73. No solution provided.

75. $(f-g)(x)=f(x)-g(x)=(2x+1)-(x^2+3)=-x^2+2x-2$

$(f-g)(1)=-(1)^2+2(1)-2=-1+2-2=-1$
$(f-g)(4)=-(4)^2+2(4)-2=-16+8-2=-10$
$(f-g)(6)=-(6)^2+2(6)-2=-36+12-2=-26$

77. $(f/g)(x)=\dfrac{f(x)}{g(x)}=\dfrac{2x+1}{x^2+3}$

$(f/g)(0)=\dfrac{2(0)+1}{(0)^2+3}=\dfrac{0+1}{0+3}=\dfrac{1}{3}$

$(f/g)(3)=\dfrac{2(3)+1}{(3)^2+3}=\dfrac{6+1}{9+3}=\dfrac{7}{12}$

$(f/g)(5)=\dfrac{2(5)+1}{(5)^2+3}=\dfrac{10+1}{25+3}=\dfrac{11}{28}$

Section 7.7 Looking Deeper: Dynamical Systems

1. All models in both examples are dynamical systems. That is, sequences of numbers of the form $A_0, A_1, A_2, \ldots$, where the value of A_{n+1} depends on the value of A_n.

3. In Example 2, the equilibrium value $\frac{5{,}000}{6}$ was stable. In Example 3, the equilibrium value $\frac{5}{3}$ was unstable.

5. $A_0 = 3$
$A_1 = 2\cdot 3 - 1 = 6 - 1 = 5$
$A_2 = 2\cdot 5 - 1 = 10 - 1 = 9$

7. $A_0 = -2$
$A_1 = -3\cdot(-2) + 4 = 6 + 4 = 10$
$A_2 = -3\cdot 10 + 4 = -30 + 4 = -26$
$A_3 = -3\cdot(-26) + 4 = 78 + 4 = 82$

9. $A_0 = 4$
$A_1 = 1.8\cdot 4 - 2 = 7.2 - 2 = 5.2$
$A_2 = 1.8\cdot 5.2 - 2 = 9.36 - 2 = 7.36$
$A_3 = 1.8\cdot 7.36 - 2 = 13.248 - 2 = 11.248$
$A_4 = 1.8\cdot 11.248 - 2 = 20.2464 - 2 = 18.2464$

11. In $A_{n+1} = 2A_n + 3$, substitute a for both A_{n+1} and A_n.
$a = 2a + 3$
$-a = 3$
$a = -3$
Since $2 > 1$, the equilibrium value is unstable.
Since $4 > 1$, the equilibrium value is unstable.

13. In $B_{n+1} = 0.25B_n + 4$, substitute a for both B_{n+1} and B_n.
$a = 0.25a + 4$
$0.75a = 4$
$a = \frac{4}{0.75} = \frac{400}{75} = \frac{16}{3}$
Since $-1 < 0.25 < 1$, the equilibrium value is stable.

15. $A_{n+1} = 1.05A_n$ where $n = 0, 1, 2, \ldots$ and $A_0 = 1{,}000$
$A_1 = 1.05\cdot 1{,}000 = 1{,}050$
$A_2 = 1.05\cdot 1{,}050 = 1{,}102.5$
You would have \$1,102.50 at the end of 2 years.

17. $P_{n+1} = [1+(0.08)\cdot(1-P_n)]\cdot P_n$ where $n = 0,1,2,...$ and $P_0 = 0.30$

$P_1 = [1+(0.08)\cdot(1-0.30)]\cdot 0.30$
$P_1 = [1+(0.08)\cdot 0.70]\cdot 0.30$
$P_1 = [1+0.056]\cdot 0.30$
$P_1 = 1.056\cdot 0.30$
$P_1 = 0.3168$

$P_2 = [1+(0.08)\cdot(1-0.3168)]\cdot 0.3168$
$P_2 = [1+(0.08)\cdot 0.6832]\cdot 0.3168$
$P_2 = [1+0.054656]\cdot 0.3168$
$P_2 = 1.054656\cdot 0.3168$
$P_2 \approx 0.3341150208$

Approximately 33.4% of the lemur population's maximum capacity will be attained at the end of two years.

19. $D_{n+1} = 0.60\cdot D_n + 250$ where $n = 0,1,2,...$ and $D_0 = 0$

$D_1 = 0.60\cdot 0 + 250 = 0 + 250 = 250$
$D_2 = 0.60\cdot 250 + 250 = 150 + 250 = 400$
$D_3 = 0.60\cdot 400 + 250 = 240 + 250 = 490$

490 milligrams of antibiotic will be in your bloodstream after three doses.

21. There are equations relating A_1 to A_0, A_2 to A_1, A_3 to A_2, and so on.

23. If $-1 < m < 1$, then the equilibrium value will be stable.

25. – 27. No solution provided.

29. $C_k = (0.99988)^k \cdot C_0 = 0.99988^k$ since $C_0 = 1$ where $k = 0, 1, 2, ...$

$$0.60 = 0.99988^k$$
$$\log 0.60 = \log 0.99988^k$$
$$\log 0.60 = k\cdot \log 0.99988$$
$$\frac{\log 0.60}{\log 0.99988} = k$$
$$k \approx \frac{-0.22184875}{-0.00005212} \approx 4,256.50$$

The fossilized plant is approximately 4,300 years old.

31. $P_0 = \frac{750}{1000} = 0.75$ and 100 turkeys represents $\frac{100}{1000} = 0.10$ or 10% of the maximum population.

$P_1 = [1+(0.10)\cdot(1-0.75)]\cdot 0.75$
$P_1 = [1+(0.10)\cdot 0.25]\cdot 0.75$
$P_1 = [1+0.025]\cdot 0.75$
$P_1 = 1.025\cdot 0.75$
$P_1 = 0.76875$
$\Rightarrow 0.76875 - 0.10 = 0.66875$

$P_2 = [1+(0.10)\cdot(1-0.66875)]\cdot 0.66875$
$P_2 = [1+(0.10)\cdot 0.33125]\cdot 0.66875$
$P_2 = [1+0.033125]\cdot 0.66875$
$P_2 = 1.033125\cdot 0.66875$
$P_2 \approx 0.69090234375$
$\Rightarrow 0.69090234375 - 0.10$
$= 0.59090234375$

31. (continued)

$$P_3 \approx [1+(0.10)\cdot(1-0.59090234375)]\cdot 0.59090234375$$
$$P_3 \approx [1+(0.10)\cdot 0.40909765625]\cdot 0.59090234375$$
$$P_3 \approx [1+0.040909765625]\cdot 0.59090234375$$
$$P_3 \approx 1.040909765625\cdot 0.59090234375$$
$$P_3 \approx 0.61507602014$$
$$\Rightarrow 0.61507602014-0.10=0.51507602014$$

There would be approximately 515 turkeys at the end of the three years.

Chapter Review Exercises

1. a)

$$\frac{2}{3}x+2=\frac{1}{6}x+4$$
$$6\cdot\left[\frac{2}{3}x+2\right]=\left[\frac{1}{6}x+4\right]\cdot 6$$
$$4x+12=x+24$$
$$3x+12=24$$
$$3x=12$$
$$x=4$$

b)

$$0.3x-2=3.5x-0.4$$
$$10\cdot[0.3x-2]=[3.5x-0.4]\cdot 10$$
$$3x-20=35x-4$$
$$-32x-20=-4$$
$$-32x=16$$
$$x=\frac{16}{-32}=-\frac{1}{2}$$

3. a) Let h be the number of hours Darryl works.
Let s be Darryl's salary for the week.

$$s=40\cdot 5+(h-40)(5)(2)$$
$$s=200+10(h-40)$$

b) $s=200+10(46-40)=200+10\cdot 6=200+60=\$260$

5. $m=\dfrac{y_2-y_1}{x_2-x_1}=\dfrac{8-5}{6-2}=\dfrac{3}{4}$

7. Let m be the number of months.

$$240+39m=350+28m$$
$$39m=110+28m$$
$$11m=110$$
$$m=10$$

after ten months

9. $m=\dfrac{y_2-y_1}{x_2-x_1}=\dfrac{9-4}{6-3}=\dfrac{5}{3}$

$$y=mx+b$$
$$y=\frac{5}{3}x+b$$
$$4=\frac{5}{3}\cdot 3+b$$
$$4=5+b$$
$$-1=b$$
$$y=\frac{5}{3}x-1$$

11. a technique by which we find a linear equation that bests models a set of data

13. We need to solve the equation $2x^2+7x=4$ or $2x^2+7x-4=0$ by using the quadratic formula, we obtain

$$x=\frac{-(7)\pm\sqrt{(7)^2-4(2)(-4)}}{2(2)}=\frac{-7\pm\sqrt{49+32}}{4}=\frac{-7\pm\sqrt{81}}{4}=\frac{-7\pm 9}{4}$$

$$x=\frac{-7-9}{4}=\frac{-16}{4}=-4 \text{ or } x=\frac{-7+9}{4}=\frac{2}{4}=\frac{1}{2}$$

15. a technique by which we find a quadratic equation that bests models a set of data

17. $A=10,000(1+0.048)^5=10,000\cdot 1.048^5\approx \$12,641.72$

19. $P_{3+1}=[1+(0.03)\cdot(1-0.50)]\cdot 0.50$

$P_4=[1+(0.03)\cdot 0.50]\cdot 0.50$

$P_4=[1+0.015]\cdot 0.50$

$P_4=1.015\cdot 0.50$

$P_4=0.5075$

21. In a logistic model we adjust the rate of growth as the population grows. The larger the population, the smaller the rate of growth.

23. Your proportion could be set up differently.

$$\frac{3.5}{840}=\frac{x}{1,500}$$

$3.5\cdot 1,500=840\cdot x$

$5,250=840x$

$\frac{5,250}{840}=x$

$x=6.25$ gal

25. $y=\frac{k}{x}$

$14=\frac{k}{3}$

$14\cdot 3=k$

$k=42$

$y=\frac{42}{x}$

$y=\frac{42}{18}$

$y=\frac{7}{3}$

27. The strength of the beam stays the same.

$$s=k\frac{(2w)\cdot d^2}{2l}=k\frac{w\cdot d^2}{l}$$

29. a) $\{1,2,3,4,5,7\}$

b) $\{-2,-1,4,6,8\}$

c) -1

d) 3 and 7

31. a) Not a function; it fails vertical line test.

b) Function; it passes vertical line test.

33. $A_0=6$

$A_1=4\cdot 6+3=24+3=27$

$A_2=4\cdot 27+3=108+3=111$

$A_3=4\cdot 111+3=444+3=447$

35. $D_{n+1}=0.50\cdot D_n+800$

$D_0=800$

$D_1=0.50\cdot 800+800=400+800=1200$

$D_2=0.50\cdot 1200+800=600+800=1400$

There will be 1400 mg of the drug in your system after 3 doses.

Chapter Test

1. a)
$$\frac{3}{4}x+5=\frac{2}{3}x+4$$
$$12\cdot\left[\frac{3}{4}x+5\right]=\left[\frac{2}{3}x+4\right]\cdot 12$$
$$9x+60=8x+48$$
$$x+60=48$$
$$x=-12$$

b)
$$0.25x+4=1.5x-0.2$$
$$100\cdot[0.25x+4]=[1.5x-0.2]\cdot 100$$
$$25x+400=150x-20$$
$$400=125x-20$$
$$420=125x$$
$$x=\frac{420}{125}=3.36$$

3. $$m=\frac{y_2-y_1}{x_2-x_1}=\frac{12-5}{8-3}=\frac{7}{5}$$

5. a) Let m be the number of minutes Brandon uses each month.
Let c be Brandon's cost for one month.
$c=30+0.045(m-1200)$

b) $c=30+0.045((1520)-1200))=30+0.045(320)=30+14.4=\$44.40$

7. 1 and b, 2 and d, 3 and a, 4 and c

9. When the rate of change between the two variables is constant.

11. Let p be the number of hours Shanaya works in the psychology lab.
Let d be the number of hours she works in the dining hall.
$8.25p+6.30d=137.70$

13. Let t be the number of years after 2002.
Let n be the number (in millions) of the U.S. population that live in poverty.
The given information corresponds to the two ordered pairs (0, 34.7) and (6, 36.5).
(Note: The number living in poverty was up 1.8 million from 2002, so there are 36.5 – 1.8 = 34.7 million in 2002)

$$m=\frac{y_2-y_1}{x_2-x_1}=\frac{36.5-34.7}{6-0}=\frac{1.8}{6}=0.3$$
$$n=mt+b$$
$$n=0.3t+b$$
$$34.7=0.3\cdot 0+b$$
$$34.7=0+b$$
$$34.7=b$$
$$n=0.3t+34.7$$
$$y=0.3\cdot 18+34.7=5.4+34.7=40.1$$

The estimated number of people living in poverty in 2020 is 40.1 million.

15. $A_{n+1} = 4 \cdot A_n + 2$

$a = 4a + 2$

$-3a = 2$

$a = -\frac{2}{3}$

The value of a is unstable since $m = 4 > 1$.

17. $A = 5{,}000(1+0.024)^8 = 5{,}000 \cdot 1.024^8 \approx \$6{,}044.63$

19. Quadratic regression.

21. a) $\{1,2,4,5,7,8\}$

b) $\{-3,-2,4,6,8\}$

c) –3

d) 2 and 7

23. $D_{n+1} = 0.35 \cdot D_n + 400$

$D_0 = 400$

$D_1 = 0.35 \cdot 400 + 400 = 140 + 400 = 540$

$D_2 = 0.35 \cdot 540 + 400 = 189 + 400 = 589$

$D_3 = 0.35 \cdot 589 + 400 = 206.15 + 400$

$= 606.15$

25. As the population increases, the environment cannot sustain the same growth rate, so the growth rate must be reduced.

27. a only

29. Your proportion could be set up differently.

$\frac{x}{90} = \frac{4}{35}$

$35 \cdot x = 4 \cdot 90$

$35x = 360$

$x = \frac{360}{35} \approx 10.3$

Curt Shilling could get 10 or 11 hits.

31.

$y = \frac{k}{x}$

$8 = \frac{k}{2/3}$

$8 \cdot \left(\frac{2}{3}\right) = k$

$k = \frac{16}{3}$

$y = \frac{16/3}{x} = \frac{16}{3x}$

$y = \frac{16}{3(4)}$

$y = \frac{16}{12} = \frac{4}{3}$

33. a) not a function; $\{0,5\},\{0,6\}$.

b) function

c) not a function; $\{-1,-1\},\{-1,1\}$

d) function

Chapter 8

Modeling With Systems of Linear Equations and Inequalities: What's the Best Way to Do It?

Section 8.1 Systems of Linear Equations

1. The lines can intersect in a single point as in Example 1; they can be parallel, with no points in common, as in Example 2; they can be the same line with infinitely many points in common, as in Example 3.

3. The numbers are easier to work with.

5. They replace curved surfaces with planar surfaces that are easier to compute.

7. The solution is $(4,2)$.

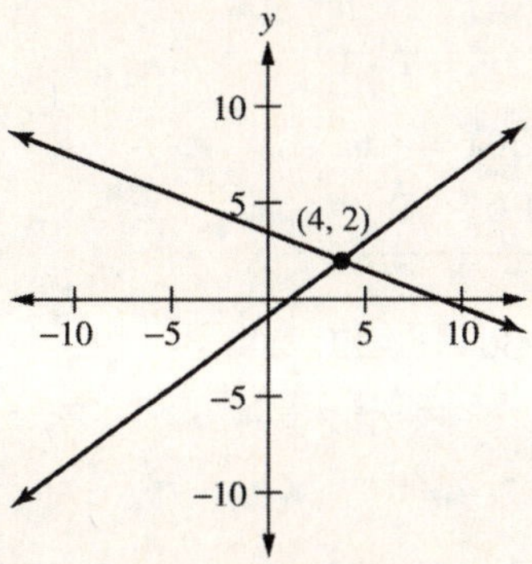

$2\cdot4+5\cdot2=8+10=18$

$-3\cdot4+4\cdot2=-12+8=-4$

9. The solution is $(3,0)$.

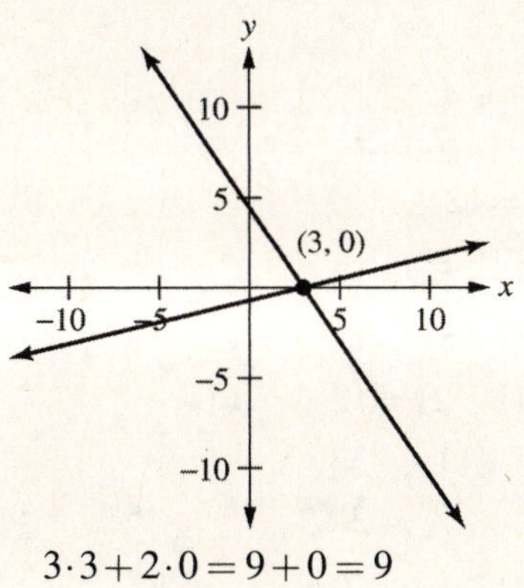

$3\cdot3+2\cdot0=9+0=9$

$-1\cdot3+2\cdot0=-3+0=-3$

11. The solution is $(4,-2)$.

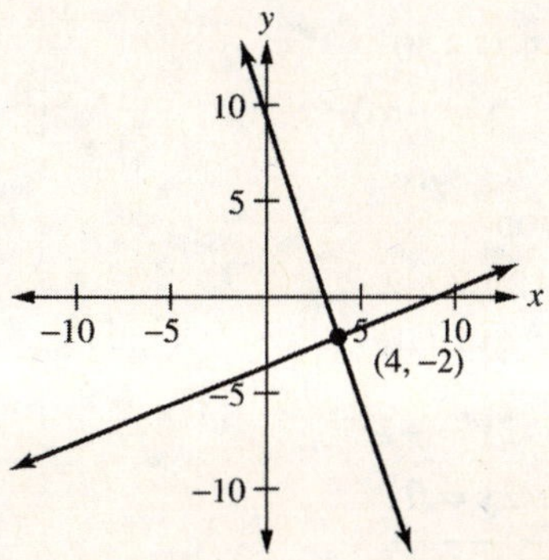

$2\cdot4-5\cdot(-2)=8-(-10)=18$

$3\cdot4+1\cdot(-2)=12+(-2)=10$

13. The solution is $(1,-1)$.

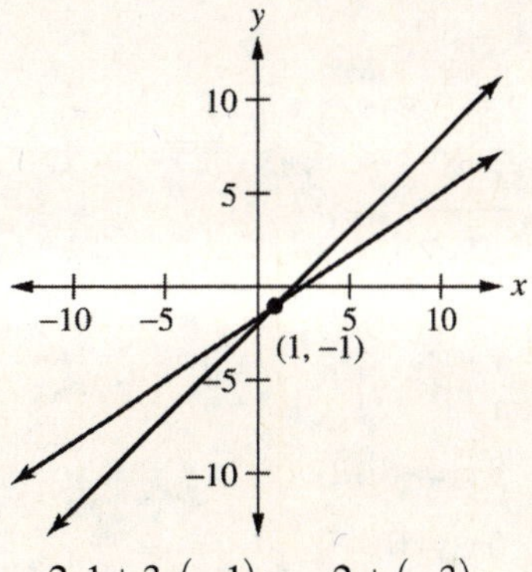

$-2\cdot1+3\cdot(-1)=-2+(-3)=-5$

$-1\cdot1+1\cdot(-1)=-1+(-1)=-2$

15. $(1,4)$

$2\cdot[-3x+2y]=[5]\cdot 2$

$6x+4y=22$

$-6x+4y=10$

$6x+4y=22$

$0+8y=32$

$y=4$

$-3x+2\cdot 4=5$

$-3x+8=5$

$-3x=-3$

$x=1$

17. $(-6,-2)$

$3x-2y=-14$

$2\cdot[-4x+y]=[22]\cdot 2$

$3x-2y=-14$

$-8x+2y=44$

$-5x+0=30$

$x=-6$

$-6\cdot 3-2y=-14$

$-18-2y=-14$

$-2y=4$

$y=-2$

19. no solution

$8x-2y=-2$

$2\cdot[-4x+y]=[3]\cdot 2$

$8x-2y=-2$

$-8x+2y=6$

$0+0=4$

21. infinite number of solutions

$4\cdot[x-y]=[-2]\cdot 4$

$-4x+4y=8$

$4x-4y=-8$

$-4x+4y=8$

$0+0=0$

23. no solution

$2\cdot[6x-9y]=[8]\cdot 2$

$3\cdot[-4x+6y]=[10]\cdot 3$

$12x-18y=16$

$-12x+18y=30$

$0+0=46$

25. $\left(\frac{1}{4},\frac{1}{2}\right)$

$12x-8y=-1$

$6\cdot[-2x+5y]=[2]\cdot 6$

$12x-8y=-1$

$-12x+30y=12$

$0+22y=11$

$y=\frac{11}{22}=\frac{1}{2}$

$12x-8\cdot\frac{1}{2}=-1$

$12x-4=-1$

$12x=3$

$x=\frac{3}{12}=\frac{1}{4}$

27. $\left(\frac{1}{5},\frac{2}{5}\right)$

$4\cdot[3x-9y]=[-3]\cdot 4$

$3\cdot[-4x+2y]=[0]\cdot 3$

$12x-36y=-12$

$-12x+6y=0$

$0-30y=-12$

$y=\frac{-12}{-30}=\frac{2}{5}$

$-4x+2\cdot\frac{2}{5}=0$

$5\cdot\left[-4x+\frac{4}{5}\right]=[0]\cdot 5$

$-20x+4=0$

$-20x=-4$

$x=\frac{-4}{-20}=\frac{1}{5}$

29. $(-1,3)$

$x-2y=-7\Rightarrow x=2y-7$

$3x+5y=12$

$3(2y-7)+5y=12$

$6y-21+5y=12$

$11y-21=12$

$11y=33$

$y=3$

$x=2\cdot 3-7=6-7=-1$

31. $(-1,4)$

$$-2x+y=6 \Rightarrow y=2x+6$$
$$-2x+3y=14$$
$$-2x+3(2x+6)=14$$
$$-2x+6x+18=14$$
$$4x+18=14$$
$$4x=-4$$
$$x=-1$$
$$y=2\cdot(-1)+6=-2+6=4$$

33. no solution

$$x+y=6$$
$$2x=8-2y \Rightarrow x=4-y$$
$$(4-y)+y=6$$
$$4-y+y=6$$
$$4=6 \text{ This is a false statement.}$$

35. infinite number of solutions

$$x-y=4$$
$$3y=3x-12 \Rightarrow y=x-4$$
$$x-(x-4)=4$$
$$x-x+4=4$$
$$4=4 \text{ This is a true statement.}$$

37. Let w be the number of games won.
Let l be the number of games lost.

$$w+l=162$$
$$w=l+30$$

$$\begin{aligned} w+l&=162 \\ \underline{w-l}&\underline{=30} \\ 2w+0&=192 \\ w&=96 \end{aligned} \qquad \begin{aligned} 96+l&=162 \\ l&=66 \end{aligned}$$

96 wins and 66 losses

39. 4 bagels and 5 ounces of cream cheese

Nutrition	Amount per Bagel (b)	Amount per Ounce of Cream Cheese (c)	Desired
Calcium	30 mg	25 mg	245 mg
Iron	2 mg	0.4 mg	10 mg

$$\begin{aligned} 30b+25c&=245 \\ 10\cdot[2b+0.4c]&=[10]\cdot 10 \end{aligned} \Rightarrow \begin{aligned} 2\cdot[30b+25c]&=[245]\cdot 2 \\ (-3)\cdot[20b+4c]&=[100]\cdot(-3) \end{aligned} \Rightarrow$$

$$\begin{aligned} 60b+50c&=490 \\ -60b-12c&=-300 \end{aligned} \Rightarrow 38c=190 \Rightarrow c=5$$

$$2b+0.4\cdot 5=10 \Rightarrow 2b+2=10 \Rightarrow 2b=8 \Rightarrow b=4$$

41. Let m be the number months.
Let c be the total cost of each system.
$c=179+17.5m$ WorldCom
$c=135+21.5m$ Satellite, Inc

$$\begin{aligned} (-1)\cdot[-17.5m+c]&=[179]\cdot(-1) \\ -21.5m+c&=135 \end{aligned} \Rightarrow \begin{aligned} 17.5m-c&=-179 \\ -21.5m+c&=135 \end{aligned} \Rightarrow -4m=-44 \Rightarrow m=11$$

After 11 months, Maneka pays $c=179+17.5(11)=\$371.50$ with either plan. After that, Worldcom is the better deal.

43. Let w be the number of games won.
Let l be the number of games lost.

$$\begin{aligned} w+l&=34 \\ \frac{w}{2}&=\frac{l}{2}+3 \end{aligned} \Rightarrow \begin{aligned} w+l&=34 \\ (2)\cdot\left[\frac{w}{2}\right]&=\left[\frac{l}{2}+3\right]\cdot(2) \end{aligned} \Rightarrow \begin{aligned} w+l&=34 \\ w&=l+6 \end{aligned} \Rightarrow \begin{aligned} w+l&=34 \\ w-l&=6 \end{aligned} \Rightarrow 2w=40 \Rightarrow w=20,$$

thus $20+l=34 \Rightarrow l=14$

20 wins and 14 losses

45. Let A be the number of passengers at Atlanta's Hartsfield International (in millions).
Let C be the number of passengers at Chicago's O'Hare International (in millions).

$$\begin{aligned} A+C&=162 \\ A&=C+8 \end{aligned} \Rightarrow \begin{aligned} A+C&=162 \\ A-C&=8 \end{aligned} \Rightarrow 2A=170 \Rightarrow A=85 \text{ thus } 85+C=162 \Rightarrow C=77$$

Atlanta's Hartsfield International has 85 million and Chicago's O'Hare International has 77 million.

47. In $2000+8=2008$.

Japan

Year	Motor Vehicle Production (in millions)
0	10
5	11

US

Year	Motor Vehicle Production (in millions)
0	13
5	12

Japan:

$$\text{slope}=\frac{11-10}{5-0}=\frac{1}{5}$$

$y=mx+b$

$y=\frac{1}{5}x+b$

$10=\frac{1}{5}\cdot 0+b$

$b=10$

$y=\frac{1}{5}x+10 \Rightarrow -x+5y=50$

US:

$$\text{slope}=\frac{12-13}{5-0}=\frac{-1}{5}=-\frac{1}{5}$$

$y=mx+b$

$y=-\frac{1}{5}x+b$

$13=-\frac{1}{5}\cdot 0+b$

$b=13$

$y=-\frac{1}{5}x+13 \Rightarrow x+5y=65$

We must now solve the following system.

$$\begin{aligned} -x+5y&=50 \quad \text{(Japan)} \\ x+5y&=65 \quad \text{(US)} \end{aligned}$$

$$\begin{aligned} -x+5y&=550 \\ x+5y&=65 \end{aligned} \Rightarrow 10y=115 \Rightarrow y=\frac{115}{10}=11.5 \Rightarrow x+5(11.5)=65 \Rightarrow x=7.5, \text{ after 8 years.}$$

49. \$11 per hour

Supply		Demand	
Price	Tutors Supplied	Price	Tutors Demanded
8	9	8	30
15	37	15	9

Supply:

$$\text{slope} = \frac{37-9}{15-8} = \frac{28}{7} = 4$$

$$y = mx + b$$
$$y = 4x + b$$
$$9 = 4 \cdot 8 + b$$
$$9 = 32 + b \Rightarrow b = -23$$
$$y = 4x - 23 \Rightarrow -4x + y = -23$$

Demand:

$$\text{slope} = \frac{9-30}{15-8} = \frac{-21}{7} = -3$$

$$y = mx + b$$
$$y = -3x + b$$
$$30 = -3 \cdot 8 + b$$
$$30 = -24 + b \Rightarrow b = 54$$
$$y = -3x + 54 \Rightarrow 3x + y = 54$$

We must now solve the following system.

$$-4x + y = -23 \quad \text{(supply)}$$
$$3x + y = 54 \quad \text{(demand)}$$

$$\begin{array}{c} -4x + y = -23 \\ (-1) \cdot [3x + y] = [54] \cdot (-1) \end{array} \Rightarrow \begin{array}{c} -4x + y = -23 \\ -3x - y = -54 \end{array} \Rightarrow -7x = -77 \Rightarrow x = 11$$

51. \$34 per book

Supply		Demand	
Price	Books Supplied	Price	Books Demanded
30	60	30	95
42	120	42	50

Supply:

$$\text{slope} = \frac{120-60}{42-30} = \frac{60}{12} = 5$$

$$y = mx + b$$
$$y = 5x + b$$
$$60 = 5 \cdot 30 + b$$
$$60 = 150 + b \Rightarrow b = -90$$
$$y = 5x - 90 \Rightarrow -5x + y = -90$$

Demand:

$$\text{slope} = \frac{50-95}{42-30} = \frac{-45}{12} = -\frac{15}{4}$$

$$y = mx + b$$
$$y = -\tfrac{15}{4}x + b$$
$$95 = -\tfrac{15}{4} \cdot 30 + b$$
$$95 = -\tfrac{225}{2} + b \Rightarrow b = \tfrac{415}{2}$$
$$y = -\tfrac{15}{4}x + \tfrac{415}{2} \Rightarrow 15x + 4y = 830$$

We must now solve the following system.

$$-5x + y = -90 \quad \text{(supply)}$$
$$15x + 4y = 830 \quad \text{(demand)}$$

$$\begin{array}{c} (-4) \cdot [-5x + y] = [-90\ ] \cdot (-4) \\ 15x + 4y = 830 \end{array} \Rightarrow \begin{array}{c} 20x - 4y = 360 \\ 15x + 4y = 830 \end{array} \Rightarrow 35x = 1190 \Rightarrow x = 34$$

53. When solving, there is a unique value for *x* or *y*.

55. When solving, there is a true statement, such as 0 + 0 = 0.

57. – 59. No solution provided.

61. a) $\begin{array}{r}(-2)\cdot[2x+3y+z]=[10]\cdot(-2)\\ x-y+2z=7\end{array} \Rightarrow \begin{array}{r}-4x-6y-2z=-20\\ x-y+2z=7\end{array} \Rightarrow -3x-7y-13$

b) $\begin{array}{r}2x+3y+z=10\\ 3x+y-z=4\end{array} \Rightarrow 5x+4y=14$

63. The first and the second equation:

$$\begin{array}{r}x-3y+2z=6\\ x+y-2z=2\end{array} \Rightarrow \tfrac{1}{2}\cdot[2x-2y]=[8]\cdot\tfrac{1}{2} \Rightarrow x-y=4$$

The first and the third equation:

$$\begin{array}{r}x-3y+2z=6\\ (-2)[3x+2y+z]=[13]\cdot(-2)\end{array} \Rightarrow \begin{array}{r}x-3y+2z=6\\ -6x-4y-2z=-26\end{array} \Rightarrow -5x-7y=-20$$

Solving for *x* and *y*:

$$\begin{array}{r}5\cdot[x-y]=[4]\cdot5\\ -5x-7y=-20\end{array} \Rightarrow \begin{array}{r}5x-5y=20\\ -5x-7y=-20\end{array} \Rightarrow -12y=0 \Rightarrow y=0$$

$$x-0=4 \Rightarrow x=4$$

Solving for *z*:

$$4-3\cdot0+2z=6 \Rightarrow 4-0+2z=6 \Rightarrow 4+2z=6 \Rightarrow 2z=2 \Rightarrow z=1$$

The solution to the system is therefore $(4,0,1)$.

65. The first and the second equation:

$$\begin{array}{r}x-y-z=2\\ 4x+y+2z=24\end{array} \Rightarrow 5x+z=26$$

The first and the third equation:

Since we eliminated *y* above, we should do the same in the next grouping.

$$\begin{array}{r}x-y-z=2\\ x+y-z=6\end{array} \Rightarrow \tfrac{1}{2}\cdot[2x-2z]=[8]\cdot\tfrac{1}{2} \Rightarrow x-z=4$$

Solving for *x* and *z*:

$$\begin{array}{r}5x+z=26\\ x-z=4\end{array} \Rightarrow 6x=30 \Rightarrow x=5$$

$$5-z=4 \Rightarrow -z=-1 \Rightarrow z=1$$

Solving for *y*:

$5-y-1=2 \Rightarrow 4-y=2 \Rightarrow -y=-2 \Rightarrow y=2$; The solution to the system is therefore $(5,2,1)$.

Section 8.2 Systems of Linear Inequalities

1. It is a half-plane.

3. There were an infinite number of solutions. There are an infinite number of servings of shrimp and broccoli that satisfy the nutritional requirements.

5. a and d

a) $3\cdot 3+4\cdot 5\overset{?}{\geq}2 \Rightarrow 9+20\overset{?}{\geq}2 \Rightarrow 29\geq 2$; True

b) $3\cdot 1+4\cdot(-2)\overset{?}{\geq}2 \Rightarrow 3-8\overset{?}{\geq}2 \Rightarrow -5\geq 2$; False

c) $3\cdot 0+4\cdot 0\overset{?}{\geq}2 \Rightarrow 0+0\overset{?}{\geq}2 \Rightarrow 0\geq 2$; False

d) $3\cdot(-4)+4\cdot 6\overset{?}{\geq}2 \Rightarrow -12+24\overset{?}{\geq}2 \Rightarrow 12\geq 2$; True

7. c and d

a) $5\cdot 3-3\cdot 2\overset{?}{<}2 \Rightarrow 15-6\overset{?}{<}2 \Rightarrow 9<2$; False

b) $5\cdot 1-3\cdot 1\overset{?}{<}2 \Rightarrow 5-3\overset{?}{<}2 \Rightarrow 2<2$; False

c) $5\cdot 0-3\cdot 0\overset{?}{<}2 \Rightarrow 0-0\overset{?}{<}2 \Rightarrow 0<2$; True

d) $5\cdot(-2)-3\cdot(-3)\overset{?}{<}2 \Rightarrow -10+9\overset{?}{<}2 \Rightarrow -1<2$; True

9. $3\cdot 0+4\cdot 0\overset{?}{\geq}12 \Rightarrow 0+0\overset{?}{\geq}12 \Rightarrow 0\geq 12$; False

Shade the side that does not contain $(0,0)$.

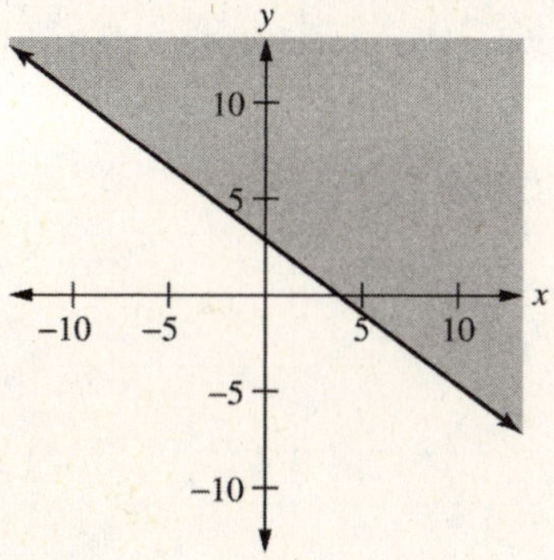

11. $2\cdot 0-4\cdot 0\overset{?}{<}12 \Rightarrow 0-0\overset{?}{<}12 \Rightarrow 0<12$; True

Shade the side that contains $(0,0)$.

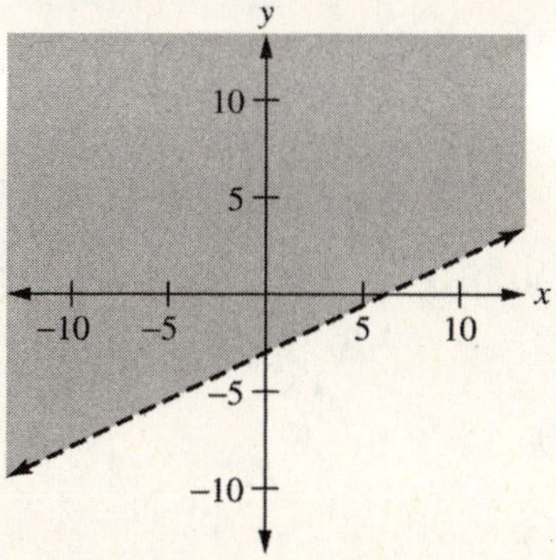

13. $0\overset{?}{\geq}3\cdot 0-9 \Rightarrow 0\overset{?}{\geq}0-9 \Rightarrow 0\geq -9$; True

Shade the side that contains $(0,0)$.

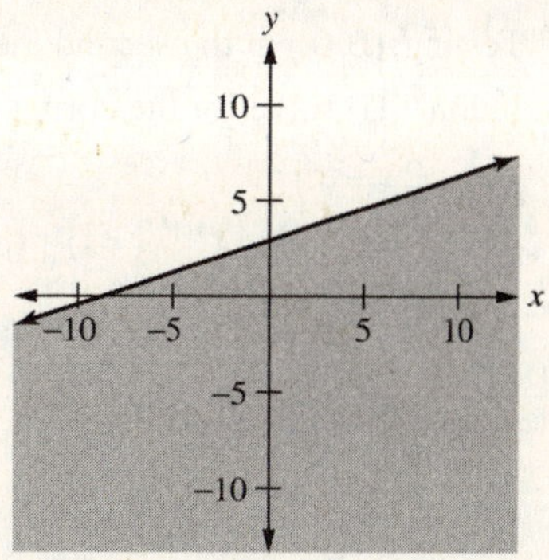

15. $4\cdot 0-8\overset{?}{<}2\cdot 0 \Rightarrow 0-8\overset{?}{<}0 \Rightarrow -8<0$; True

Shade the side that contains $(0,0)$.

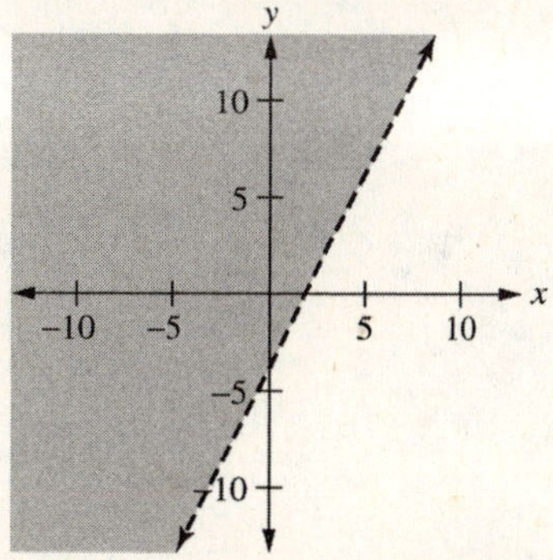

17. Testing $(0,0)$ in the first inequality, we get $2\cdot 0-3\cdot 0=0$ which is greater than -5, so $(0,0)$ is not a solution. Testing $(0,0)$ in the second inequality, we get $0-2\cdot 0=0$ which is greater than -8, so $(0,0)$ is not a solution. To solve for the corner point, we solve the system.

$$\begin{array}{c} 2x-3y=-5 \\ x-2y=-8 \end{array} \Rightarrow \begin{array}{c} 2x-3y=-5 \\ (-2)\cdot[x-2y]=[-8]\cdot(-2) \end{array} \Rightarrow \begin{array}{c} 2x-3y=-5 \\ -2x+4y=16 \end{array} \Rightarrow y=11$$

$$2x-3\cdot 11=-5 \Rightarrow 2x-33=-5 \Rightarrow 2x=28 \Rightarrow x=14$$

Thus the corner point is $(14,11)$.

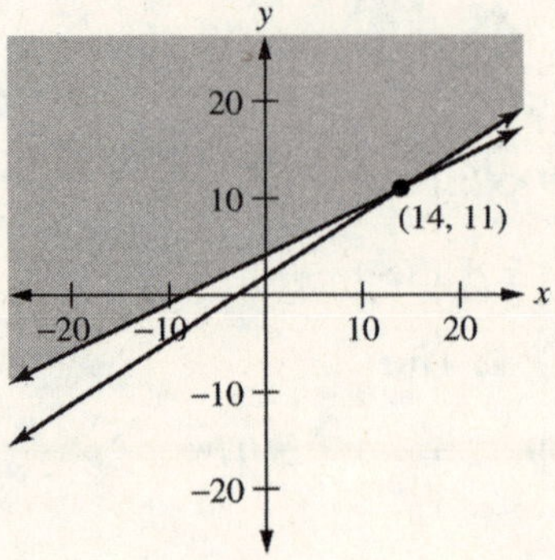

19. Testing $(0,0)$ in the first inequality, we get $2\cdot 0-0=0$ which is less than 3, so $(0,0)$ is not a solution. Testing $(0,0)$ in the second inequality, we get $0-0=0$ which is greater than -1, so $(0,0)$ is not a solution. To solve for the corner point, we solve the system.

$$\begin{array}{c} 2x-y=3 \\ x-y=-1 \end{array} \Rightarrow \begin{array}{c} 2x-y=3 \\ (-1)\cdot[x-y]=[-1]\cdot(-1) \end{array} \Rightarrow \begin{array}{c} 2x-y=3 \\ -x+y=1 \end{array} \Rightarrow x=4$$

$$4-y=-1 \Rightarrow -y=-5 \Rightarrow y=5$$

Thus the corner point is $(4,5)$.

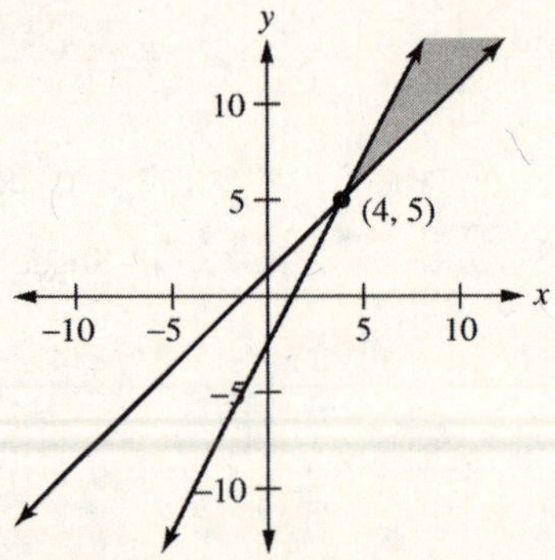

21. Testing $(0,0)$ in the first inequality, we get $-4\cdot 0+3\cdot 0=0$ which is less than 23, so $(0,0)$ is a solution. Testing $(0,0)$ in the second inequality, we get $3\cdot 0\overset{?}{>}19-5\cdot 0\Rightarrow 0\overset{?}{>}19-0\Rightarrow 0>19$ which is false, so $(0,0)$ is not a solution. To solve for the corner point, we solve the system.

$$\begin{array}{r} -4x+3y=23 \\ 3x=19-5y \end{array} \Rightarrow \begin{array}{r} 3\cdot[-4x+3y]=[23]\cdot 3 \\ 4\cdot[3x+5y]=[19]\cdot 4 \end{array} \Rightarrow \begin{array}{r} -12x+9y=69 \\ 12x+20y=76 \end{array} \Rightarrow 29y=145\Rightarrow y=5$$

$$3x=19-5\cdot 5\Rightarrow 3x=19-25\Rightarrow 3x=-6\Rightarrow x=-2$$

Thus the corner point is $(-2,5)$.

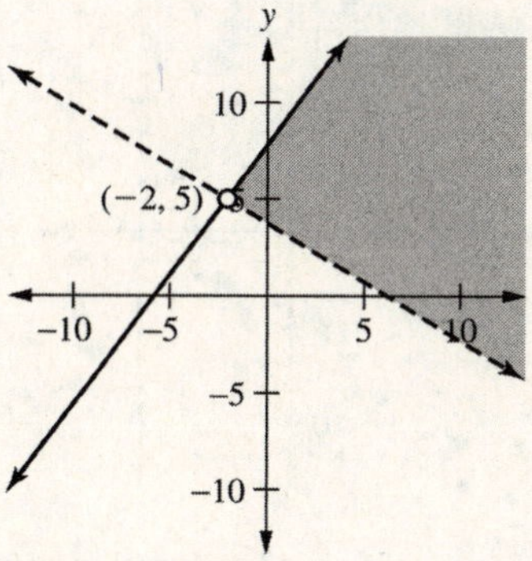

23. Testing $(0,0)$ in the first inequality, we get $3\cdot 0+5\cdot 0=0$ which is less than 32, so $(0,0)$ is a solution. Testing $(0,0)$ in the second inequality, we get $0\geq 4$ which is false, so $(0,0)$ is not a solution. To solve for the corner point, we solve the system.

$$\begin{array}{c} 3x+5y=32 \\ y=4 \end{array}$$

$$3x+5\cdot 4=32\Rightarrow 3x+20=32\Rightarrow 3x=12\Rightarrow x=4$$

Thus the corner point is $(4,4)$.

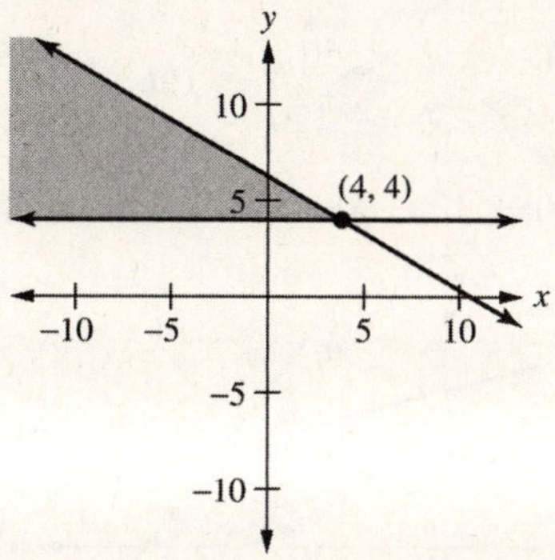

25. Testing $(0,0)$ in the first inequality, we get $2\cdot 0+5\cdot 0=0$ which is less than 26, so $(0,0)$ is not a solution. Testing $(0,0)$ in the second inequality, we get $0<8$ which is true, so $(0,0)$ is a solution. To solve for the corner point, we solve the system.

$$2x+5y=26$$
$$x=8$$
$$2\cdot 8+5y=26 \Rightarrow 16+5y=26 \Rightarrow 5y=10 \Rightarrow y=2$$

Thus the corner point is $(8,2)$.

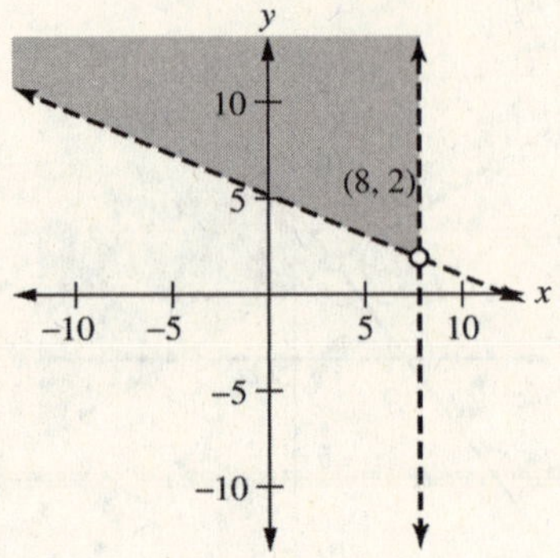

27. Since the number of entertainment centers cannot be negative, we will consider restrictions in the first quadrant given by the following table.

	Athens (a)	Barcelona (b)	Available	Inequalities
fancy molding in feet	4	15	360	$4a+15b\le 360$
time in hours	4	3	120	$4a+3b\le 120$

The origin, $(0,0)$, an a-intercept, $(30,0)$, and a b-intercept, $(0,24)$, are all corner points. To find the fourth corner point we solve the following system.

$$4a+15b=360$$
$$4a+3b=120$$

$$\begin{array}{c} 4a+15b=360 \\ (-1)\cdot[4a+3b]=[120]\cdot(-1) \end{array} \Rightarrow \begin{array}{c} 4a+15b=360 \\ -4a-3b=-120 \end{array} \Rightarrow 12b=240 \Rightarrow b=20$$

$4a+3\cdot 20=120 \Rightarrow 4a+60=120 \Rightarrow 4a=60 \Rightarrow a=15$ The fourth corner point is $(15,20)$.

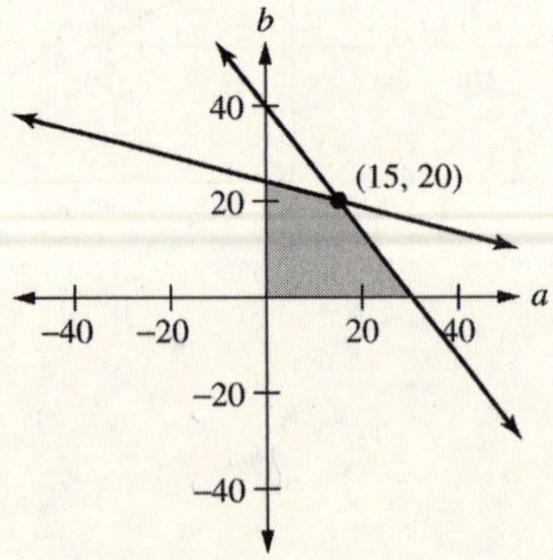

29. Let h be the number of hats.
Let t be the number of T-shirts.

Since the number of items cannot be negative, we will consider restrictions in the first quadrant given by the following.

$$30h + 20t \le 600$$

The origin, $(0,0)$, the h-intercept, $(20,0)$, and the t-intercept, $(0,30)$, are all corner points.

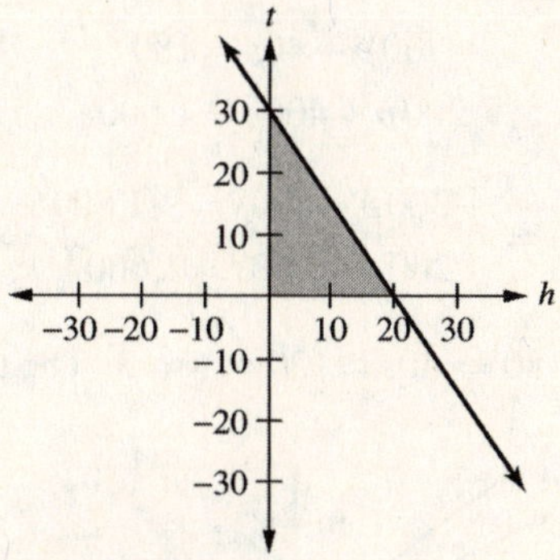

31. Let b be the amount invested in bonds in dollars.
Let m be the amount invested in mutual funds in dollars.

Since the amount invested cannot be negative, we will consider restrictions in the first quadrant given by the following.

$$b + m \le 18{,}000$$
$$b \ge 2m$$

The origin, $(0,0)$, and a b-intercept, $(18000,0)$, are corner points. To find the third corner point we solve the following system.

$$b + m = 18{,}000$$
$$b = 2m$$

$2m + m = 18{,}000 \Rightarrow 3m = 18{,}000 \Rightarrow m = 6{,}000$

$b = 2 \cdot 6{,}000 = 12{,}000$ The third corner point is $(12000, 6000)$.

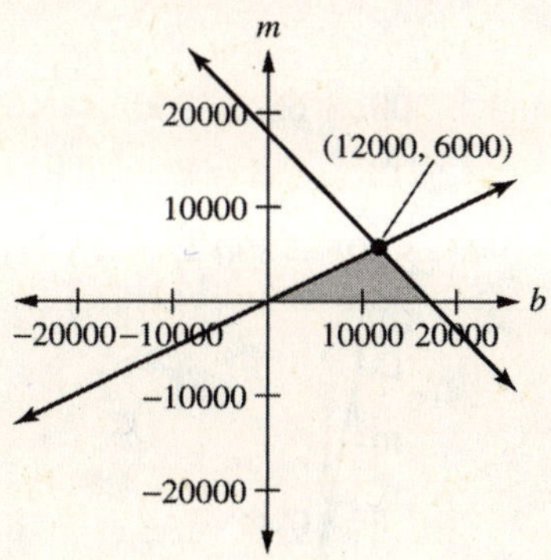

33. Since the amount of nutritional supplements cannot be negative, we will consider restrictions in the first quadrant given by the following table.

	PowerUp (p)	StressTabs (s)	Required	Inequalities
Niacin in mg	30	40	180	$30p+40s \geq 180$
Vitamin C in mg	200	400	1,600	$200p+400s \geq 1,600$

A p-intercept, $(8,0)$, and an s-intercept, $\left(0,4\frac{1}{2}\right)$, are both corner points. To find the third corner point we solve the following system.

$$30p+40s=180$$
$$200p+400s=1,600$$

$$\begin{array}{c}(-10)\cdot[30p+40s]=[180]\cdot(-10)\\ 200p+400s=1,600\end{array} \Rightarrow \begin{array}{c}-300p-400s=-1,800\\ 200p+400s=1,600\end{array} \Rightarrow -100p=-200 \Rightarrow p=2$$

$30\cdot 2+40s=180 \Rightarrow 60+40s=180 \Rightarrow 40s=120 \Rightarrow s=3$ The third corner point is $(2,3)$.

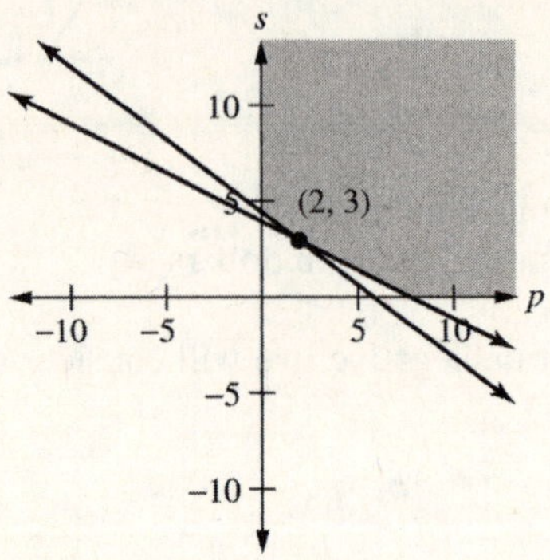

35. Let s be the number of sandwiches.
Let c be the number of slices of cheese.

Since the number of sandwiches or slices of cheese cannot be negative, we will consider restrictions in the first quadrant given by the following.

$$300s+60c \leq 900$$
$$s \geq 2$$

Two s-intercepts, $(2,0)$ and $(3,0)$, are corner points. To find the third corner point we solve the following system.

$$300s+60c=900$$
$$s=2$$

$300\cdot 2+60c=900 \Rightarrow 600+60c=900 \Rightarrow 60c=300 \Rightarrow c=5$ The third corner point is $(2,5)$.

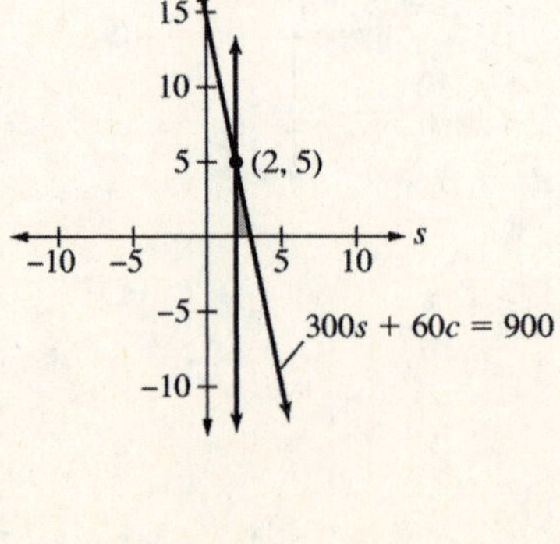

37. Let r be the amount spent on radio advertising in dollars.
Let n be the amount spent on newspaper advertising in dollars.

Since the amount spent on advertising cannot be negative, we will consider restrictions in the first quadrant given by the following.

$$\begin{aligned} 100r + 300n &\leq 2,400 \\ r &\geq 3n \end{aligned}$$

The origin, $(0,0)$, and an r-intercept, $(24,0)$, are corner points. To find the third corner point we solve the following system.

$$\begin{aligned} 100r + 300n &= 2,400 \\ r &= 3n \end{aligned}$$

$100 \cdot 3n + 300n = 2,400 \Rightarrow 300n + 300n = 2,400 \Rightarrow 600n = 2,400 \Rightarrow n = 4$

$r = 3 \cdot 4 = 12$ The third corner point is $(12,4)$.

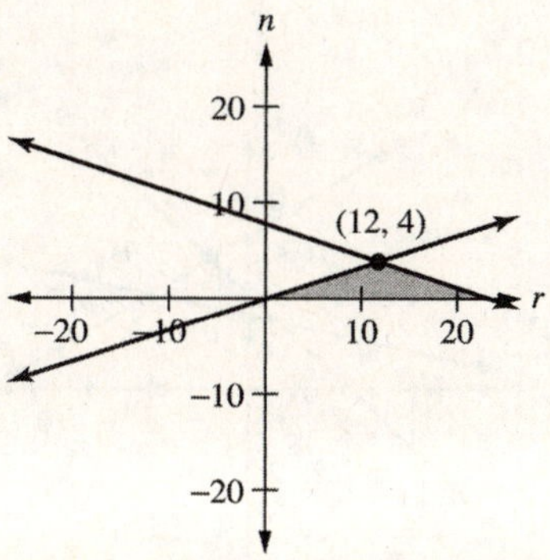

39. To determine which half-plane contains solutions to a linear inequality.

41. The x- and y-coordinates may represent real quantities that could not be negative.

43. There are no points in which the regions given by the two inequalities intersect.

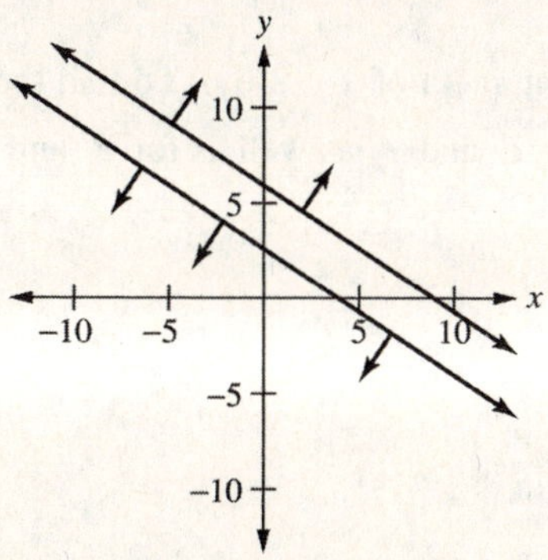

45. There are no points in which the regions given by the two inequalities intersect.

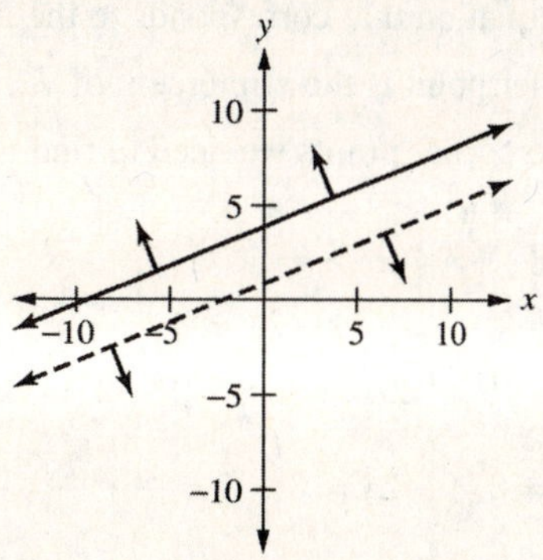

47. No solution provided.

49. The constraints $x \geq 0$ and $y \geq 0$ imply that we are limited to quadrant I.
Let the equation that corresponds to the inequality $2x+3y \leq 25$ be ℓ_1.
Let the equation that corresponds to the inequality $5y \geq 20+x$ be ℓ_2.
Let the equation that corresponds to the inequality $y \leq x+5$ be ℓ_3.

Two of the corner points are y-intercepts, namely $(0,4)$ and $(0,5)$. To find the other two corner points we need to find the point of intersection for ℓ_1 and ℓ_2 as well as for ℓ_1 and ℓ_3.

ℓ_1 and ℓ_2:

$$\begin{aligned}2x+3y&=25\\5y&=20+x\end{aligned} \Rightarrow \begin{aligned}2x+3y&=25\\2\cdot[-x+5y]&=[20]\cdot 2\end{aligned} \Rightarrow \begin{aligned}2x+3y&=25\\-2x+10y&=40\end{aligned} \Rightarrow 13y=65 \Rightarrow y=5$$

$5\cdot 5=20+x \Rightarrow 25=20+x \Rightarrow 5=x$ A corner point is $(5,5)$.

ℓ_1 and ℓ_3:

$$\begin{aligned}2x+3y&=25\\y&=x+5\end{aligned} \Rightarrow 2x+3(x+5)=25 \Rightarrow 2x+3x+15=25 \Rightarrow 5x+15=25 \Rightarrow 5x=10 \Rightarrow x=2$$

$y=2+5=7$ A corner point is $(2,7)$.

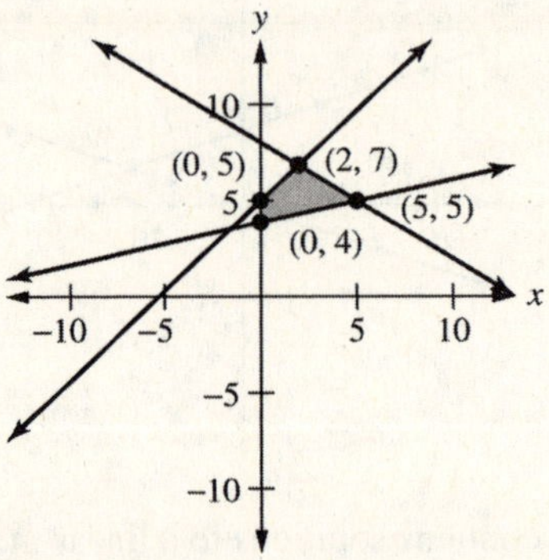

51. The constraints $x \geq 0$ and $y \geq 0$ imply that we are limited to quadrant I.

Let the equation that corresponds to the inequality $x-2y \leq 8$ be ℓ_1.
Let the equation that corresponds to the inequality $2x+y \leq 19$ be ℓ_2.
Let the equation that corresponds to the inequality $x-y \geq 5$ be ℓ_3.
One corner point is the x-intercept of ℓ_3, $(5,0)$. Another is the x-intercept of ℓ_1, $(8,0)$. To find the other two corner points we need to find the point of intersection for ℓ_1 and ℓ_2 as well as for ℓ_2 and ℓ_3.

ℓ_1 and ℓ_2:

$$\begin{aligned}x-2y&=8\\2x+y&=19\end{aligned} \Rightarrow \begin{aligned}x-2y&=8\\2\cdot[2x+y]&=[19]\cdot 2\end{aligned} \Rightarrow \begin{aligned}x-2y&=8\\4x+2y&=38\end{aligned} \Rightarrow 5x=46 \Rightarrow x=\tfrac{46}{5}$$

$\frac{46}{5}-2y=8 \Rightarrow -2y=8-\frac{46}{5}=\frac{40}{5}-\frac{46}{5}=-\frac{6}{5} \Rightarrow y=\frac{3}{5}$ A corner point is $\left(\frac{46}{5},\frac{3}{5}\right)$.

ℓ_2 and ℓ_3:

$$\begin{aligned}2x+y&=19\\x-y&=5\end{aligned} \Rightarrow 3x=24 \Rightarrow x=8$$

$8-y=5 \Rightarrow -y=-3 \Rightarrow y=3$ A corner point is $(8,3)$.

51. (continued)

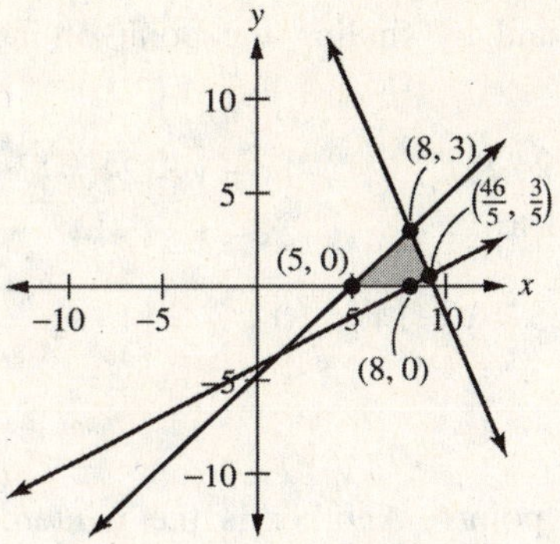

For Exercises 53 – 55, we consider how the origin $(0,0)$ relates in each of the equations. We will replace the blank with either $\leq$ or $\geq$, depending on whether we want the statement to be true or false.

Case A: For $3y = 43 - 5x$ we have $3\cdot 0 \overset{?}{\_} 43 - 5\cdot 0$ or $0 \overset{?}{\_} 43 - 0$ or $0 \overset{?}{\_} 43$.

Case B: For $11y = 27 - 2x$ we have $11\cdot 0 \overset{?}{\_} 27 - 2\cdot 0$ or $0 \overset{?}{\_} 27 - 0$ or $0 \overset{?}{\_} 27$.

Case C: For $3y = 1 + 2x$ we have $3\cdot 0 \overset{?}{\_} 1 + 2\cdot 0$ or $0 \overset{?}{\_} 1 + 0$ or $0 \overset{?}{\_} 1$.

53. In Region 1, when we look at the location of the origin, $(0,0)$, Case A should be true, Case B should be false and Case C should be false. This yields the following set of inequalities.

$$3y \leq 43 - 5x$$
$$11y \geq 27 - 2x$$
$$3y \geq 1 + 2x$$

55. In Region 3, when we look at the location of the origin, $(0,0)$, Case A should be true, Case B should be false and Case C should be true. This yields the following set of inequalities.

$$3y \leq 43 - 5x$$
$$11y \geq 27 - 2x$$
$$3y \leq 1 + 2x$$

Section 8.3 Looking Deeper: Linear Programming

1. In Example 1, the maximum value of the quantity $2x - 5y$ within the solution set for the system of inequalities is being found.

3. The constraints are on copper, $2s + 16c \leq 160$, and on labor, $1s + 3c \leq 40$.

5. The origin, $(0,0)$, is a corner point. Another is the y-intercept of $x + 2y = 16$, namely $(0,8)$. Another is the x-intercept of $4x + y = 24$, namely $(6,0)$.

   A fourth corner point can be found by finding the point of intersection between $x + 2y = 16$ and $2x + y = 14$.

$$\begin{array}{c} x + 2y = 16 \\ (-2)\cdot[2x + y] = [14]\cdot(-2) \end{array} \Rightarrow \begin{array}{c} x + 2y = 16 \\ -4x - 2y = -28 \end{array} \Rightarrow -3x = -12 \Rightarrow x = 4$$

$$2\cdot 4 + y = 14 \Rightarrow 8 + y = 14 \Rightarrow y = 6$$

   This corner point is $(4,6)$.

5. (continued)

A fifth corner point can be found by finding the point of intersection between $4x+y=24$ and $2x+y=14$.

$$\begin{matrix} 4x+y=24 \\ (-1)\cdot[2x+y]=[14]\cdot(-1) \end{matrix} \Rightarrow \begin{matrix} 4x+y=24 \\ -2x-y=-14 \end{matrix} \Rightarrow 2x=10 \Rightarrow x=5$$

$$2\cdot5+y=14 \Rightarrow 10+y=14 \Rightarrow y=4$$

This corner point is $(5,4)$.

7. The origin, $(0,0)$, is a corner point. Another is the y-intercept of $x+4y=24$, namely $(0,6)$. Another is the x-intercept of $3x+2y=22$, namely $\left(\frac{22}{3},0\right)$.

A fourth corner point can be found by finding the point of intersection between $x+y=8$ and $x+4y=24$.

$$\begin{matrix} x+y=8 \\ (-1)[x+4y]=[24]\cdot(-1) \end{matrix} \Rightarrow \begin{matrix} x+y=8 \\ -x-4y=-24 \end{matrix} \Rightarrow -3y=-16 \Rightarrow y=\tfrac{16}{3}$$

$$x+\tfrac{16}{3}=8 \Rightarrow x=8-\tfrac{16}{3}=\tfrac{24}{3}-\tfrac{16}{3}=\tfrac{8}{3}$$

This corner point is $\left(\frac{8}{3},\frac{16}{3}\right)$.

A fifth corner point can be found by finding the point of intersection between $x+y=8$ and $3x+2y=22$.

$$\begin{matrix} (-2)\cdot[x+y]=[8]\cdot(-2) \\ 3x+2y=22 \end{matrix} \Rightarrow \begin{matrix} -2x-2y=-16 \\ 3x+2y=22 \end{matrix} \Rightarrow x=6$$

$$6+y=8 \Rightarrow y=2$$

This corner point is $(6,2)$.

9. The origin, $(0,0)$, is a corner point. Another is the y-intercept of $x+2y=16$, namely $(0,8)$. Another is the x-intercept of $2x+y=16$, namely $(8,0)$.

A fourth corner point can be found by finding the point of intersection between $x+y=10$ and $x+2y=16$.

$$\begin{matrix} x+y=10 \\ (-1)[x+2y]=[16]\cdot(-1) \end{matrix} \Rightarrow \begin{matrix} x+y=10 \\ -x-2y=-16 \end{matrix} \Rightarrow -y=-6 \Rightarrow y=6$$

$$x+6=10 \Rightarrow x=4$$

This corner point is $(4,6)$.

A fifth corner point can be found by finding the point of intersection between $x+y=10$ and $2x+y=16$.

$$\begin{matrix} x+y=10 \\ (-1)[2x+y]=[16]\cdot(-1) \end{matrix} \Rightarrow \begin{matrix} x+y=10 \\ -2x-y=-16 \end{matrix} \Rightarrow -x=-6 \Rightarrow x=6$$

$$6+y=10 \Rightarrow y=4$$

This corner point is $(6,4)$.

9. (continued)

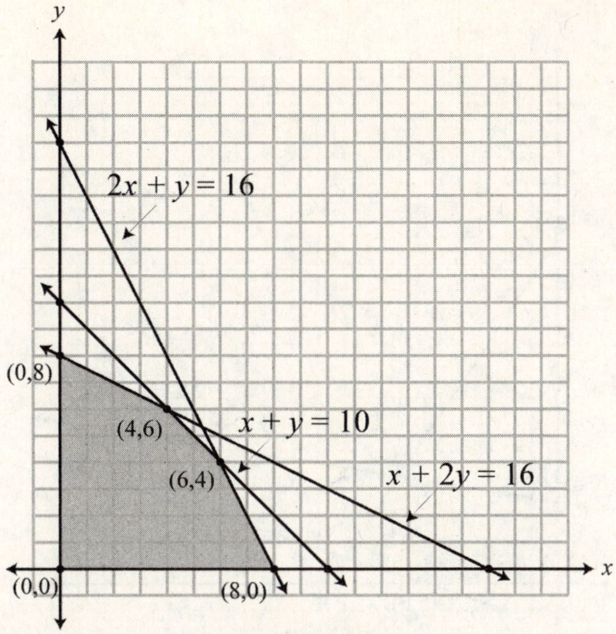

Corner point	Value of $P = 3x + 4y$	
$(0,0)$	$3\cdot0+4\cdot0=0+0=0$	
$(0,8)$	$3\cdot0+4\cdot8=0+32=32$	
$(8,0)$	$3\cdot8+4\cdot0=24+0=24$	
$(4,6)$	$3\cdot4+4\cdot6=12+24=36$	$\Leftarrow$ Maximum
$(6,4)$	$3\cdot6+4\cdot4=18+16=34$	

Maximum is 36 at $(4,6)$.

11. The origin, $(0,0)$, is a corner point. Another is the y-intercept of $x-6y=-18$, namely $(0,3)$. Another is the x-intercept of $x-3y=5$, namely $(5,0)$.

A fourth corner point can be found by finding the point of intersection between $3x+2y=26$ and $x-3y=5$.

$$\begin{matrix} 3x+2y=26 \\ (-3)[x-3y]=[5]\cdot(-3) \end{matrix} \Rightarrow \begin{matrix} 3x+2y=26 \\ -3x+9y=-15 \end{matrix} \Rightarrow 11y=11 \Rightarrow y=1$$

$$x-3\cdot1=5 \Rightarrow x-3=5 \Rightarrow x=8$$

This corner point is $(8,1)$.

A fifth corner point can be found by finding the point of intersection between $3x+2y=26$ and $x-6y=-18$.

$$\begin{matrix} 3x+2y=26 \\ (-3)\cdot[x-6y]=[-18]\cdot(-3) \end{matrix} \Rightarrow \begin{matrix} 3x+2y=26 \\ -3x+18y=54 \end{matrix} \Rightarrow 20y=80 \Rightarrow y=4$$

$$x-6\cdot4=-18 \Rightarrow x-24=-18 \Rightarrow x=6$$

This corner point is $(6,4)$.

11. (continued)

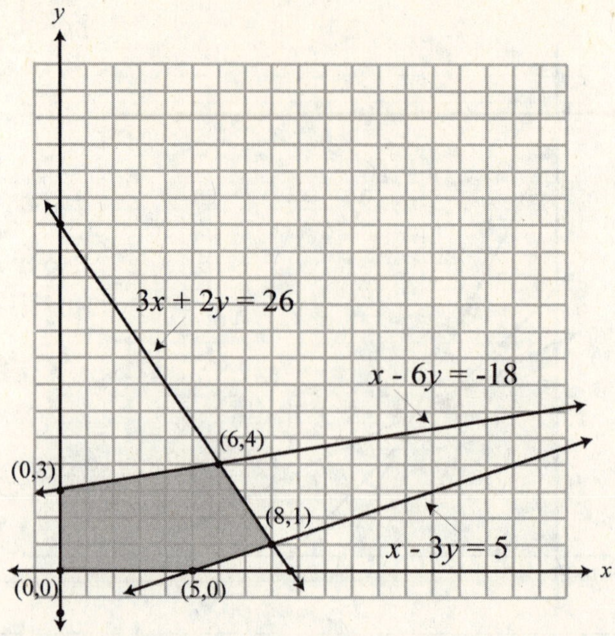

Corner point	Value of $P = 5x - y$	
$(0,0)$	$5\cdot 0 - 0 = 0 - 0 = 0$	
$(0,3)$	$5\cdot 0 - 3 = 0 - 3 = -3$	$\Leftarrow$ Minimum
$(5,0)$	$5\cdot 5 - 0 = 25 - 0 = 25$	
$(8,1)$	$5\cdot 8 - 1 = 40 - 1 = 39$	
$(6,4)$	$5\cdot 6 - 4 = 30 - 4 = 26$	

Minimum is -3 at $(0,3)$.

13. A corner point can be found by finding the point of intersection between $y = 3$ and $x - y = -1$.

$$x - 3 = -1 \Rightarrow x = 2$$

This corner point is $(2,3)$.

A second corner point can be found by finding the point of intersection between $y = 3$ and $2x + y = 21$.

$$2x + 3 = 21 \Rightarrow 2x = 18 \Rightarrow x = 9$$

This corner point is $(9,3)$.

A third corner point can be found by finding the point of intersection between $y = 5$ and $x - y = -1$.

$$x - 5 = -1 \Rightarrow x = 4$$

This corner point is $(4,5)$.

A fourth corner point can be found by finding the point of intersection between $y = 5$ and $2x + y = 21$.

$$2x + 5 = 21 \Rightarrow 2x = 16 \Rightarrow x = 8$$

This corner point is $(8,5)$.

13. (continued)

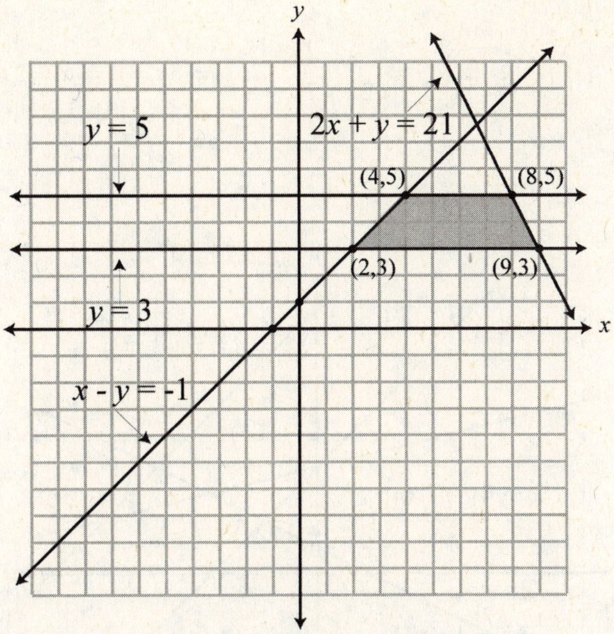

Corner point	Value of $P = 20x - 12y$	
$(2,3)$	$20 \cdot 2 - 12 \cdot 3 = 40 - 36 = 4$	⇐ Minimum
$(9,3)$	$20 \cdot 9 - 12 \cdot 3 = 180 - 36 = 144$	
$(4,5)$	$20 \cdot 4 - 12 \cdot 5 = 80 - 60 = 20$	
$(8,5)$	$20 \cdot 8 - 12 \cdot 5 = 160 - 60 = 100$	

Minimum is 4 at $(2,3)$.

15. The corner point can be found by finding the point of intersection between $x-2y=-2$ and $x+y=4$.

$$\begin{array}{c} x-2y=-2 \\ 2\cdot[x+y]=[4]\cdot 2 \end{array} \Rightarrow \begin{array}{c} x-2y=-2 \\ 2x+2y=8 \end{array} \Rightarrow 3x=6 \Rightarrow x=2$$

$$2+y=4 \Rightarrow y=2$$

The corner point is $(2,2)$.

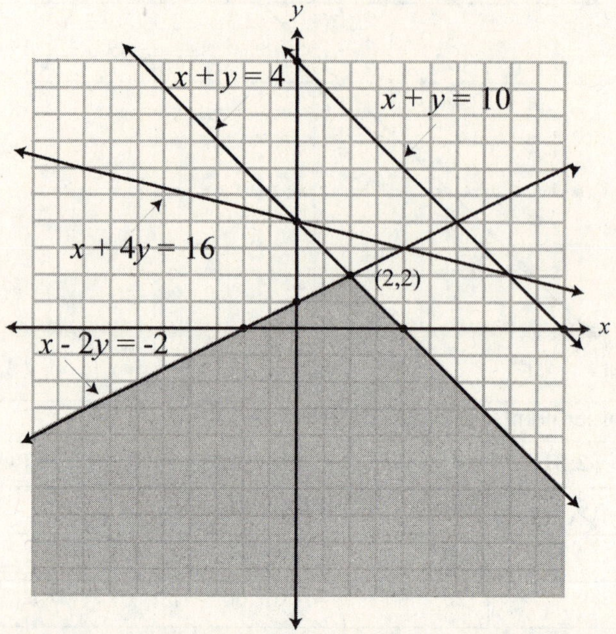

Corner point	Value of $P=5x+12y$	
$(2,2)$	$5\cdot 2+12\cdot 2=10+24=34$	$\Leftarrow$ Maximum

Maximum is 34 at $(2,2)$.

17. A corner point can be found by finding the point of intersection between $6x-7y=-6$ and $3x+4y=42$.

$$\begin{array}{c} 6x-7y=-6 \\ (-2)\cdot[3x+4y]=[42]\cdot(-2) \end{array} \Rightarrow \begin{array}{c} 6x-7y=-6 \\ -6x-8y=-84 \end{array} \Rightarrow -15y=-90 \Rightarrow y=6$$

$$3x+4\cdot 6=42 \Rightarrow 3x+24=42 \Rightarrow 3x=18 \Rightarrow x=6$$

This corner point is $(6,6)$.

17. (continued)

A second corner point can be found by finding the point of intersection between $6x-7y=-6$ and $6x-22y=-51$.

$$\begin{array}{c} 6x-7y=-6 \\ (-1)\cdot[6x-22y]=[-51]\cdot(-1) \end{array} \Rightarrow \begin{array}{c} 6x-7y=-6 \\ -6x+22y=51 \end{array} \Rightarrow 15y=45 \Rightarrow y=3$$

$$6x-7\cdot 3=-6 \Rightarrow 6x-21=-6 \Rightarrow 6x=15 \Rightarrow x=\tfrac{15}{6}=\tfrac{5}{2}$$

This corner point is $\left(\frac{5}{2},3\right)$.

A third corner point can be found by finding the point of intersection between $6x-22y=-51$ and $3x+4y=42$.

$$\begin{array}{c} 6x-22y=-51 \\ (-2)\cdot[3x+4y]=[42]\cdot(-2) \end{array} \Rightarrow \begin{array}{c} 6x-22y=-51 \\ -6x-8y=-84 \end{array} \Rightarrow -30y=-135 \Rightarrow y=\tfrac{135}{30}=\tfrac{9}{2}$$

$$3x+4\cdot\tfrac{9}{2}=42 \Rightarrow 3x+18=42 \Rightarrow 3x=24 \Rightarrow x=8$$

This corner point is $\left(8,\frac{9}{2}\right)$.

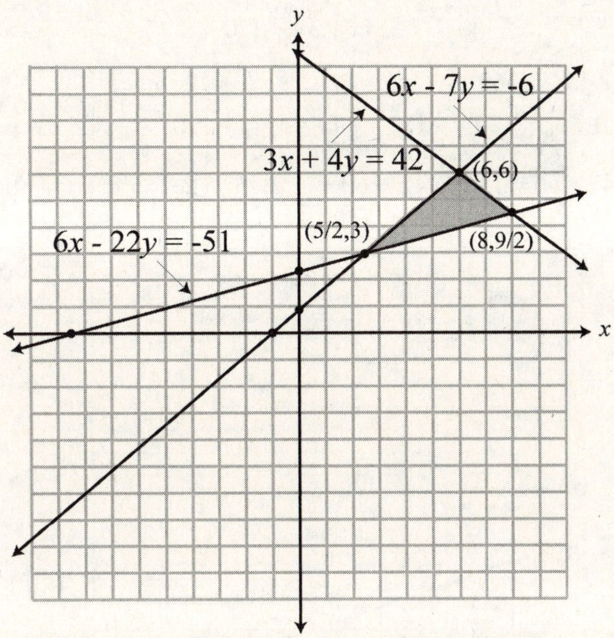

Corner point	Value of $P=4x+2y$	
$(6,6)$	$4\cdot 6+2\cdot 6=24+12=36$	
$\left(\frac{5}{2},3\right)$	$4\cdot\frac{5}{2}+2\cdot 3=10+6=16$	⇐ Minimum
$\left(8,\frac{9}{2}\right)$	$4\cdot 8+2\cdot\frac{9}{2}=32+9=41$	

Minimum is 16 at $\left(\frac{5}{2},3\right)$.

19. Let x be the number of Athens entertainment centers.
Let y be the number of Barcelona entertainment centers.

Since the number of entertainment centers cannot be negative, we will consider restrictions in the first quadrant given by the following table.

	Athens (x)	Barcelona (y)	Available	Inequalities
Fancy molding in feet	4	15	360	$4x+15y \le 360$
Time in hours	4	3	120	$4x+3y \le 120$

The constraints-objective are therefore as follows.

Maximize $P = 9x+12y$ subject to the constraints

$$4x+15y \le 360$$
$$4x+3y \le 120$$
$$x \ge 0,\ y \ge 0$$

The origin, $(0,0)$, an x-intercept, $(30,0)$, and a y-intercept, $(0,24)$, are all corner points. To find the fourth corner point we solve the following system.

$$4x+15y=360$$
$$4x+3y=120$$

$$\begin{array}{c} 4x+15y=360 \\ (-1)\cdot[4x+3y]=[120]\cdot(-1) \end{array} \Rightarrow \begin{array}{c} 4x+15y=360 \\ -4x-3y=-120 \end{array} \Rightarrow 12y=240 \Rightarrow y=20$$

$4x+3\cdot 20=120 \Rightarrow 4x+60=120 \Rightarrow 4x=60 \Rightarrow x=15$ The fourth corner point is $(15,20)$

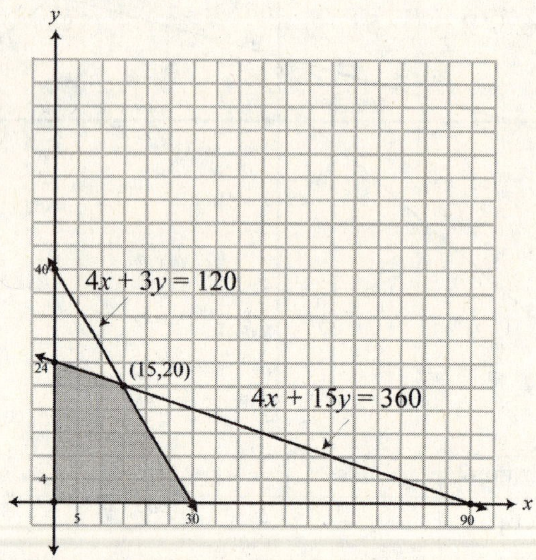

Corner point	Value of $P=9x+12y$	
$(0,0)$	$9\cdot 0+12\cdot 0=0+0=0$	
$(30,0)$	$9\cdot 30+12\cdot 0=270+0=270$	
$(0,24)$	$9\cdot 0+12\cdot 24=0+288=288$	
$(15,20)$	$9\cdot 15+12\cdot 20=135+240=375$	$\Leftarrow$ Maximum

Scott should manufacture 15 Athens and 20 Barcelona entertainment centers to produce a maximum profit of \$375.

21. Let x be the number of hats.
Let y be the number of T-shirts.

The constraints-objective are therefore as follows.

Maximize $P = 4x + 3y$ subject to the constraints

$$10x + 9y \leq 140$$
$$y \geq 2x$$
$$x \geq 0, y \geq 0$$

The origin, $(0,0)$, and a y-intercept, $(0, \frac{140}{9})$, are corner points. To find the third corner point we solve the following system.

$$10x + 9y = 140$$
$$y = 2x$$

$$10x + 9(2x) = 140 \Rightarrow 10x + 18x = 140 \Rightarrow 28x = 140 \Rightarrow x = 5$$

$y = 2 \cdot 5 = 10$ The third corner point is $(5,10)$.

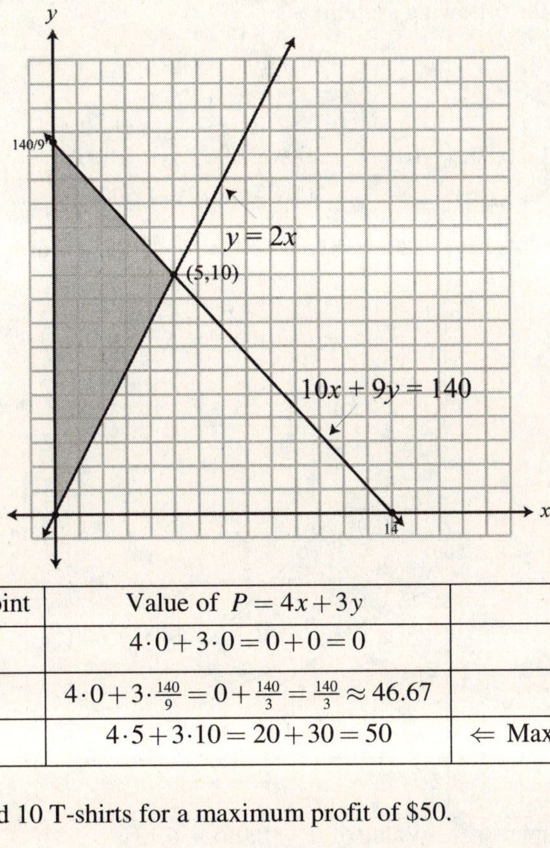

Corner point	Value of $P = 4x + 3y$	
$(0,0)$	$4 \cdot 0 + 3 \cdot 0 = 0 + 0 = 0$	
$(0, \frac{140}{9})$	$4 \cdot 0 + 3 \cdot \frac{140}{9} = 0 + \frac{140}{3} = \frac{140}{3} \approx 46.67$	
$(5,10)$	$4 \cdot 5 + 3 \cdot 10 = 20 + 30 = 50$	$\Leftarrow$ Maximum

Nicole should make 5 hats and 10 T-shirts for a maximum profit of $50.

Note: Although in this solution we considered a corner point of $(0, \frac{140}{9})$, in a real-life situation we would only consider whole numbers.

23. Let x be the number of PowerUp tablets.
Let y be the number of StressTabs tablets.

Since the amount of nutritional supplements cannot be negative, we will consider restrictions in the first quadrant given by the following table.

	PowerUp (x)	StressTabs (y)	Required	Inequalities
Niacin in mg	30	40	180	$30x+40y \geq 180$
Vitamin C in mg	200	400	1,600	$200x+400y \geq 1,600$

The constraints-objective are therefore as follows.

Minimize $P = 0.09x + 0.13y$ subject to the constraints

$$30x+40y \geq 180$$
$$200x+400y \geq 1,600$$
$$x+y \leq 8$$
$$x \geq 0, y \geq 0$$

An x-intercept, $(8,0)$, and two y-intercept, $(0,4\frac{1}{2})$ and $(0,8)$, are all corner points. To find the fourth corner point we solve the following system.

$$30x+40y=180$$
$$200x+400y=1,600$$

$$\begin{matrix}(-10)\cdot[30x+40y]=[180]\cdot(-10) \\ 200x+400y=1,600\end{matrix} \Rightarrow \begin{matrix}-300x-400y=-1,800 \\ 200x+400y=1,600\end{matrix} \Rightarrow -100x=-200 \Rightarrow x=2$$

$30\cdot 2+40y=180 \Rightarrow 60+40y=180 \Rightarrow 40y=120 \Rightarrow y=3$ The fourth corner point is $(2,3)$.

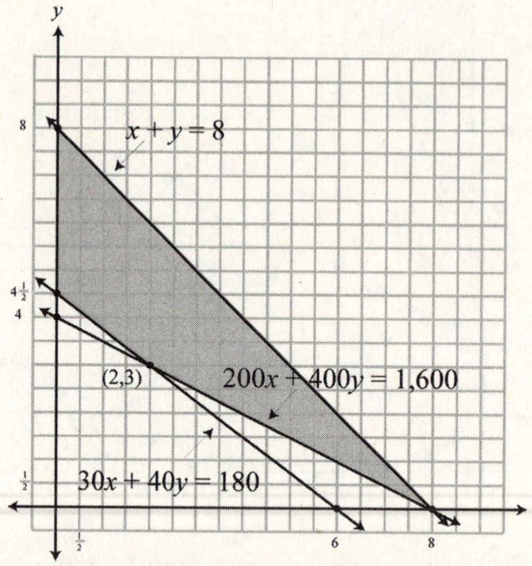

Corner point	Value of $P=0.09x+0.13y$	
$(8,0)$	$0.09\cdot 8+0.13\cdot 0=0.72+0=0.72$	
$(0,4\frac{1}{2})$	$0.09\cdot 0+0.13\cdot\frac{9}{2}=0+0.585\approx 0.59$	
$(0,8)$	$0.09\cdot 0+0.13\cdot 8=0+1.04=1.04$	
$(2,3)$	$0.09\cdot 2+0.13\cdot 3=0.18+0.39=0.57$	$\Leftarrow$ Minimum

Raphael should take 2 PowerUp tablets and 3 StressTabs for a minimum cost of \$0.57.

25. Let x be the number of lizards.
Let y be the number of frogs.

Since the number of animals cannot be negative, we will consider restrictions in the first quadrant given by the following table.

	Lizards (x)	Frogs (y)	Available	Inequalities
Living space in square feet	2	2	400	$2x+2y\le 400$
Cost to feed in dollars	6	1	600	$6x+y\le 600$

The constraints-objective are therefore as follows.

Maximize $P=17x+6y$ subject to the constraints

$$2x+2y\le 400$$
$$6x+y\le 600$$
$$x\ge 0,\ y\ge 0$$

The origin, $(0,0)$, an x-intercept, $(100,0)$, and a y-intercept, $(0,200)$, are all corner points. To find the fourth corner point we solve the following system.

$$2x+2y=400$$
$$6x+y=600$$

$$\begin{array}{c}2x+2y=400\\(-2)\cdot[6x+y]=[600]\cdot(-2)\end{array}\Rightarrow\begin{array}{c}2x+2y=400\\-12x-2y=-1,200\end{array}\Rightarrow -10x=-800\Rightarrow x=80$$

$6\cdot 80+y=600\Rightarrow 480+y=600\Rightarrow y=120$ The fourth corner point is $(80,120)$.

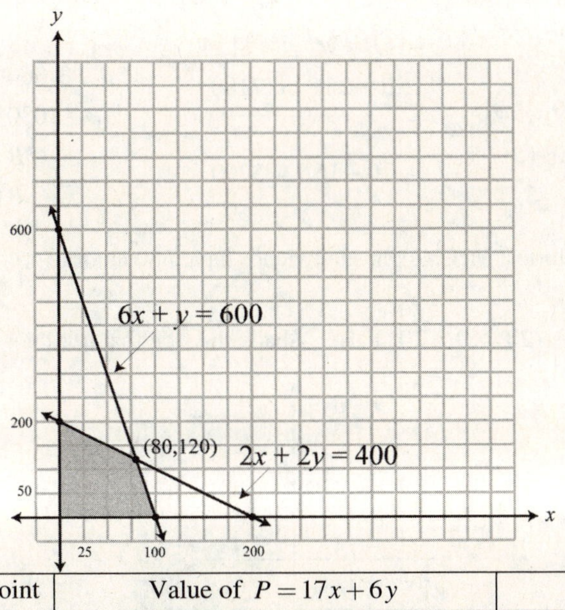

Corner point	Value of $P=17x+6y$	
$(0,0)$	$17\cdot 0+6\cdot 0=0+0=0$	
$(100,0)$	$17\cdot 100+6\cdot 0=1,700+0=1,700$	
$(0,200)$	$17\cdot 0+6\cdot 200=0+1,200=1,200$	
$(80,120)$	$17\cdot 80+6\cdot 120=1,360+720=2,080$	$\Leftarrow$ Maximum

The Reptile Farm should host 80 lizards and 120 frogs for a maximum profit of \$2,080.

27. The variables, the constraints, and the objective function.

29. First, graph the system of constraints, second, find the corner points; third, evaluate the objective function at the corner points.

31. – 35. No solution provided.

Chapter Review Exercises

1. a) $(3,-5)$

$$4x-3y=27$$
$$(-4)\cdot[x+2y]=[-7]\cdot(-4)$$

$$\begin{array}{r} 4x-3y=27 \\ -4x-8y=28 \\ \hline 0-11y=55 \\ y=-5 \end{array} \qquad \begin{array}{r} x+2\cdot(-5)=-7 \\ x-10=-7 \\ x=3 \end{array}$$

b) The lines intersect at a single point.

3. a) no solution

$$(3)\cdot[-4x+5y]=[3]\cdot(3)$$
$$12x-15y=-10$$

$$\begin{array}{r} -12x+15y=9 \\ 12x-15y=-10 \\ \hline 0=-1 \end{array}$$

b) The lines are parallel.

5. Let B be the number of Big Macs.
Let C be the number of Chicken Sandwiches.
$5B+3C=3780$ Actual Meal
$3B+5C=3420$ Optional Meal

$$\begin{array}{c} (-3)\cdot[5B+3C]=[3780]\cdot(-3) \\ (5)\cdot[3B+5C]=[3420]\cdot(5) \end{array} \Rightarrow \begin{array}{r} -15B-9C=-11340 \\ 15B+25C=17100 \\ \hline 0+16B=5760 \\ y=360 \end{array} \Rightarrow \begin{array}{r} 5B+3\cdot 360=3780 \\ 5B+1080=3780 \\ 5B=2700 \\ B=540 \end{array}$$

Big Mac's have 540 calories; McChicken sandwiches have 360 calories.

7. $6\cdot 0+5\cdot 0 \overset{?}{>} 20 \Rightarrow 0+0 \overset{?}{>} 20 \Rightarrow 0>20$; False. Shade the side that does not contain $(0,0)$.

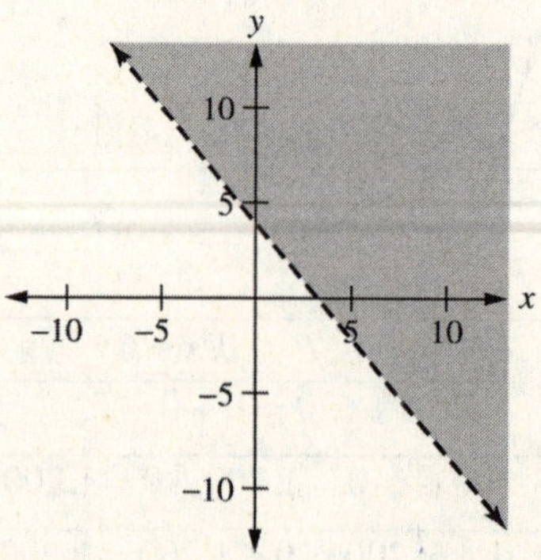

9. Let t be the number of T-shirts.
Let h be the number of hats.

Since the number of hats and T-shirts cannot be negative, we will consider restrictions in the first quadrant given by the following.

$$\begin{aligned} 4t+2h &\le 180 \\ h &\ge t+15 \end{aligned}$$

The h-intercepts, $(0,90)$ and $(0,15)$, are corner points. To find the third corner point we solve the following system.

$$\begin{aligned} 4t+2h &= 180 \\ h &= t+15 \end{aligned}$$

$$\begin{aligned} 4t+2(t+15) &= 180 \\ 4t+2t+30 &= 180 \\ 6t+30 &= 180 \\ 6t &= 150 \\ t &= 25 \end{aligned}$$

$h=25+15=40$

The third corner point is $(25,40)$.

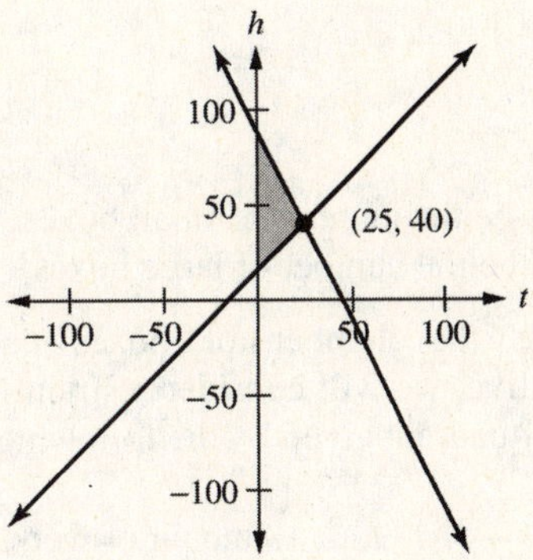

Chapter Test

1. a) no solution

$(4)\cdot[6x-2y]=[-4]\cdot(4)$

$-24x+8y=10$

$$\begin{aligned} 24x-8y &= -16 \\ -24x+8y &= 10 \\ \hline 0 &= -6 \end{aligned}$$

b) The lines are parallel.

3. $(2,-3)$

$(3)\cdot[3x+4y]=[-6]\cdot(3)$

$(4)\cdot[2x-3y]=[13]\cdot(4)$

$$\begin{aligned} 9x+12y &= -18 \\ 8x-12y &= 52 \\ \hline 17x+0 &= 34 \\ x &= 2 \end{aligned}$$

$$\begin{aligned} 3(2)+4y &= -6 \\ 6+4y &= -6 \\ 4y &= -12 \\ y &= -3 \end{aligned}$$

5. Let f be the number of hours Shandra worked at the fast food restaurant.
Let s be the number of hours she worked giving swimming lessons.

$$\begin{aligned} s &= f+5 \\ 5.65f+8.50s &= 212.30 \end{aligned} \Rightarrow \begin{aligned} s &= f+5 \\ 100\cdot[5.65f+8.50s] &= [212.30]\cdot 100 \end{aligned} \Rightarrow \begin{aligned} s &= f+5 \\ 565f+850s &= 21,230 \end{aligned}$$

Using substitution, $565f+850(f+5)=21,230 \Rightarrow 565f+850f+4,250=21,230 \Rightarrow$

$1,415f+4,250=21,230 \Rightarrow 1,415f=16,980 \Rightarrow f=12$

Shandra worked 12 hours at the restaurant and $s=12+5=17$ hours giving swimming lessons.

7. $4 \cdot 0 - 3 \cdot 0 \overset{?}{\geq} 8 \Rightarrow 0 - 0 \overset{?}{\geq} 8 \Rightarrow 0 \geq 8$; False. Shade the side that does not contain $(0,0)$.

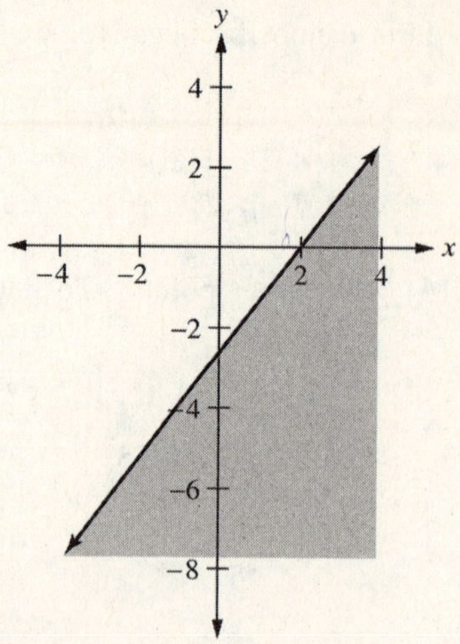

9. Let s be the number of small boxes.
Let l be the number of large boxes.

Since the number of boxes cannot be negative, we will consider restrictions in the first quadrant given by the following.

$$2s + 6l \leq 90 \quad \text{(Amount of wood)}$$
$$s + 2l \leq 40 \quad \text{(Amount of time)}$$

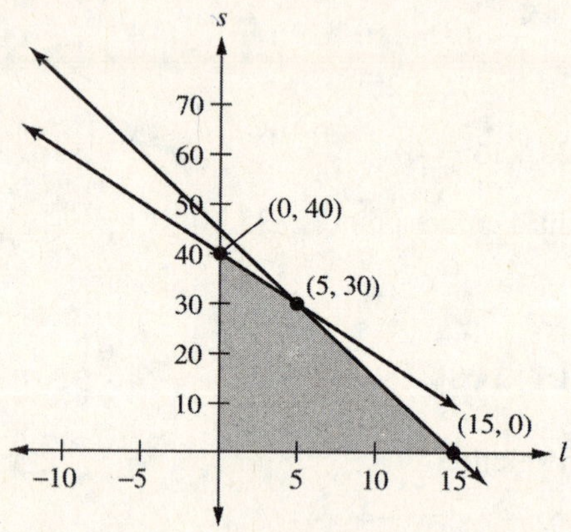

The points, $(40,0)$ and $(0,15)$, are corner points. To find the third corner point we solve the following system.

$$2s + 6l = 90$$
$$-2 \cdot [s + 2l] = 40 \cdot -2$$

$$\begin{array}{r} 2s + 6l = 90 \\ -2s - 4l = -80 \\ \hline 2l = 10 \\ l = 5 \end{array} \qquad \begin{array}{l} s + 2 \cdot 5 = 40 \\ s + 10 = 40 \\ s = 30 \end{array}$$

The third corner point is $(5,30)$.

Chapter 9

Consumer Mathematics: The Mathematics of Everyday Life

Section 9.1 Percent Change and Taxes

1. We represented 19% as 19 hundredths, or 0.19, then placed the 32 in the two decimal places to the right of 0.19.

3. We found the difference between the selling price and the dealer's cost and then divided this value by the dealer's cost.

5. 0.78

7. 0.08

9. 0.2735

11. 0.0035

13. 43%

15. 36.5%

17. 145%

19. 0.2%

21. $\frac{3}{4} = 0.75 = 75\%$

23. $\frac{5}{16} = 0.3125 = 31.25\%$

25. $\frac{5}{2} = 2.5 = 250\%$

27. $\frac{4}{250} = 0.016 = 1.6\%$

29. $percent \times 80 = 12$

$$percent = \frac{12}{80} = 0.15 = 15\%$$

31. $0.22 \times base = 77$

$$base = \frac{77}{0.22} = 350$$

33. $0.1225 \times 160 = amount$

$$19.6 = amount$$

35. $percent \times 48 = 8.4$

$$percent = \frac{8.4}{48} = 0.175 = 17.5\%$$

37. $0.2325 \times base = 29.76$

$$base = \frac{29.76}{0.2325} = 128$$

39. $\frac{294.6}{3{,}124} \approx 0.094 = 9.4\%$

41. The average price of a new home in the Northeast in 2003 represents 100% of the 2005 price plus the 7.88% increase or $100\% + 7.88\% = 107.88\%$.

$$1.0788 \times base = 428{,}000 \Rightarrow base = \frac{428{,}000}{1.0788} \approx \$397{,}000$$

The average price for a new home in the Northeast in 2005 was approximately $397,000.

43. $\dfrac{502}{502+476+476+250+222}=\dfrac{502}{1{,}926}\approx 0.2606=26.06\%$

45. $\dfrac{476-222}{222}=\dfrac{254}{222}\approx 1.1441=114.41\%$

114.41% larger

47. $\dfrac{448}{448+354+290+317+388}=\dfrac{448}{1{,}797}\approx 0.2493=24.93\%$

49. $\dfrac{448-317}{317}=\dfrac{131}{317}\approx 0.4132=41.32\%$

51. The 81 authentic autographs represent 6% of all autographs in circulation.

$$0.06\times base=81\Rightarrow base=\frac{81}{0.06}\approx 1{,}350$$

There are $1{,}350-81=1{,}269$ non-authentic Beatles autographs in circulation.

53. $\dfrac{301.6-281}{281}=\dfrac{20.6}{281}\approx 0.073=7.3\%$

55. The amount in 2006 represents 100% of the amount in 2005 plus the 103.3% increase or $100\%+103.3\%=203.3\%$.

$$2.033\times 13.6=27.6$$

Approximately 27.6 million albums were downloaded in 2006.

57. a) $\dfrac{16{,}065-14{,}875}{14{,}875}=\dfrac{1{,}190}{14{,}875}=0.08=8\%$

b) She divided by 16,065 instead of 14,875.

$$\frac{16{,}065-14{,}875}{16{,}065}=\frac{1{,}190}{16{,}065}=0.074=7.4\%$$

59. Amount of increase is $0.35\times 28{,}000=\$9{,}800$. Her new salary will be $28{,}000+9{,}800=\$37{,}800$.

61. The sale price represents $100\%-15\%=85\%$ of the original price. $0.85\times Base=\$578$. The original price was $Base=\dfrac{\$578}{0.85}=\$680$.

63. $\dfrac{12{,}711-11{,}400}{11{,}400}=\dfrac{1{,}311}{11{,}400}=0.115=11.5\%$

65. The amount that the fund is worth this quarter represents 100% of the amount the fund was worth last quarter minus the 12% decrease or $100\%-12\%=88\%$.

$$0.88\times base=11{,}264\Rightarrow base=\frac{11{,}264}{0.88}=\$12{,}800$$

The fund was worth $12,800 last quarter.

67. Using line 4, the Federal Income tax is $35,771.50

$$15{,}107.50+0.28(148{,}000-74{,}200)=15{,}107.50+0.28(73{,}800)=$$
$$15{,}107.50+20{,}664=35{,}771.5$$

$35,771.50 is owed.

69. Using line 3, the Federal Income tax is $8,507.50

$$4{,}220+0.25(47{,}800-30{,}650)=4{,}220+0.25(17{,}150)=$$
$$4{,}220+4{,}287.5=8{,}507.5$$

$8,507.50 is owed.

71. – 73. No solution provided.

75. per hundred

77. We could have used the dealer's cost for the base, the markup as the amount, and then solved for percent.

79. We must calculate the value for the 1st, 2nd, 3rd, then 4th year.

Value at the start of year		Depreciation	Value at the end of year
1	$18,000	$0.12\times 18{,}000=\$2{,}160$	$18{,}000-2{,}160=\$15{,}840$
2	$15,840	$0.12\times 15{,}840=\$1{,}900.80$	$15{,}840-1{,}900.80=\$13{,}939.20$
3	$13,939.20	$0.12\times 13{,}939.20\approx\$1{,}672.70$	$13{,}939.20-1{,}672.70=\$12{,}266.50$
4	$12,266.50	$0.12\times 12{,}266.50=\$1{,}471.98$	$12{,}266.50-1{,}471.98=\$10{,}794.52$

81. No; The price after the reduction will be less than the original price. Examples will vary.

Consider an original price of $1,000. In general, after an increase of $x\%$, we have a new price of

$$1{,}000+0.01\cdot x\times 1{,}000=1{,}000+10x.$$

After a decrease of $x\%$, we have a new price of

$$(1{,}000+10x)-0.01\cdot x\times(1{,}000+10x)=$$
$$1{,}000+10x-10x-0.1\cdot x^2=$$
$$1{,}000-0.1x^2=$$
$$1{,}000-0.0001\cdot x^2\times 1{,}000$$

or

The final price is $(0.01\cdot x^2)\%$ less than the original price.

83. No; It is the same as an increase of 32%. Examples will vary.

One way to approach this problem is to start with an amount such as 10,000 and determine what this amount after the two increases.

After the 10% increase: $10{,}000+0.10\times 10{,}000=10{,}000+1{,}000=11{,}000$

After the 20% increase: $11{,}000+0.20\times 11{,}000=11{,}000+2{,}200=13{,}200$

A single increase of 32% would result in the amount.

$10{,}000+0.32\times 10{,}000=10{,}000+3{,}200=\$13{,}200$

Section 9.2 Interest

1. We substituted values for A, r, and t in the equation $A = P(1+rt)$, and then solved for P.

3. $\log(3^x) = x\log(3)$

5. $I = P \times r \times t$
$I = 1{,}000 \times 0.08 \times 3$
$I = \$240$

7. $I = P \times r \times t$
$700 = 3{,}500 \times r \times 4$
$700 = 14{,}000r$
$0.05 = r$
$r = 5\%$

9. $A = P(1+rt)$
$A = 2{,}500(1+(0.08)3)$
$A = 2{,}500(1+0.24)$
$A = 2{,}500(1.24)$
$A = 3{,}100$
$A = \$3{,}100$

11. $A = P(1+rt)$
$1{,}770 = P(1+(0.06)3)$
$1{,}770 = P(1+0.18)$
$1{,}770 = P(1.18)$
$1{,}500 = P$
$P = \$1{,}500$

13. $A = P(1+rt)$
$1{,}400 = 1{,}250(1+(r)2)$
$1{,}400 = 1{,}250(1+2r)$
$1{,}400 = 1{,}250 + 2500r$
$150 = 2500r$
$0.06 = r$
$r = 6\%$

15. $\frac{r}{m} = \frac{0.18}{12} = 0.015 = 1.5\%$

17. $\frac{r}{m} = \frac{0.12}{365} = 12/365\%$

19. $A = P\left(1+\frac{r}{m}\right)^n$
$A = 5{,}000\left(1+\frac{0.05}{1}\right)^{1\cdot 5}$
$A = 5{,}000(1+0.05)^5$
$A = 5{,}000(1.05)^5$
$A \approx 6{,}381.41$
$A \approx \$6{,}381.41$

21. $A = P\left(1+\frac{r}{m}\right)^n$
$A = 4{,}000\left(1+\frac{0.08}{4}\right)^{4\cdot 2}$
$A = 4{,}000(1+0.02)^8$
$A = 4{,}000(1.02)^8$
$A \approx 4{,}686.64$
$A \approx \$4{,}686.64$

23. $A = P\left(1+\frac{r}{m}\right)^n$
$A = 20{,}000\left(1+\frac{0.08}{12}\right)^{12\cdot 2}$
$A = 20{,}000\left(1+\frac{0.08}{12}\right)^{24}$
$A \approx 23{,}457.76$
$A \approx \$23{,}457.76$

25. $A = P\left(1+\frac{r}{m}\right)^n$
$A = 4{,}000\left(1+\frac{0.10}{365}\right)^{365\cdot 2}$
$A = 4{,}000\left(1+\frac{0.10}{365}\right)^{730}$
$A \approx 4{,}885.48$
$A \approx \$4{,}885.48$

27. $\left(1+\dfrac{0.075}{12}\right)^{12}-1\approx 0.0776=7.76\%$

29. $\left(1+\dfrac{0.06}{4}\right)^{4}-1\approx 0.0614=6.14\%$

31. You must compare $\left(1+\dfrac{r}{m}\right)^{n}$ for each case where m is the number of times the money compounds in a single year.

5% compounded yearly:

$$\left(1+\frac{r}{m}\right)^{m}=\left(1+\frac{0.05}{1}\right)^{1}$$
$$=1+0.05$$
$$=1.05$$

4.95% compounded quarterly:

$$\left(1+\frac{r}{m}\right)^{m}=\left(1+\frac{0.0495}{4}\right)^{4}\approx 1.05043$$

4.95% compounded quarterly is better.

33.
$$A=P\left(1+\frac{r}{m}\right)^{n}$$
$$30,000=P\left(1+\frac{0.06}{4}\right)^{4\cdot 15}$$
$$30,000=P(1+0.015)^{60}$$
$$30,000=P(1.015)^{60}$$
$$\frac{30,000}{1.015^{60}}=P$$
$$P\approx \$12,278.88$$

35.
$$3^{x}=10$$
$$\log 3^{x}=\log 10$$
$$x\log 3=\log 10$$
$$x=\log 10/\log 3$$
$$x\approx 2.096$$

37.
$$(1.05)^{x}=2$$
$$\log(1.05)^{x}=\log 2$$
$$x\log(1.05)=\log 2$$
$$x=\log 2/\log(1.05)$$
$$x\approx 14.207$$

39.
$$x^{3}=10$$
$$\left(x^{3}\right)^{1/3}=(10)^{1/3}$$
$$x=10^{1/3}\approx 2.1544$$

41.
$$x^{4}=10$$
$$\left(x^{4}\right)^{1/4}=(10)^{1/4}$$
$$x=10^{1/4}\approx 1.7783$$

43.
$$A=P(1+r)^{t}$$
$$2500=2000(1+r)^{5}$$
$$1.25=(1+r)^{5}$$
$$(1.25)^{1/5}=\left((1+r)^{5}\right)^{1/5}$$
$$(1.25)^{1/5}=1+r$$
$$(1.25)^{1/5}-1=r$$
$$r\approx 0.0456=4.56\%$$

45.
$$A=P(1+r)^{t}$$
$$1,500=1,000(1+0.04)^{t}$$
$$1.5=(1.04)^{t}$$
$$\log(1.5)=\log(1.04)^{t}$$
$$\log(1.5)=t\cdot\log(1.04)$$
$$\log(1.5)/\log(1.04)=t$$
$$t\approx 10.34$$

47. a) $36\cdot 136=\$4,896$

b) $4,896-3,600=\$1,296$

49. $I=P\times r\times t$

$$I=10,000\times 0.08\times\frac{1}{12}$$
$$I=\$66.67$$

51.
$$I=P\times r\times t$$
$$300=1,200\times r\times 2$$
$$300=2,400r$$
$$0.125=r$$
$$r=12.5\%$$

53.
$$
\begin{aligned}
I &= P \times r \times t \\
425 - 400 &= 400 \times r \times \frac{1}{12} \\
25 &= 400 \times r \times \frac{1}{12} \\
25 &= \frac{100}{3} r \\
0.75 &= r \\
r &= 75\%
\end{aligned}
$$

55. $4.65(1.027)^{10} \approx \$6.07$

57. $3.25(1.028)^{10} \approx \$4.29$

59. $4.65(1+2.26)^5 = 4.65(3.26)^5 \approx \$1,712.15$

61. a) She earned $2,450 - 2,375 = \$75$ in interest.

$$
\begin{aligned}
I &= P \times r \times t \\
75 &= 2,375 \times r \times \frac{8}{12} \\
75 &= \frac{4750}{3} \times r \\
\frac{9}{190} &= r
\end{aligned}
$$

$r \approx 0.04736842$ or approximately 4.74%

b)
$$
\begin{aligned}
A &= P\left(1+\frac{r}{m}\right)^n \\
2,450 &= 2,375\left(1+\frac{r}{12}\right)^{12 \cdot \frac{8}{12}} \\
\frac{98}{95} &= \left(1+\frac{r}{12}\right)^8 \\
\left(\frac{98}{95}\right)^{1/8} &= \left(\left(1+\frac{r}{12}\right)^8\right)^{1/8} \\
\left(\frac{98}{95}\right)^{1/8} &= 1+\frac{r}{12} \\
\left(\frac{98}{95}\right)^{1/8} - 1 &= \frac{r}{12}
\end{aligned}
$$

$r \approx 4.67\%$

63.
$$
\begin{aligned}
A &= P\left(1+\frac{r}{m}\right)^n \\
20,000 &= 9,420\left(1+\frac{0.075}{12}\right)^{12t} \\
20,000 &= 9,420(1+0.00625)^{12t} \\
\frac{1,000}{471} &= (1.00625)^{12t} \\
\log\left(\frac{1,000}{471}\right) &= \log(1.00625)^{12t} \\
\log\left(\frac{1,000}{471}\right) &= 12t \cdot \log(1.00625) \\
t &\approx \frac{\log\left(\frac{1,000}{471}\right) / \log(1.00625)}{12}
\end{aligned}
$$

$t \approx 10.07$ years

65. $I = Prt$

67. A is the future amount, P is the principle, r is the yearly interest rate, m is the number of compounding periods per year, and n is the total number of compounding periods.

69. – 71. No solution provided.

73. $\left(1+\frac{0.10}{100,000}\right)^{100,000} - 1 \approx 0.105170863 = 10.5170863\%$

75. $e^r - 1 \approx 2.718281828^{0.10} - 1 \approx 0.105170918 = 10.5170918\%$

Section 9.3 Consumer Loans

1. We added the loan amount and the interest, and then divided the sum by the number of payments.

3. It shows the balance for the loan for each day of September.

5. $I = Prt = 900(0.12)(2) = 216$ Monthly Payment $= \frac{P+I}{n} = \frac{900+216}{2\cdot 12} = \frac{1{,}116}{24} = \$46.50$

7. $I = Prt = 1{,}360(0.08)(4) = 435.20$

$$\text{Monthly Payment} = \frac{P+I}{n} = \frac{1{,}360+435.20}{4\cdot 12} = \frac{1{,}795.20}{48} = \$37.40$$

9. $I = Prt = 1{,}280(0.095)(2) = \$243.20$

$$\text{Monthly Payment} = \frac{P+I}{n} = \frac{1{,}280+243.20}{2\cdot 12} = \frac{1{,}523.20}{24} \approx \$63.47$$

11. $I = Prt = 6{,}480(0.1165)(4) = \$3{,}019.68$

$$\text{Monthly Payment} = \frac{P+I}{n} = \frac{6{,}480+3{,}019.68}{4\cdot 12} = \frac{9{,}499.68}{48} = \$197.91$$

13. $I = Prt = 900(0.12)(2) = 216$ Monthly Payment $= \frac{P+I}{n} = \frac{900+216}{2\cdot 12} = \frac{1{,}116}{24} = \$46.50$

$\frac{900}{24} = \$37.50$ of the loan amount is paid each month 46.50-37.50 $= \$9$ in interest each month

$\frac{9}{37.50} = 0.24 = 24\%$ per month or $12\cdot 24\% = 288\%$ annual interest rate

15. $I = Prt = 1360(0.08)(4) = 435.20$ Monthly Payment $= \frac{P+I}{n} = \frac{1{,}360+435.20}{4\cdot 12} = \frac{1{,}795.20}{48} = \$37.40$

$\frac{1{,}360}{48} \approx \$28.33$ of the loan amount is paid each month 37.40-28.33 $= \$9.07$ in interest each month

$\frac{9.07}{28.33} \approx 0.32 = 32\%$ per month or $12\cdot 32\% = 384\%$ annual interest rate

17. a) $I = Prt = 11{,}000(0.092)(4) = 4{,}048$

$$\text{Monthly Payment} = \frac{P+I}{n} = \frac{11{,}000+4048}{4\cdot 12} = \frac{15{,}048}{48} = \$313.50$$

b) $I = Prt = 9{,}000(0.092)(4) = 3{,}312$

$$\text{Monthly Payment} = \frac{P+I}{n} = \frac{9{,}000+3{,}312}{4\cdot 12} = \frac{12{,}312}{48} = \$256.50$$

His monthly payments will be reduced by 313.50 – 256.50 = \$57.00

17. (continued)

c) Monthly Payment $= 200 = \dfrac{P+I}{48} \Rightarrow P+I = 9{,}600 = P + P(0.092)(4) = P + 0.368P = 1.368P \Rightarrow$

$9{,}600 = 1.368P \Rightarrow P \approx \$7{,}017.54$ This represents the amount financed, so

$11{,}000 - 7{,}017.54 = \$3{,}982.46$ should be the down payment.

19. a) $I = Prt = 15{,}000(0.096)(3) = 4{,}320$

Monthly Payment $= \dfrac{P+I}{n} = \dfrac{15{,}000 + 4{,}320}{3 \cdot 12} = \dfrac{19{,}320}{36} \approx \$536.67$

b) $I = Prt = 12{,}000(0.096)(3) = 3{,}456$

Monthly Payment $= \dfrac{P+I}{n} = \dfrac{12{,}000 + 3{,}456}{3 \cdot 12} = \dfrac{15{,}456}{36} \approx \$429.34$

Her monthly payments will be reduced by 536.67 – 429.34=$107.33

c) Monthly Payment $= 300 = \dfrac{P+I}{36} \Rightarrow P+I = 10{,}800 = P + P(0.096)(3) = P + 0.288P = 1.288P \Rightarrow$

$10{,}800 = 1.288P \Rightarrow P \approx \$8{,}385.09$ This represents the amount financed, so

$15{,}000 - 8{,}385.09 = \$6{,}614.91$ should be the down payment.

21.

Method	Finance Charge = (Last Month's Balance)rt	P	r	t	Finance Charge = $I=Prt$
Unpaid Balance	$475 \times (0.18/12) \approx 7.13$	Last Month's Balance + Finance – Payment + New Charges – Returned Charges = 475 + 7.13 – 225 + 180 – 145 = 292.13	18%	$\frac{1}{12}$	$292.13 \times (0.18/12)$ $\approx$ **$4.38**

23.

Method	Finance Charge = (Last Month's Balance)rt	P	r	t	Finance Charge = $I=Prt$
Unpaid Balance	$640 \times (0.165/12) = 8.80$	Last Month's Balance + Finance – Payment + New Charges = 640 + 8.80–320 + 140 + 35 + 75 = 578.80	16.5%	$\frac{1}{12}$	$578.80 \times (0.165/12)$ $\approx$ **$7.96**

25.

Method	Finance Charge = (Last Month's Balance)rt	P	r	t	Finance Charge = $I=Prt$
Unpaid Balance	$460 \times (0.188/12) \approx 7.21$	Last Month's Balance + Finance – Payment + New Charges = 460 + 7.21 – 300 + 140 + 135 + 175 = 617.21	18.8%	$\frac{1}{12}$	$617.21 \times (0.188/12)$ $\approx$ **$9.67**

27.

Day	Balance	Number of days × Balance
1,2,3,4	\$280	$4 \times 280 = 1{,}120$
5,6,7,8,9,10,11,12,13,14	\$205	$10 \times 205 = 2{,}050$
15,16,17,18,19,20	\$340	$6 \times 340 = 2{,}040$
21,22,23	\$356	$3 \times 356 = 1{,}068$
24,25,26,27,28,29,30,31	\$382	$8 \times 382 = 3{,}056$

Average daily balance $\dfrac{1{,}120+2{,}050+2{,}040+1{,}068+3{,}056}{31} = \dfrac{9{,}334}{31} \approx 301.10$

Finance charge $I = Prt = 301.10(.21)(31/365) \approx \$5.37$

29.

Day	Balance	Number of days × Balance
1,2	\$240	$2 \times 240 = 480$
3,4,5,6,7,8,9,10,11,12	\$375	$10 \times 375 = 3{,}750$
13,14,15,16,17,18,19,20,21,22	\$225	$10 \times 225 = 2{,}250$
23,24,25,26,27	\$255	$5 \times 255 = 1{,}275$
28,29,30	\$283	$3 \times 283 = 849$

Average daily balance $\dfrac{480+3{,}750+2{,}250+1{,}275+849}{30} = \dfrac{8{,}604}{30} = 286.80$

Finance charge $I = Prt = 286.80(.21)(30/365) \approx \$4.95$

31. Finance charge on unpaid balance $I = Prt = 280 \cdot 0.21 \cdot (1/12) = \$4.90$

Method	P	r	t	Finance Charge = $I = Prt$
Unpaid Balance	Last month's balance + finance charge + purchases – payment = 280 + 4.90 + 135 + 16 + 26 – 75 = 386.90	21%	$\frac{1}{12}$	$386.90 \times (0.21/12)$ $\approx$ **\$6.77**

33. Finance charge on unpaid balance $I = Prt = 240 \cdot 0.21 \cdot (1/12) = \$4.20$

Method	P	r	t	Finance Charge = $I = Prt$
Unpaid Balance	Last month's balance + finance charge + purchases – payment = 240 + 4.20 + 135 + 30 + 28– 150 = 287.20	21%	$\frac{1}{12}$	$287.20 \times (0.21/12)$ $\approx$ **\$5.03**

35. Add-on Interest: $I = Prt = 1{,}000(0.105)\left(\frac{10}{12}\right) = 87.50$

Unpaid Balance Method

Beginning of Month	Finance Charge = (Last Month's Balance)rt	Last Month's Balance + Finance Charge – Payment
2	1,000 × (0.18/12) = 15.00	1,000 + 15.00 – (100 + 15.00) = 900
3	900 × (0.18/12) = 13.50	900 + 13.50 – (100 + 13.50) = 800
4	800 × (0.18/12) = 12.00	800 + 12.00 – (100 + 12.00) = 700
5	700 × (0.18/12) = 10.50	700 + 10.50 – (100 + 10.50) = 600
6	600 × (0.18/12) = 9.00	600 + 9.00 – (100 + 9.00) = 500
7	500 × (0.18/12) = 7.50	500 + 7.50 – (100 + 7.50) = 400
8	400 × (0.18/12) = 6.00	400 + 6.00 – (100 + 6.00) = 300
9	300 × (0.18/12) = 4.50	300 + 4.50 – (100 + 4.50) = 200
10	200 × (0.18/12) = 3.00	200 + 3.00 – (100 + 3.00) = 100
11	100 × (0.18/12) = 1.50	100 + 1.50 – (100 + 1.50) = 0

The total interest paid for the unpaid balance method is $82.50, so the unpaid balance method would accumulate less finance charges.

37. Unpaid Balance Method

Beginning of Month	Finance Charge = (Last Month's Balance)r	Last Month's Balance + Finance Charge
2	1,150 × (0.0175) ≈ 20.13	1,150 + 20.13 = 1,170.13
3	1,170.13 × (0.0175) ≈ 20.48	1,170.13 + 20.48 = 1,190.61
4	1,190.61 × (0.0175) ≈ 20.84	1,190.61 + 20.84 = 1,211.45

The total interest accumulated with the unpaid balance method is $61.45 –or – By using the formula $A = P(1+r)^n$ where P=1,150, r=0.0175, and n=3 we get

$$A = 1{,}150(1+0.0175)^3 = 1{,}150 \cdot 1.0175^3 \approx 1{,}211.44$$
$$1{,}211.44 - 1{,}150 = \$61.44$$

39. The interest for the loan is "added on" before calculating the payments.

41. Charge a large purchase to your card early in the month and then pay off the debt before the end of the month so no interest is charged for that purchase.

43. – 47. No solution provided.

Section 9.4 Annuities

1. The expression $\frac{r}{m}$ is the monthly interest rate; m is the number of compounding periods per year; R is the monthly payment.

3. $\frac{x^8 - 1}{x - 1}$

Your calculations may differ slightly from ours due to rounding at intermediate steps.

5. $\frac{r}{m}=\frac{0.06}{12}=0.005$; $n=7$; $A=100\left(\frac{(1+0.005)^7-1}{0.005}\right)\approx \$710.59$

7. $\frac{r}{m}=\frac{0.03}{12}$; $n=12\cdot 8=96$; $A=200\left(\frac{\left(1+\frac{0.03}{12}\right)^{96}-1}{0.03/12}\right)\approx \$21{,}669.48$

9. $\frac{r}{m}=\frac{0.09}{12}$ and $n=12\cdot 4=48$; $A=400\left(\frac{\left(1+\frac{0.09}{12}\right)^{48}-1}{0.09/12}\right)\approx \$23{,}008.28$

11. $\frac{r}{m}=\frac{0.095}{12}$ and $n=12\cdot 8=96$; $A=600\left(\frac{\left(1+\frac{0.095}{12}\right)^{96}-1}{0.095/12}\right)\approx \$85{,}785.11$

13. $\frac{r}{m}=\frac{0.08}{4}$ and $n=4\cdot 5=20$; $A=500\left(\frac{\left(1+\frac{0.08}{4}\right)^{20}-1}{0.08/4}\right)\approx \$12{,}148.68$

15. $\frac{r}{m}=\frac{0.036}{4}$; $n=4\cdot 6=24$; $A=280\left(\frac{\left(1+\frac{0.036}{4}\right)^{24}-1}{0.036/4}\right)\approx \$7{,}463.67$

17. $\frac{r}{m}=\frac{0.06}{12}=0.005$ and $n=12\cdot 1=12$

$$2{,}000=R\left(\frac{(1+0.005)^{12}-1}{0.005}\right)$$
$$2{,}000\approx R(12.3355623729)$$
$$R\approx\frac{2{,}000}{12.3355623729}\approx \$162.14$$

19. $\frac{r}{m}=\frac{0.075}{12}=0.00625$ and $n=12\cdot 2=24$

$$5{,}000=R\left(\frac{(1+0.00625)^{24}-1}{0.00625}\right)$$
$$5{,}000\approx R(25.8067228988)$$
$$R\approx\frac{5{,}000}{25.8067228988}\approx \$193.75$$

21.
$$3^x=20$$
$$\log 3^x=\log 20$$
$$x\cdot\log 3=\log 20$$
$$x=\frac{\log 20}{\log 3}\approx 2.7268$$

23.
$$\begin{aligned} 8^x &= 10 \\ \log 8^x &= \log 10 \\ x \cdot \log 8 &= \log 10 \\ x &= \frac{\log 10}{\log 8} \approx 1.1073 \end{aligned}$$

25.
$$\begin{aligned} \frac{8^x - 5}{6} &= 20 \\ 8^x - 5 &= 120 \\ 8^x &= 125 \\ \log 8^x &= \log 125 \\ x \cdot \log 8 &= \log 125 \\ x &= \frac{\log 125}{\log 8} \approx 2.3219 \end{aligned}$$

27.
$$\begin{aligned} \frac{8^x + 2}{5} &= 12 \\ 8^x + 2 &= 60 \\ 8^x &= 58 \\ \log 8^x &= \log 58 \\ x \cdot \log 8 &= \log 58 \\ x &= \frac{\log 58}{\log 8} \approx 1.9527 \end{aligned}$$

29.
$$\begin{aligned} 10,000 &= 200\left(\frac{\left(1+\frac{0.09}{12}\right)^n - 1}{0.09/12}\right) \\ 50 &= \left(\frac{(1.0075)^n - 1}{0.0075}\right) \\ 0.375 &= 1.0075^n - 1 \\ 1.375 &= 1.0075^n \\ \log 1.375 &= \log 1.0075^n \\ \log 1.375 &= n \cdot \log 1.0075 \\ \frac{\log 1.375}{\log 1.0075} &= n \\ n &\approx 42.62 \end{aligned}$$

31.
$$\begin{aligned} 5,000 &= 150\left(\frac{\left(1+\frac{0.06}{12}\right)^n - 1}{0.06/12}\right) \\ \frac{100}{3} &= \left(\frac{(1.005)^n - 1}{0.005}\right) \\ 0.1666667 &\approx 1.005^n - 1 \\ 1.1666667 &\approx 1.005^n \\ \log 1.1666667 &\approx \log 1.005^n \\ \log 1.1666667 &\approx n \cdot \log 1.005 \\ \frac{\log 1.1666667}{\log 1.005} &\approx n \\ n &\approx 30.91 \end{aligned}$$

33.
$$\begin{aligned} 6,000 &= 250\left(\frac{\left(1+\frac{0.075}{12}\right)^n - 1}{0.075/12}\right) \\ 24 &= \left(\frac{(1.00625)^n - 1}{0.00625}\right) \\ 0.15 &= 1.00625^n - 1 \\ 1.15 &= 1.00625^n \\ \log 1.15 &= \log 1.00625^n \\ \log 1.15 &= n \cdot \log 1.00625 \\ \frac{\log 1.15}{\log 1.00625} &= n \\ n &\approx 22.43 \end{aligned}$$

35. $\frac{r}{m} = \frac{0.065}{12}$ and $n = 30$; $A = 75\left(\frac{\left(1+\frac{0.065}{12}\right)^{30} - 1}{0.065/12}\right) \approx \$2,435.99$

37. $\frac{r}{m}=\frac{0.102}{12}$ and $n=12\cdot 3=36$; $A=150\left(\frac{\left(1+\frac{0.102}{12}\right)^{36}-1}{0.102/12}\right)\approx \$6,286.36$

39. $\frac{r}{m}=\frac{0.065}{12}$ and $n=12\cdot 15=180$; $A=400\left(\frac{\left(1+\frac{0.065}{12}\right)^{180}-1}{0.065/12}\right)\approx \$121,417.91$

41. $n=12\cdot 8=96$

$$14,000=R\left(\frac{(1+0.007)^{96}-1}{0.007}\right)$$

$$14,000\approx R(136.224411498)$$

$$R\approx \frac{14,000}{136.2244114988}\approx \$102.78$$

43. $\frac{r}{m}=\frac{0.082}{12}$ and $n=6$

$$600=R\left(\frac{\left(1+\frac{0.082}{12}\right)^{6}-1}{0.082/12}\right)$$

$$600\approx R(6.10343868814)$$

$$R\approx \frac{600}{6.10343868814}\approx \$98.31$$

45. a)

tax-deferred plan

$\frac{r}{m}=\frac{0.06}{12}=0.005$, $n=12\cdot 30=360$;

$R=300$

$$A=300\left(\frac{(1.005)^{360}-1}{0.005}\right)\approx \$301,354.51$$

non-deferred plan

$\frac{r}{m}=\frac{0.06}{12}=0.005$, $n=12\cdot 30=360$;

$R=300\cdot 0.75=225$

$$A=225\left(\frac{(1.005)^{360}-1}{0.005}\right)\approx \$226,015.88$$

b)

$$I=301,354.51-300\cdot 360$$
$$=301,354.51-108,000$$
$$=\$193,354.51$$

$$I=226,015.88-225\cdot 360$$
$$=226,015.88-81,000$$
$$=\$145,015.88$$

c)

When you withdraw your money, you will have $301,354.51(0.82)=\$247,110.70$.

When you withdraw your money, you will have $81,000+145,015.88(0.82)=$ \$199,913.02.

The tax-deferred plan earns \$47,197.68 more.

47. a)

tax-deferred plan

$\frac{r}{m}=\frac{0.04}{12}$, $n=12\cdot 20=240$;

$R=400$

$$A=400\left(\frac{\left(1+\frac{0.04}{12}\right)^{240}-1}{0.04/12}\right)\approx \$146,709.85$$

non-deferred plan

$\frac{r}{m}=\frac{0.04}{12}$, $n=12\cdot 20=240$;

$R=400\cdot 0.70=280$

$$A=280\left(\frac{\left(1+\frac{0.04}{12}\right)^{240}-1}{0.04/12}\right)\approx \$102,696.90$$

b)

$I=146,709.85-400\cdot 240$
$=146,709.85-96,000$
$=\$50,709.85$

$I=102,696.90-280\cdot 240$
$=102,696.90-67,200$
$=\$35,496.90$

c)

When you withdraw your money, you will have $146,709.85(0.70)=\$102,696.90$.

When you withdraw your money, you will have $67,200+35,496.90(0.70)$ = $92,047.83.

The tax-deferred plan earns $10,649.07 more.

49. a)

tax-deferred plan

$\frac{r}{m}=\frac{0.034}{12}$, $n=12\cdot 35=420$;

$R=500$

$$A=500\left(\frac{\left(1+\frac{0.034}{12}\right)^{420}-1}{0.034/12}\right)\approx \$402,627.32$$

non-deferred plan

$\frac{r}{m}=\frac{0.034}{12}$, $n=12\cdot 35=420$;

$R=500\cdot 0.75=375$

$$A=375\left(\frac{\left(1+\frac{0.034}{12}\right)^{420}-1}{0.034/12}\right)\approx \$301,970.49$$

b)

$I=402,627.32-500\cdot 420$
$=402,627.32-210,000$
$=\$192,627.32$

$I=301,970.49-375\cdot 420$
$=301,970.49-157,500$
$=\$144,470.49$

c)

When you withdraw your money, you will have $402,627.32(0.70)=\$281,839.12$.

When you withdraw your money, you will have $157,500+144,470.49(0.70)$ = $258,629.34.

The tax-deferred plan earns $23,209.78 more.

51. $$400{,}000 = 5{,}000\left(\frac{\left(1+\frac{0.108}{12}\right)^n - 1}{0.108/12}\right)$$

$$80 = \left(\frac{(1+0.009)^n - 1}{0.009}\right)$$

$$0.72 = 1.009^n - 1$$

$$1.72 = 1.009^n$$

$$\log 1.72 = \log 1.009^n$$

$$\log 1.72 = n \cdot \log 1.009$$

$$\frac{\log 1.72}{\log 1.009} = n$$

$$n \approx 61 \text{ months}$$

53. $$30{,}000 = 550\left(\frac{\left(1+\frac{0.078}{12}\right)^n - 1}{0.078/12}\right)$$

$$\frac{600}{11} = \left(\frac{(1+0.0065)^n - 1}{0.0065}\right)$$

$$0.3545455 \approx 1.0065^n - 1$$

$$1.3545455 \approx 1.0065^n$$

$$\log 1.3545455 \approx \log 1.0065^n$$

$$\log 1.3545455 \approx n \cdot \log 1.0065$$

$$\frac{\log 1.3545455}{\log 1.0065} \approx n$$

$$n \approx 47 \text{ months}$$

55. The exponent property; $\log a^x = x \log a$.

57. No solution provided.

59. $\frac{r}{m} = \frac{0.0935}{12}$ and $n = 12 \cdot 20 = 240$

$$98{,}695.16 = R\left(\frac{\left(1+\frac{0.0935}{12}\right)^{240} - 1}{0.0935/12}\right)$$

$$98{,}695.16 \approx R(698.366876471)$$

$$R \approx \frac{98{,}695.16}{698.366876471} \approx \$141.33$$

61. $\frac{r}{m} = \frac{0.06}{12} = 0.005$ and $n = 12 \cdot 10 = 120$

$$A = 200\left(\frac{(1+0.005)^{120} - 1}{0.005}\right) \approx \$32{,}775.87$$

$$A = P\left(1+\frac{r}{m}\right)^n$$

$$A = 32{,}775.87\left(1+\frac{0.06}{12}\right)^{12\cdot 30}$$

$$A = 32{,}775.87(1+0.005)^{360}$$

$$A = 32{,}775.87(1.005)^{360}$$

$$A \approx 197{,}395.14$$

$$A \approx \$197{,}395.14$$

63. No solution provided.

Section 9.5 Amortization

1. The left side represents the amount owed on the loan after 48 months of compounding. The right side represents the amount that must be in the annuity to pay off the loan.

3. $\frac{r}{m}=\frac{0.10}{12}$ and $n=12\cdot 4=48$

$$5{,}000\left(1+\frac{0.10}{12}\right)^{48}=R\left(\frac{\left(1+\frac{0.10}{12}\right)^{48}-1}{0.10/12}\right)$$

$$7{,}446.77049303\approx R(58.7224918326)$$

$$R\approx\frac{7{,}446.77049303}{58.722491826}\approx \$126.82$$

5. $\frac{r}{m}=\frac{0.075}{12}=0.00625$ and $n=12\cdot 4=48$

$$8{,}000(1+0.00625)^{48}=R\left(\frac{(1+0.00625)^{48}-1}{0.00625}\right)$$

$$10{,}788.7932107\approx R(55.7758642148)$$

$$R\approx\frac{10{,}788.7932107}{55.7758642148}\approx \$193.44$$

7. $\frac{r}{m}=\frac{0.0825}{12}=0.006875$ and $n=12\cdot 4=48$

$$12{,}500(1+0.006875)^{48}=R\left(\frac{(1+0.006875)^{48}-1}{0.006875}\right)$$

$$17{,}367.4791291\approx R(56.6397571391)$$

$$R\approx\frac{17{,}367.4791291}{56.6397571391}\approx \$306.64$$

9. $\frac{r}{m}=\frac{0.08}{12}$

$$1{,}900\left(1+\frac{0.08}{12}\right)^{18}=R\left(\frac{\left(1+\frac{0.08}{12}\right)^{18}-1}{0.08/12}\right)$$

$$2{,}141.39107973\approx R(19.0571905052)$$

$$R\approx\frac{2{,}141.39107973}{19.0571905052}\approx \$112.37$$

11.

Payment Number	Amount of Payment	Interest Payment	Applied to Principal	Balance
				\$5,000.00
1	\$126.82	\$41.67	\$85.15	\$4,914.85
2	\$126.82	\$40.96	\$85.86	\$4,828.99
3	\$126.82	\$40.24	\$86.58	\$4,742.41

13.

Payment Number	Amount of Payment	Interest Payment	Applied to Principal	Balance
				\$12,500.00
1	\$306.64	\$85.94	\$220.70	\$12,279.30
2	\$306.64	\$84.42	\$222.22	\$12,057.08
3	\$306.64	\$82.89	\$223.75	\$11,833.33

15. a) $\frac{r}{m} = \frac{0.07}{12}$ and $n = 12 \cdot 30 = 360$

$$100,000\left(1+\frac{0.07}{12}\right)^{360} = R\left(\frac{\left(1+\frac{0.07}{12}\right)^{360}-1}{0.07/12}\right) \qquad R \approx \frac{811,649.747526}{1,219.97099576} \approx \$665.31$$

$$811,649.747526 \approx R(1,219.97099576)$$

Payment Number	Amount of Payment	Interest Payment	Applied to Principal	Balance
				$100,000.00
1	$665.31	$583.33	$81.98	$99,918.02
2	$665.31	$582.86	$82.45	$99,835.57
3	$665.31	$582.37	$82.94	$99,752.63

b)

Payment Number	Amount of Payment	Interest Payment	Applied to Principal	Balance
				$100,000.00
1	$765.31	$583.33	$181.98	$99,818.02
2	$765.31	$582.27	$183.04	$99,634.98
3	$765.31	$581.20	$184.11	$99,450.87

c) The 4th payment line for a payment of $665.31 is

4	$665.31	$581.89	$83.42	$99,669.21

The 4th payment line for a payment of $765.31 is

4	$765.31	$580.13	$185.18	$99,265.69

The difference is $581.89 - 580.13 = \$1.76$.

17. a) $\frac{r}{m} = \frac{0.072}{12} = 0.006$ and $n = 12 \cdot 4 = 48$

$$11,500(1+0.006)^{48} = R\left(\frac{(1+0.006)^{48}-1}{0.006}\right) \qquad 15,325.0152341 \approx R(55.4350033931)$$

$$R \approx \frac{15,325.0152341}{55.4350033931} \approx \$276.46$$

b) $48 \cdot 276.46 = 13,270.08$ and $13,270.08 - 11,500.00 = \$1,770.08$

This represents the approximate amount of interest paid. The actual amount is slightly lower due to the fact that we round our payments up to the next penny and our final payment is slightly less than the other 47 payments.

19. a) $\frac{r}{m}=\frac{0.18}{12}=0.015$ and $n=5$

$$275(1+0.015)^5=R\left(\frac{(1+0.015)^5-1}{0.015}\right)$$

$$296.253101068\approx R(5.15226692563)$$

$$R\approx\frac{296.253101068}{5.15226692563}\approx \$57.50$$

b) $5\cdot 57.50=287.50$ and $287.50-275.00=\$12.50$

This represents the approximate amount of interest paid. The actual amount is slightly lower due to the fact that we round our payments up to the next penny and our final payment is slightly less than the other 4 payments.

21. a) 1st year: $$R=\frac{200000\left(1+\frac{0.04}{12}\right)^{360}}{\frac{\left(\left(1+\frac{0.04}{12}\right)^{360}-1\right)}{0.04/12}}=\frac{662699.6029}{694.0494044}\approx \$954.83$$

b) In year 5: $r=0.04+0.02+0.02+0.02+0.02=0.12$

$$R=\frac{200000\left(1+\frac{0.12}{12}\right)^{360}}{\frac{\left(\left(1+\frac{0.12}{12}\right)^{360}-1\right)}{0.12/12}}=\frac{7189928.266}{3494.964133}\approx \$2,057.23$$

23. a) 1st year: $$R=\frac{220000\left(1+\frac{0.044}{12}\right)^{360}}{\frac{\left(\left(1+\frac{0.044}{12}\right)^{360}-1\right)}{0.044/12}}=\frac{821566.9617}{745.7441674}\approx \$1101.68$$

b) In year 5: $r=0.044+0.02+0.02+0.01+0.018=0.112$

$$R=\frac{220000\left(1+\frac{0.112}{12}\right)^{360}}{\frac{\left(\left(1+\frac{0.112}{12}\right)^{360}-1\right)}{0.112/12}}=\frac{6235689.472}{2929.718899}\approx \$2,128.43$$

25. $A = 50{,}000\left(\frac{(1+0.10)^{20}-1}{0.10}\right) \approx \$2{,}863{,}749.97$

$$A = P\left(1+\frac{r}{m}\right)^{n}$$
$$2{,}863{,}749.97 = P(1+0.10)^{20}$$
$$2{,}863{,}749.97 = P(1.10)^{20}$$
$$\frac{2{,}863{,}749.97}{1.10^{20}} = P$$
$$P \approx \$425{,}678.19$$

The present value of the annual payments is slightly better than the lump sum of \$425,000.

27. $\frac{r}{m} = \frac{0.108}{12} = 0.009$ and $n = 12 \cdot 4 = 48$

$$P(1+0.009)^{48} = 350\left(\frac{(1+0.009)^{48}-1}{0.009}\right)$$
$$P(1.53736142389) \approx 20{,}897.3887067$$
$$P \approx \frac{20{,}897.3887067}{1.53736142389} \approx \$13{,}593.02$$

29. $\frac{r}{m} = \frac{0.09}{12} = 0.0075$ and $n = 12 \cdot 20 = 240$

$$A = 350\left(\frac{(1+0.0075)^{240}-1}{0.0075}\right) \approx \$233{,}760.40$$
$$A = P\left(1+\frac{r}{n}\right)^{nt}$$
$$233{,}760.40 = P(1+0.0075)^{240}$$
$$233{,}760.40 = P(1.0075)^{240}$$
$$\frac{233{,}760.40}{1.0075^{240}} = P$$
$$P \approx \$38{,}900.73$$

The present value of the annual payments is slightly worse than the lump sum of \$40,000.

31. We need to find the monthly payments first, where $\frac{r}{m} = \frac{0.085}{12}$ and $n = 12 \cdot 5 = 60$.

$$12{,}000\left(1+\frac{0.085}{12}\right)^{60} = R\left(\frac{\left(1+\frac{0.085}{12}\right)^{60}-1}{0.085/12}\right)$$
$$118{,}327.6071744 \approx R(74.4424373453)$$
$$R \approx \frac{118{,}327.6071744}{74.4424373453} \approx \$246.20$$

Now find the unpaid balance where $n = 12 \cdot 3 = 36$.

$$U = 12{,}000\left(1+\frac{0.085}{12}\right)^{36} - 246.20\left(\frac{\left(1+\frac{0.085}{12}\right)^{36}-1}{0.085/12}\right) \approx 15{,}471.63 - 10{,}055.46 = \$5{,}416.17$$

33. We need to find the monthly payments first where $\frac{r}{m}=\frac{0.07}{12}$ and $n=12\cdot 30=360$.

$$120{,}000\left(1+\frac{0.07}{12}\right)^{360}=R\left(\frac{\left(1+\frac{0.07}{12}\right)^{360}-1}{0.07/12}\right)$$

$$973{,}979.697032\approx R(1{,}219.97099576)$$

$$R\approx\frac{973{,}979.697032}{1{,}219.97099576}\approx \$798.37$$

Now find the unpaid balance where $n=12\cdot 8=96$.

$$U=120{,}000\left(1+\frac{0.07}{12}\right)^{96}-798.37\left(\frac{\left(1+\frac{0.07}{12}\right)^{96}-1}{0.07/12}\right)\approx 209{,}739.17-102{,}350.09=\$107{,}389.08$$

35. We need to find the monthly payments first where $\frac{r}{m}=\frac{0.08}{12}$ and $n=12\cdot 2=24$.

$$8{,}000\left(1+\frac{0.08}{12}\right)^{24}=R\left(\frac{\left(1+\frac{0.08}{12}\right)^{24}-1}{0.08/12}\right)$$

$$9{,}383.10345397\approx R(25.9331897619)$$

$$R\approx\frac{9{,}383.10345397}{25.9331897619}\approx \$361.82$$

Now find the unpaid balance where $n=12$.

$$U=8{,}000\left(1+\frac{0.08}{12}\right)^{12}-361.82\left(\frac{\left(1+\frac{0.08}{12}\right)^{12}-1}{0.08/12}\right)\approx 8{,}664.00-4{,}504.63=\$4{,}159.37$$

37. a) We need to find the unpaid balance where $\frac{r}{m}=\frac{0.08}{12}$.

$$U=10{,}000\left(1+\frac{0.08}{12}\right)^{24}-244.13\left(\frac{\left(1+\frac{0.08}{12}\right)^{24}-1}{0.08/12}\right)\approx 11{,}728.88-6{,}331.07=\$5{,}397.81$$

Next we need to determine the new monthly payments where $\frac{r}{m}=\frac{0.065}{12}$.

$$5{,}397.81\left(1+\frac{0.065}{12}\right)^{24}=R\left(\frac{\left(1+\frac{0.065}{12}\right)^{24}-1}{0.065/12}\right)$$

$$6{,}145.02307866\approx R(25.5561107016)$$

$$R\approx\frac{6{,}145.02307866}{25.5561107016}\approx \$240.46$$

b) $(244.13-240.46)\cdot 24=\$88.08$

39. a) We need to find the unpaid balance where $\frac{r}{m} = \frac{0.08}{12}$.

$$U = 100{,}000\left(1+\frac{0.08}{12}\right)^{60} - 836.45\left(\frac{\left(1+\frac{0.08}{12}\right)^{60}-1}{0.08/12}\right) \approx 148{,}984.57 - 61{,}459.72 = \$87{,}524.85$$

Next we need to determine the new monthly payments where $\frac{r}{n} = \frac{0.07}{12}$.

$$87{,}524.85\left(1+\frac{0.07}{12}\right)^{180} = R\left(\frac{\left(1+\frac{0.07}{12}\right)^{180}-1}{0.07/12}\right)$$

$$249{,}353.635276 \approx R(316.962296718) \Rightarrow R \approx \frac{249{,}353.635276}{316.962296718} \approx \$786.70$$

b) $(836.45 - 786.70)\cdot 180 = \$8{,}955.00$

41. a) We need to find the unpaid balance where $\frac{r}{m} = \frac{0.08}{12}$.

$$U = 40{,}000\left(1+\frac{0.08}{12}\right)^{24} - 976.52\left(\frac{\left(1+\frac{0.08}{12}\right)^{24}-1}{0.08/12}\right) \approx 46{,}915.52 - 25{,}324.28 = \$21{,}591.24$$

Next we need to determine the new monthly payments where $\frac{r}{m} = \frac{0.075}{12} = 0.00625$.

$$21{,}591.24(1+0.00625)^{24} = R\left(\frac{(1+0.00625)^{24}-1}{0.00625}\right)$$

$$25{,}073.7346733 \approx R(25.8067228988) \Rightarrow R \approx \frac{25{,}073.7346733}{25.8067228988} \approx \$971.60$$

b) $(976.52 - 971.60)\cdot 24 = \$118.08$

43. We saw the left expression in calculating compound interest and the right expression in computing the future value of annuities.

45. No solution provided.

47. – 49. No solution provided.

Section 9.6 Looking Deeper: Annual Percentage Rate

1. You had paid \$300 on an outstanding balance of \$1,000, so the interest rate was $\frac{\$300}{\$1{,}000} = 0.30$ or 30%.

3. $I = P \times r \times t$
$I = 6{,}000 \times 0.08 \times 3$
$I = \$1{,}440.00$

$6{,}000 \times a \times 1 + 4{,}000 \times a \times 1 + 2{,}000 \times a \times 1 = 1{,}440$
$12{,}000a = 1{,}440$
$a = \dfrac{1{,}440}{12{,}000} = 0.12 = 12\%$

5. $\dfrac{\text{finance charge}}{\text{amount borrowed}} \times 100 = \dfrac{270}{1{,}800} \times 100 = \$15.00$

7. $\dfrac{\text{finance charge}}{\text{amount borrowed}} \times 100 = \dfrac{260}{2{,}000} \times 100 = \$13.00$

9. $\dfrac{\text{finance charge}}{\text{amount borrowed}} \times 100 = \dfrac{420}{3{,}000} \times 100 = \$14.00$
From Table 9.6, we see that for a 24-payment loan, this corresponds to an APR of roughly 13%.

11. $\dfrac{\text{finance charge}}{\text{amount borrowed}} \times 100 = \dfrac{165}{4{,}000} \times 100 \approx \$4.13$
From Table 9.6, we see that for a 6-payment loan, this corresponds to an APR of roughly 14%.

13. 12.45%

$\text{APR} \approx \dfrac{2 \times 36 \times 0.064}{36+1} = \dfrac{4.608}{37} = 0.1245$

15. 13.67%

$\text{APR} \approx \dfrac{2 \times 42 \times 0.07}{42+1} = \dfrac{5.88}{43} = 0.1367$

17. Finance charge = $24 \cdot 485 - 10{,}000 = 11{,}640 - 10{,}000 = 1{,}640$

$\dfrac{\text{finance charge}}{\text{amount borrowed}} \times 100 = \dfrac{1{,}640}{10{,}000} \times 100 = \$16.40$
From Table 9.6, we see that for a 24-payment loan, this corresponds to an APR of roughly 15%.

19. Finance charge = $48 \cdot 116.50 - 4{,}500 = 5{,}592 - 4{,}500 = 1{,}092$

$\dfrac{\text{finance charge}}{\text{amount borrowed}} \times 100 = \dfrac{1{,}092}{4{,}500} \times 100 \approx \$24.27$
From Table 9.6, we see that for a 48-payment loan, this corresponds to an APR of roughly 11%.

21. $I = Prt = 2{,}000 \times 0.08 \times 2 = 320$

$\dfrac{\text{finance charge}}{\text{amount borrowed}} \times 100 = \dfrac{320}{2{,}000} \times 100 = \$16.00$
From Table 9.6, we see that for a 24-payment loan, this corresponds to an APR of roughly 15%.

23. $I = Prt = P \times 0.082 \times 3 = 0.246P$

$\dfrac{\text{finance charge}}{\text{amount borrowed}} \times 100 = \dfrac{0.246P}{P} \times 100 = \$24.60$
From Table 9.6, we see that for a 36-payment loan, this corresponds to an APR of roughly 15%.

25. b) is better

An add-on interest loan at 8.4% for three years:

$I = Prt = 5{,}000 \times 0.084 \times 3 = 1{,}260$

$$\frac{\text{finance charge}}{\text{amount borrowed}} \times 100 = \frac{1{,}260}{5{,}000} \times 100 = \$25.20$$

From Table 9.6, we see that for a 36-payment loan, this corresponds to an APR of roughly 15%.

24 payments of $230:

Finance charge = $24 \cdot 230 - 5{,}000 = 5{,}520 - 5{,}000 = 520$

$$\frac{\text{finance charge}}{\text{amount borrowed}} \times 100 = \frac{520}{5{,}000} \times 100 = \$10.40$$

From Table 9.6, we see that for a 24-payment loan, this corresponds to an APR of roughly 10%.

27. b) is better

An add-on interest loan at 8.9% for four years:

$I = Prt = 5{,}000 \times 0.089 \times 4 = 1{,}780$

$$\frac{\text{finance charge}}{\text{amount borrowed}} \times 100 = \frac{1{,}780}{5{,}000} \times 100 = \$35.60$$

From Table 9.6, we see that for a 48-payment loan, this corresponds to an APR of roughly 16%.

36 payments of $165:

Finance charge = $36 \cdot 165 - 5{,}000 = 5{,}940 - 5{,}000 = 940$

$$\frac{\text{finance charge}}{\text{amount borrowed}} \times 100 = \frac{940}{5{,}000} \times 100 = \$18.80$$

From Table 9.6, we see that for a 36-payment loan, this corresponds to an APR of roughly 12%.

29. 24 payments of $18.75:

Finance charge = $24 \cdot 18.75 - 375 = 450 - 375 = 75$

$$\frac{\text{finance charge}}{\text{amount borrowed}} \times 100 = \frac{75}{375} \times 100 = \$20.00$$

From Table 9.6, we see that for a 24-payment loan, this corresponds to an APR of over 16%.

31. First calculate the finance charge per $100 of the loan. Next, use the line corresponding to the number of loan payment to locate the amount that is closest to the finance charge per $100. The percent shown at the top of that column is the approximate APR for the loan.

33. – 35. No solution provided.

Chapter Review Exercises

1. 12.45%

3. $\frac{11}{16} = 0.6875 = 68.75\%$

5. The $238.4 million represents 13.2% of the total spent on chocolate candy in 2007.

$$0.132 \times base = 238.4 \Rightarrow base = \frac{238.4}{0.132} \approx 1{,}806$$

Total amount spent on chocolate candy in 2007 was $1,806 million.

7. $I = P \times r \times t$
$I = 1{,}500 \times 0.09 \times 2$
$I = 270$
$1{,}500 + 270 = \$1{,}770$

9. $A = P\left(1 + \frac{r}{m}\right)^n$

$A = 11{,}400\left(1 + \frac{0.18}{12}\right)^{12\left(\frac{1}{2}\right)}$

$A = 11{,}400(1 + 0.015)^6$

$A \approx 12{,}465.25$

$A \approx \$12{,}465.25$

11. $A = P\left(1 + \frac{r}{m}\right)^n$

$2{,}000 = 1{,}000\left(1 + \frac{0.064}{12}\right)^n$

$2 \approx (1.00533333333)^n$

$\log(2) \approx \log(1.00533333333)^n$

$\log(2) \approx n \cdot \log(1.00533333333)$

$n \approx \frac{\log(2)}{\log(1.00533333333)} \approx 130.311$

$n \approx 131$ months

13. $I = Prt = 1{,}320(0.0825)(3) = \$326.70$

$\text{Monthly Payment} = \frac{P + I}{n} = \frac{1{,}320 + 326.70}{3 \cdot 12} = \frac{1{,}646.70}{36} \approx \$45.75$

15.

Day	Balance	Number of days × Balance
1,2,3,4,5	$275	5 × 275 = 1,375
6,7,8,9,10,11	$200	6 × 200 = 1,200
12,13,14,15,16,17,18	$315	7 × 315 = 2,205
19,20,21,22,23	$335	5 × 335 = 1,675
24,25,26,27,28,29,30,31	$351	8 × 351 = 2,808

Average daily balance $\frac{1{,}375 + 1{,}200 + 2{,}205 + 1{,}675 + 2{,}808}{31} = \frac{9{,}263}{31} \approx 298.81$

Finance charge $I = Prt = 298.81(0.18)(31/365) \approx \$4.57$

17. $r = \frac{0.06}{12} = 0.005$

$2{,}000 = R\left(\frac{(1 + 0.005)^{36} - 1}{0.005}\right)$

$2{,}000 \approx R(39.3361049646)$

$R \approx \frac{2{,}000}{39.3361049646} \approx \$50.85$

19. $\frac{r}{m} = \frac{0.09}{12} = 0.0075$

$10{,}000 = 300\left(\frac{(1 + 0.0075)^n - 1}{0.0075}\right)$

$\frac{100}{3} = \left(\frac{(1.0075)^n - 1}{0.0075}\right)$

$0.25 = 1.0075^n - 1$

$1.25 = 1.0075^n$

$\log 1.25 = \log 1.0075^n$

$\log 1.25 = n \cdot \log 1.0075$

$\frac{\log 1.25}{\log 1.0075} = n$

$n \approx 30$ months

21. $\frac{r}{n} = \frac{0.10}{12}$ and $n = 12 \cdot 4 = 48$

$$5,000\left(1+\frac{0.10}{12}\right)^{48} = R\left(\frac{\left(1+\frac{0.10}{12}\right)^{48}-1}{0.10/12}\right)$$

$$7,446.77049303 \approx R(58.7224918326)$$

$$R \approx \frac{7,446.77049303}{58.722491826} \approx \$126.82$$

23. $A = 50,000\left(\frac{(1+0.08)^{20}-1}{0.08}\right) \approx \$2,288,098.21$

$$A = P\left(1+\frac{r}{m}\right)^{n}$$

$$2,288,098.21 = P(1+0.08)^{20}$$

$$2,288,098.21 \approx P(4.66095714385)$$

$$\frac{2,288,098.21}{4.66095714385} \approx P$$

$$P \approx \$490,907.37$$

The present value is \$490,907.37; the lump sum of \$500,000 is a better deal.

25. For this exercise, we will use the same method as described in the footnote for Exercises 21 – 23 in section 9.5.

a) 1st year: $R = \frac{180000\left(1+\frac{0.045}{12}\right)^{360}}{\frac{\left(\left(1+\frac{0.045}{12}\right)^{360}-1\right)}{0.045/12}} = \frac{692585.649}{759.3861467} \approx \$912.04$

b) In year 5: $r = 0.125$

$$R = \frac{180000\left(1+\frac{0.125}{12}\right)^{360}}{\frac{\left(\left(1+\frac{0.125}{12}\right)^{360}-1\right)}{0.125/12}} = \frac{7506767.182}{3907.609164} \approx \$1,921.07$$

27. 15.2%

$$\text{APR} \approx \frac{2\times 20\times 0.08}{20+1} = \frac{3.2}{21} = 0.152$$

Chapter Test

1. 36.24%

3. $\frac{7}{16} = 0.4375 = 43.75\%$

5. $$\begin{aligned} I &= P \times r \times t \\ I &= 3,400 \times 0.025 \times 3 \\ I &= 255 \\ 3,400 + 255 &= \$3,655 \end{aligned}$$

7. The amount of decrease in price is $\$169.99 - \$149.99 = \$20$.
The percent reduction in price is $\frac{\$20}{\$169.99} \approx 0.1177 = 11.77\%$.

9. $I = 162.50 \cdot 24 - 3,000 = 900$

$$\begin{aligned} I &= P \times r \times t \\ 900 &= 3,000 \times r \times 2 \\ 900 &= 6,000r \\ 0.15 &= r \\ r &= 15\% \end{aligned}$$

11. $$\begin{aligned} A &= P\left(1 + \frac{r}{n}\right)^n \\ 2,000 &= 1,000\left(1 + \frac{0.048}{12}\right)^n \\ 2 &\approx (1.004)^n \\ \log(2) &\approx \log(1.004)^n \\ \log(2) &\approx n \cdot \log(1.004) \\ n &\approx \frac{\log(2)}{\log(1.004)} \approx 173.633 \\ nt &\approx 174 \text{ months} \end{aligned}$$

13. $$\begin{aligned} A &= P\left(1 + \frac{r}{m}\right)^n \\ 15,000 &= P\left(1 + \frac{0.042}{12}\right)^{12 \cdot 18} \\ 15,000 &= P(1 + 0.0035)^{216} \\ 15,000 &= P(1.0035)^{216} \\ \frac{15,000}{1.0035^{216}} &= P \\ P &\approx \$7,052.42 \end{aligned}$$

15. $$\begin{aligned} A &= P(1 + r)^t \\ 2,400 &= 2,100(1 + r)^3 \\ \frac{8}{7} &= (1 + r)^3 \\ \left(\frac{8}{7}\right)^{1/3} &= \left((1 + r)^3\right)^{1/3} \\ \left(\frac{8}{7}\right)^{1/3} &= 1 + r \\ r &= \left(\frac{8}{7}\right)^{1/3} - 1 \approx 0.0455 = 4.55\% \end{aligned}$$

17.

Day	Balance	Number of days × Balance
1,2,3	$425	3 × 425 = 1,275
4,5,6,7,8,9	$340	6 × 340 = 2,040
10,11,12,13,14	$365	5 × 365 = 1,825
15,16,17,18,19,20,21,22,23,24	$380	10 × 380 = 3,800
25,26,27,28,29,30	$460	6 × 460 = 2,760

Average daily balance $\dfrac{1,275+2,040+1,825+3,800+2,760}{30}=\dfrac{11,700}{30}=390.00$

Finance charge $I=Prt=390.00(0.21)(30/365)\approx \$6.74$

19. $\dfrac{5^x-4}{3}=10$

$$5^x-4=30$$
$$5^x=34$$
$$\log 5^x=\log 34$$
$$x\cdot\log 5=\log 34$$
$$x=\frac{\log 34}{\log 5}\approx 2.1911$$

21. $r=\dfrac{0.04}{12}=0.0033333333$

$$1,800=R\left(\frac{(1+0.0033333333)^{36}-1}{0.0033333333}\right)$$
$$1,800\approx R(38.181562)$$
$$R\approx\frac{1,800}{38.181562}\approx \$47.15$$

23. $\dfrac{r}{m}=\dfrac{0.0375}{12}=0.003125$

$$9,000=450\left(\frac{(1+0.003125)^n-1}{0.003125}\right)$$
$$20=\left(\frac{(1.003125)^n-1}{0.003125}\right)$$
$$0.0625=1.003125^n-1$$
$$1.0625=1.003125^n$$
$$\log 1.0625=\log 1.003125^n$$
$$\log 1.0625=n\cdot\log 1.003125$$
$$\frac{\log 1.0625}{\log 1.003125}=n$$
$$n\approx 20 \text{ months}$$

25. $A=100,000\left(\dfrac{(1+0.04)^{10}-1}{0.04}\right)\approx \$1,200,610.71$

$$A=P\left(1+\frac{r}{m}\right)^n$$
$$1,200,610.71=P(1+0.04)^{10}$$
$$1,200,610.71\approx P(1.480244285)$$
$$\frac{1,200,610.71}{1.480244285}\approx P$$
$$P\approx \$811,089.58$$

The annuity is the better deal because its present value is $811,089.58.

27. We need to find the monthly payments first.

$$\frac{r}{m} = \frac{0.096}{12} = 0.008 \text{ and } n = 12 \cdot 5 = 60$$

$$16{,}500(1+0.008)^{60} = R\left(\frac{(1+0.008)^{60} - 1}{0.008}\right)$$

$$26{,}614.350421808 \approx R(76.6238668)$$

$$R \approx \frac{26{,}614.350421808}{76.6238668} \approx \$347.34$$

Now find the unpaid balance where $n = 12 \cdot 2 = 24$.

$$U = 16{,}500(1+0.008)^{24} - 347.34\left(\frac{(1+0.008)^{24} - 1}{0.008}\right)$$

$$U \approx 19{,}977.30 - 9{,}150.03 = \$10{,}827.27$$

Chapter 10

Geometry: Ancient And Modern Mathematics Embrace

Section 10.1 Lines, Angles, and Circles

1. a) acute   b) right   c) obtuse   d) straight

3. The length of arc AB is proportional to the measure of central angle $\angle ACB$.

5. 7 and 9; 8 and 10; 2 and 5; 1 and 6; 3 and 4

7. 3 and 7

9. 10; 8

11. 9 and 10; 7 and 10; 7 and 8; 8 and 9; 3 and 10; 4 and 10; 4 and 8; 3 and 8

13. true

15. false; Complementary angles are two angles whose sum of the measures is 90 degrees. The two angles do not each have to measure 45 degrees.

17. true

19. false; For example, a $30°$ angle is the complement of a $60°$ angle and the supplement of a $150°$ angle.

21. false; For example, the supplement of a $30°$ angle is a $150°$ angle.

23. e and d

25. b and f; c and g

27. complement: $90° - 30° = 60°$; supplement: $180° - 30° = 150°$

29. complement: none; supplement: $180° - 120° = 60°$

31. complement: $90° - 51.2° = 38.8°$; supplement: $180° - 51.2° = 128.8°$

33. $m\angle a = 180° - 36° = 144°$; $m\angle b = 36°$; $m\angle c = 144°$

35. $m\angle a = 45°$; $m\angle b = 180° - 45° = 135°$; $m\angle c = 45°$

37. $m\angle a = 90° - 38° = 52°$; $m\angle b = 90°$; $m\angle c = 180° - 52° = 128°$

39. $$\frac{\text{measure of central angle}}{360°} = \frac{\text{arc length}}{\text{circumference}}$$
$$\frac{90°}{360°} = \frac{a}{24 \text{ feet}}$$
$$90 \cdot 24 = 360a$$
$$2{,}160 = 360a$$
$$a = \frac{2{,}160}{360} = 6$$

Arc *AB* has length 6 feet.

41. $$\frac{\text{measure of central angle}}{360°} = \frac{\text{arc length}}{\text{circumference}}$$
$$\frac{m}{360°} = \frac{4 \text{ m}}{12 \text{ m}}$$
$$12m = 4 \cdot 360$$
$$12m = 1{,}440$$
$$m = 120$$

The measure of the central angle, $\angle ACB$, is $120°$.

43. $$\frac{\text{measure of central angle}}{360°} = \frac{\text{arc length}}{\text{circumference}}$$
$$\frac{30°}{360°} = \frac{100 \text{ mm}}{c}$$
$$30c = 360 \cdot 100$$
$$30c = 36{,}000$$
$$c = 1{,}200$$

The circumference of the circle is 1,200 millimeters.

45. $$\frac{\text{measure of central angle}}{360°} = \frac{\text{arc length}}{\text{circumference}}$$
$$\frac{180° - 60°}{360°} = \frac{a}{18 \text{ ft}}$$
$$\frac{120°}{360°} = \frac{a}{18 \text{ ft}}$$
$$360a = 120 \cdot 18$$
$$360a = 2{,}160$$
$$a = 6$$

The arc length is 6 feet.

47. $$\frac{\text{measure of central angle}}{360°} = \frac{\text{arc length}}{\text{circumference}}$$
$$\frac{m}{360°} = \frac{15 \text{ in} - 3 \text{ in}}{30 \text{ in}}$$
$$\frac{m}{360°} = \frac{12 \text{ in}}{30 \text{ in}}$$
$$30m = 360 \cdot 12$$
$$30m = 4{,}320$$
$$m = 144°$$

$m\angle BCD$ is 144°.

49. $$2x + x = 180°$$
$$3x = 180°$$
$$x = 60°$$

51. $$3x + 2x = 90°$$
$$5x = 90°$$
$$x = 18°$$

53. 12

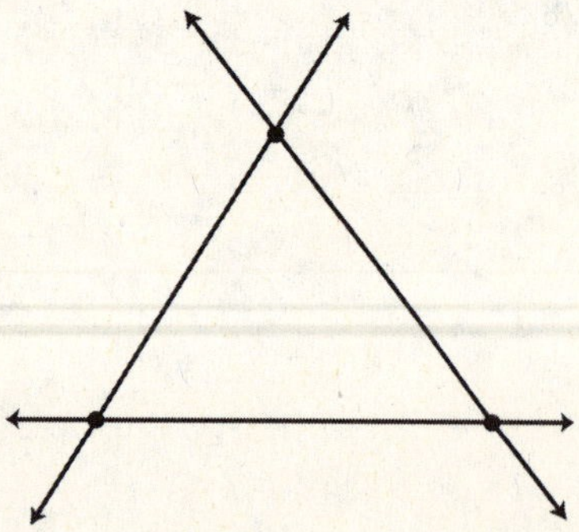

55. $$\frac{\text{measure of central angle}}{360°} = \frac{\text{arc length}}{\text{circumference}}$$
$$\frac{m}{360°} = \frac{1{,}000 \text{ miles}}{25{,}000 \text{ miles}}$$
$$25{,}000m = 360 \cdot 1{,}000$$
$$25{,}000m = 360{,}000$$
$$m = 14.4$$

The measure of the angle should have been 14.4°.

57. The sum of supplementary angles is 180°; the sum of complementary angles is 90°.

59. "Alternate" means that the angles are on opposite sides of the transversal. "Exterior" means that the angles are outside the parallel lines.

61. This is possible. If the two lines are perpendicular, then all four angles are right angles. Each angle measures $90°$.

63. No more than two could be obtuse. Because vertical angles are equal, if there were a third obtuse angle then the fourth would also have to be obtuse, giving a sum of more than $360°$.

65. Yes, if the angles both measure $45°$.

67. No solution provided.

69.

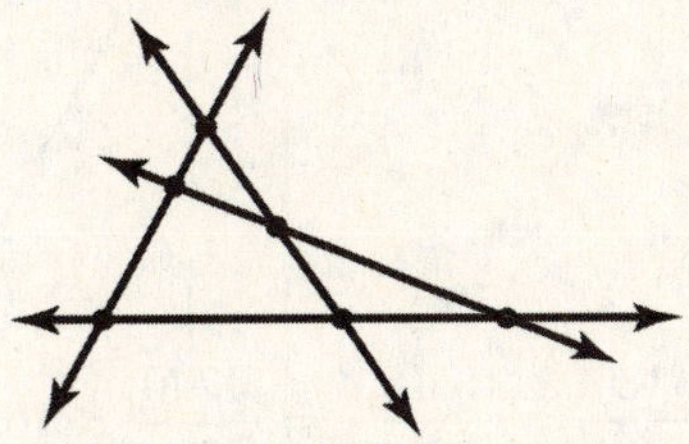

71.

Number of lines	Most number of angles formed
2	$4 = 4 \cdot 1$
3	$12 = 4 \cdot 3 = 4 \cdot (1+2)$
4	$24 = 4 \cdot 6 = 4 \cdot (1+2+3)$
5	$40 = 4 \cdot 10 = 4 \cdot (1+2+3+4)$
. .	. .
10	$4 \cdot (1+2+3+4+5+6+7+8+9) = 4 \cdot 45 = 180$

Section 10.2 Polygons

1. Alternate interior angles are equal; $m\angle 1 = m\angle 4$, $m\angle 3 = m\angle 5$.

3. It was the supplement of $\angle WVZ$, which had a measure of $108°$ and $180° - 108° = 72°$.

5. false; A trapezoid has at least one pair of parallel sides where a parallelogram must have two pairs of parallel sides.

7. true

9. false; Consider a rhombus and a square.

11. polygon

13. not a polygon; not made of line segments

15.
$$m\angle A + m\angle B + m\angle C = 180°$$
$$m\angle A + 2m\angle A + 3m\angle A = 180°$$
$$6m\angle A = 180°$$
$$m\angle A = 30°$$

$$m\angle B = 2m\angle A = 2 \cdot 30° = 60°$$
$$m\angle C = 3m\angle A = 3 \cdot 30° = 90°$$

17.
$$m\angle A + m\angle B + m\angle C = 180°$$
$$m\angle A + (m\angle A + 10) + (2m\angle A - 10) = 180°$$
$$4m\angle A = 180°$$
$$m\angle A = 45°$$

$$m\angle B = m\angle A + 10 = 45 + 10 = 55°$$
$$m\angle C = 2m\angle A - 10 = 2 \cdot 45 - 10 = 90 - 10 = 80°$$

19. We can divide the hexagon into four triangles. The angle sum is $4 \cdot 180° = 720°$.

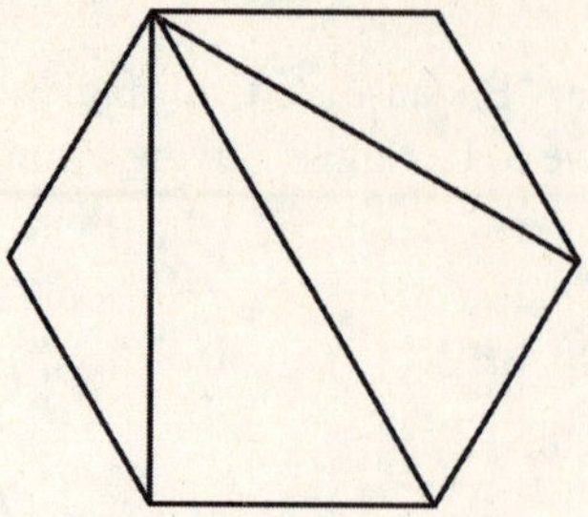

21. $\dfrac{(20-2)\cdot 180°}{20} = \dfrac{18\cdot 180°}{20} = \dfrac{3,240°}{20} = 162°$

23. $160° = \dfrac{(n-2)\cdot 180°}{n}$

$160n = (n-2)\cdot 180$

$160n = 180n - 360$

$-20n = -360$

$n = \dfrac{-360}{-20} = 18$

25. $\dfrac{x}{4} = \dfrac{10}{24}$

$24x = 10\cdot 4$

$24x = 40$

$x = \dfrac{40}{24} = \dfrac{5}{3}$ in.

27. $\dfrac{x}{3} = \dfrac{10}{6}$

$6x = 10\cdot 3$

$6x = 30$

$x = 5$ in.

29. $m\angle E = 40°$

$\dfrac{EH}{8} = \dfrac{21}{24}$

$24(EH) = 21\cdot 8$

$24(EH) = 168$

$EH = \dfrac{168}{24} = 7$

31. $m\angle I = 35°$

$\dfrac{HI}{120} = \dfrac{105}{150}$

$150(HI) = 105\cdot 120$

$150(HI) = 12,600$

$HI = \dfrac{12,600}{150} = 84$

33.

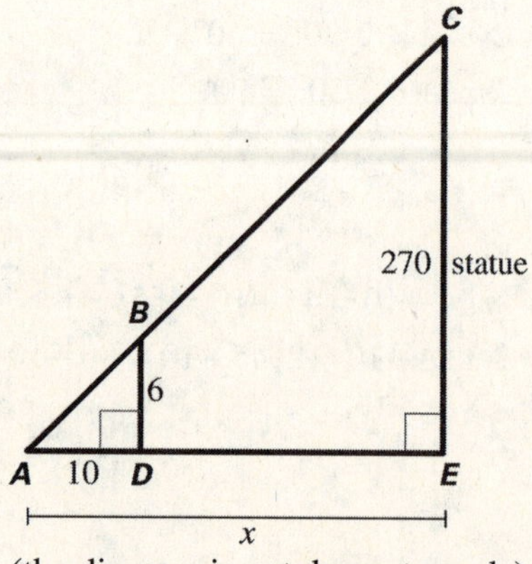

(the diagram is not drawn to scale)

Triangles ABD and ACE are similar. Therefore $\dfrac{BD}{AD} = \dfrac{CE}{AE}$, so $\dfrac{6}{10} = \dfrac{270}{x} \Rightarrow 6x = 2700 \Rightarrow$ $x = 450$.

The length of the shadow of the statue is 450 ft.

35. Let x be the width of the river.

$$\frac{x}{200}=\frac{3}{8}$$
$$8x=200\cdot 3$$
$$8x=600$$
$$x=75 \text{ ft}$$

37. Let x be the length of the pond.

$$\frac{x}{270+60}=\frac{35}{60}$$
$$\frac{x}{330}=\frac{35}{60}$$
$$60x=330\cdot 35$$
$$60x=11{,}550$$
$$x=\frac{11{,}550}{60}=192.5 \text{ feet}$$

39.

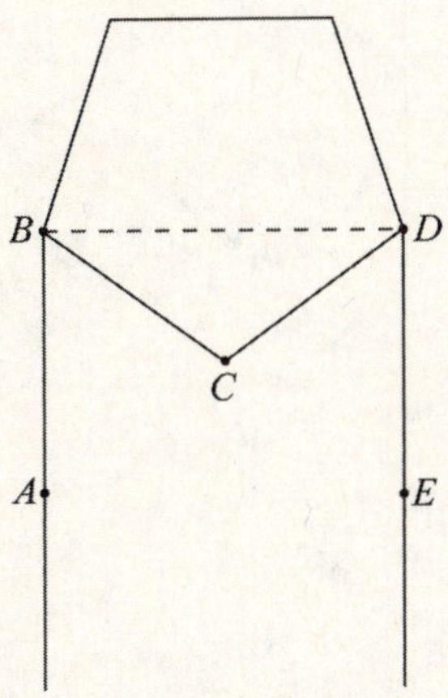

Label the points as shown. Draw a line connecting B to D. Angles BDE and ABD will be right angles. Find the measure of angle BCD.

$$m\angle BCD=\frac{(5-2)\cdot 180°}{5}=\frac{3\cdot 180°}{5}=\frac{540°}{5}=108°$$

Since triangle BCD is isosceles, $m\angle BDC=\frac{180°-108°}{2}=\frac{72°}{2}=36°$. Also since $\angle BDC$ and $\angle CDE$ are complementary, the desired angles each measure $90°-36°=54°$.

In Exercise 41, the triangles formed by the sides and supports of the smaller and larger gazebos are similar, so the corresponding sides are proportional.

41. Let x be the length of the long support

$$\frac{x}{12}=\frac{17.4}{10}$$
$$10x=12\cdot 17.4$$
$$10x=208.8$$
$$x=\frac{208.8}{10}=20.88 \text{ ft}$$

43. Isosceles triangles have two equal sides; equilateral triangles have 3 equal sides.

45. No. A scalene triangle has no sides equal.

47. Both have four equal sides. A rhombus does not have to have right interior angles.

49. No solution provided.

51. It is possible to construct. Additional answers may vary.

53. It is not possible to construct. The angle sum would be greater than 180°.

55. No solution provided.

57. The measure of the interior angles gets larger. For very large n, the interior angles have measures close to 180°. For example, evaluate the expression $\frac{(n-2)\cdot 180°}{n}$ with n=10, 100, 1000, 10000, etc.

n	$\frac{[(n-2)\cdot 180°]}{n}$	simplified
10	$\frac{(10\text{-}2)\cdot 180°}{10}$	144°
100	$\frac{(100-2)\cdot 180°}{100}$	176.4°
1,000	$\frac{(1{,}000-2)\cdot 180°}{1{,}000}$	179.64°
10,000	$\frac{(10{,}000-2)\cdot 180°}{10{,}000}$	179.964°
⋮	⋮	⋮

59. The interior angle of the regular n-gon is $\frac{(n-2)\cdot 180°}{n}$.

Since the triangle formed on the bottom of the n-gon is isosceles, the desired angles each measure

$$90° - \frac{180° - \frac{(n-2)\cdot 180°}{n}}{2} = 90° - 90° + \frac{(n-2)\cdot 180°}{2n} = \frac{(n-2)\cdot 90°}{n}.$$

Section 10.3 Perimeter and Area

1. In both cases, the area is the height times the base.

3. The area of the trapezoid is the sum of the areas of two triangles.

5. $A = l\cdot w = 16\cdot 10 = 160 \text{ ft}^2$

7. $A = h\cdot b = 7\cdot 20 = 140 \text{ in}^2$

9. $A = \frac{1}{2}\cdot(b_1 + b_2)\cdot h = \frac{1}{2}\cdot(22+14)\cdot 6 = \frac{1}{2}\cdot(36)\cdot 6 = 18\cdot 6 = 108 \text{ cm}^2$

11. $A = \frac{1}{2}\cdot h\cdot b = \frac{1}{2}\cdot 6\cdot 24 = 3\cdot 24 = 72 \text{ yd}^2$

13. $A = \pi r^2 \approx 3.14\cdot 5^2 = 3.14\cdot 25 = 78.5 \text{ cm}^2$

15. $r = \frac{d}{2} = \frac{8}{2} = 4 \text{ m}$

$A = \pi r^2 \approx 3.14\cdot 4^2 = 3.14\cdot 16 = 50.24 \text{ m}^2$

17. Entire area: $A = l \cdot w = 18 \cdot 8 = 144 \text{ yd}^2$

Unshaded area: $A = \frac{1}{2} \cdot h \cdot b = \frac{1}{2} \cdot 4 \cdot 18 = 2 \cdot 18 = 36 \text{ yd}^2$

$A = \frac{1}{2} \cdot h \cdot b = \frac{1}{2} \cdot 4 \cdot 9 = 2 \cdot 9 = 18 \text{ yd}^2$

$A = \frac{1}{2} \cdot h \cdot b = \frac{1}{2} \cdot 4 \cdot 9 = 2 \cdot 9 = 18 \text{ yd}^2$

Shaded area: $144 - 36 - 18 - 18 = 72 \text{ yd}^2$

19. Entire area: $A = l \cdot w = 8 \cdot (2 + 2) = 8 \cdot 4 = 32 \text{ m}^2$

Unshaded area: $A = \pi r^2 \approx 3.14 \cdot 2^2 = 3.14 \cdot 4 = 12.56 \text{ m}^2$

$A = \pi r^2 \approx 3.14 \cdot 2^2 = 3.14 \cdot 4 = 12.56 \text{ m}^2$

Shaded area: $32 - 12.56 - 12.56 = 6.88 \text{ m}^2$

21. Entire area: $A = l \cdot w = 2 \cdot 2 = 4 \text{ m}^2$

Unshaded area: $A = \pi r^2 \approx 3.14 \cdot \left(\frac{2}{2}\right)^2 = 3.14 \cdot 1 = 3.14 \text{ m}^2$

Shaded area: $4 - 3.14 = 0.86 \text{ m}^2$

23. We need to find the height of the parallelogram. Since the area of a parallelogram is height times base, the height is therefore 6 inches.

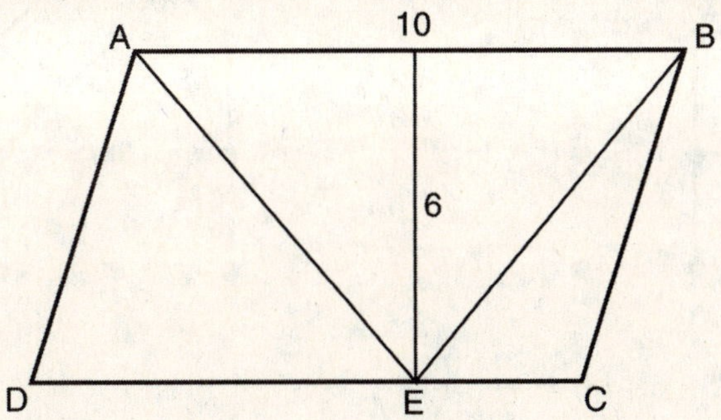

It is given that the area of ΔBEC is 6 square inches. The area of ΔABE is $A = \frac{1}{2} \cdot h \cdot b = \frac{1}{2} \cdot 6 \cdot 10 = 30$ square inches. Thus, the area of ΔADE is $60 - 6 - 30 = 24$ square inches.

25. We need to find the areas ΔEDB and ΔAEF. Since the area of ΔBAE is 30 square yards, ΔEDB also has an area of 30 square yards. This is a result of the right angles that yields these two triangles congruent. Since the area of trapezoid $ABDF$ is 66 square yards, the area of ΔAEF is $66-30-30=6$ square yards. Also, lengths AB and DE are 10 yards each because they must satisfy the relation $30=\frac{1}{2}\cdot 6\cdot b \Rightarrow 30=3b \Rightarrow b=10$.

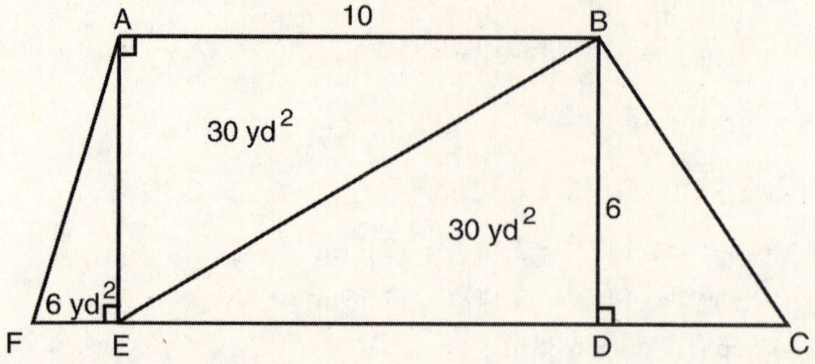

Since the area of ΔBDC is twice ΔAEF, the area of ΔBDC is $2\cdot 6=12$ square yards.

27. 7 in^2

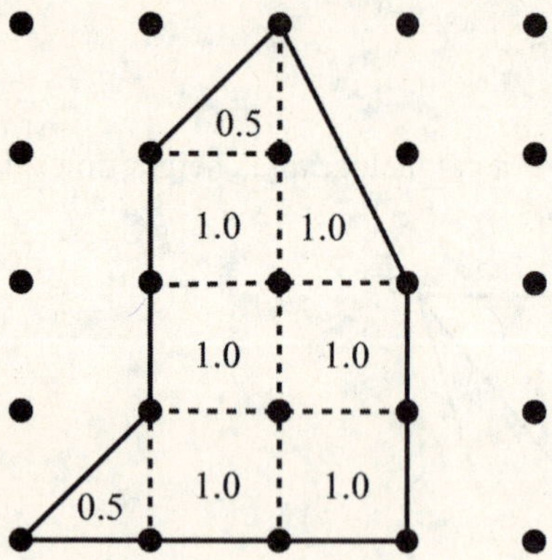

29. 3.5 in^2

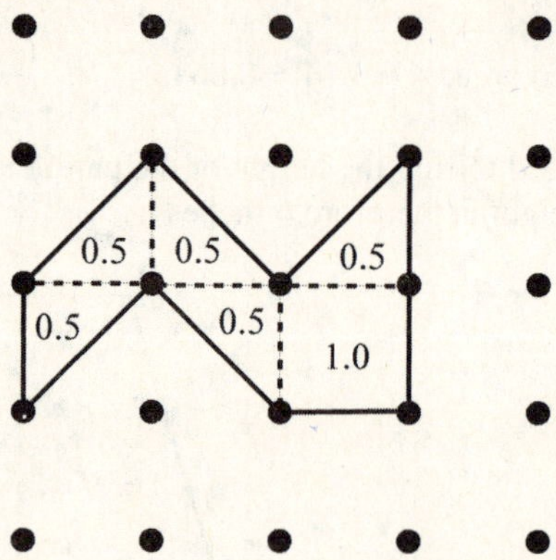

31. $s=\frac{1}{2}(a+b+c)=\frac{1}{2}(15+6+18)=\frac{39}{2}$

$$A=\sqrt{s(s-a)(s-b)(s-c)}=\sqrt{\frac{39}{2}\left(\frac{39}{2}-15\right)\left(\frac{39}{2}-6\right)\left(\frac{39}{2}-18\right)}=\sqrt{\frac{39}{2}\cdot\frac{9}{2}\cdot\frac{27}{2}\cdot\frac{3}{2}}=\sqrt{\frac{28,431}{16}}$$

$A\approx 42.15\ \text{cm}^2$

33. $s=\frac{1}{2}(a+b+c)=\frac{1}{2}(10+15+12)=\frac{37}{2}$

$$A=\sqrt{s(s-a)(s-b)(s-c)}=\sqrt{\frac{37}{2}\left(\frac{37}{2}-10\right)\left(\frac{37}{2}-15\right)\left(\frac{37}{2}-12\right)}=\sqrt{\frac{37}{2}\cdot\frac{17}{2}\cdot\frac{7}{2}\cdot\frac{13}{2}}=\sqrt{\frac{57,239}{16}}$$

$A\approx 59.81\ \text{m}^2$

35. $s=\frac{1}{2}(a+b+c)=\frac{1}{2}(19+7+18)=\frac{44}{2}=22$

$$A=\sqrt{s(s-a)(s-b)(s-c)}=\sqrt{22(22-19)(22-7)(22-18)}=\sqrt{22\cdot 3\cdot 15\cdot 4}=\sqrt{3{,}960}\approx 62.93\text{ cm}^2$$

$$A=\frac{1}{2}\cdot h\cdot b$$

$$\sqrt{3{,}960}=\frac{1}{2}\cdot h\cdot 19$$

$$h=\frac{2\sqrt{3{,}960}}{19}\approx 6.62\text{ cm}$$

37. $a^2+b^2=c^2$

$$5^2+12^2=x^2$$

$$25+144=x^2$$

$$169=x^2$$

$$x=\sqrt{169}=13\text{ m}$$

39. $a^2+b^2=c^2$

$$x^2+11^2=13^2$$

$$x^2+121=169$$

$$x^2=48$$

$$x=\sqrt{48}\approx 6.93\text{ yd}$$

41. We need to first find the height, h, of the triangle using the Pythagorean theorem.

$$8^2+h^2=10^2$$

$$64+h^2=100$$

$$h^2=36$$

$$h=\sqrt{36}=6\text{ m}$$

$$A=\frac{1}{2}\cdot h\cdot b$$

$$A=\frac{1}{2}\cdot 6\cdot(8+15)$$

$$A=\frac{1}{2}\cdot 6\cdot 23=3\cdot 23=69\text{ m}^2$$

43. area

45. perimeter

47. The distance from home plate to second base represents the hypotenuse of an isosceles right triangle. Let d represent this distance.

$$90^2+90^2=d^2$$

$$8{,}100+8{,}100=d^2$$

$$16{,}200=d^2$$

$$d=\sqrt{16{,}200}\approx 127.28\text{ ft}$$

49. AB is 2 m.

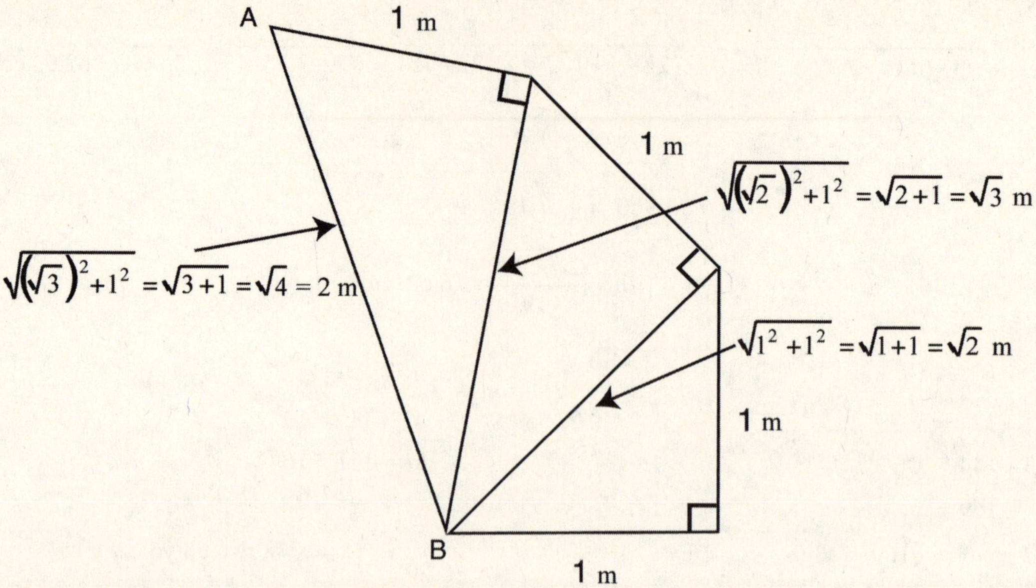

51. We need to sum the areas of a rectangle and a semi-circle (half of a circle).

Area of rectangle: $A=h\cdot b=12\cdot 6=72$ ft^2

Area of semi-circle: $A=\frac{1}{2}\pi\ r^2\approx\frac{1}{2}\cdot 3.14\cdot\left(\frac{6}{2}\right)^2=\frac{1}{2}\cdot 3.14\cdot 3^2=\frac{1}{2}\cdot 3.14\cdot 9=14.13\ \text{ft}^2$

Total area: 72 + 14.13= 86.13 ft^2

53. We need to determine the cost per square inch in order to compare the value of the two pizzas.

12-inch pizza: $A=\pi\ r^2$

$$A\approx 3.14\cdot\left(\frac{12}{2}\right)^2=3.14\cdot 6^2=3.14\cdot 36=113.04$$

The cost per square inch is approximately $\frac{5.99}{113.04}\approx 0.053$ or 5.3 cents.

16-inch pizza: $A=\pi\ r^2$

$$A\approx 3.14\cdot\left(\frac{16}{2}\right)^2=3.14\cdot 8^2=3.14\cdot 64=200.96$$

The cost per square inch is approximately $\frac{8.99}{200.96}\approx 0.045$ or 4.5 cents.

The 16-inch pizza is a better deal. The cost per square inch is less.

55. There are two rectangular pieces. Each of these has an area of $4 \cdot 100 = 400 \text{ m}^2$.
The two ends combined would form a washer-like figure in which we need to find the area of the shaded region.

Outer Area: $A_0 = \pi r^2$

$A_0 \approx 3.14 \cdot \left(\frac{20}{2}\right)^2$

$A_0 \approx 3.14 \cdot 10^2$

$A_0 \approx 3.14 \cdot 100$

$A_0 \approx 314 \text{ m}^2$

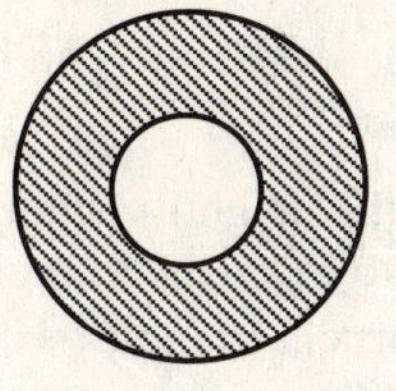

Inner Area: $A_I = \pi r^2$

$A_I \approx 3.14 \left(\frac{20-4-4}{2}\right)^2$

$A_I \approx 3.14 \cdot 6^2$

$A_I \approx 3.14 \cdot 36$

$A_I \approx 113.04 \text{ m}^2$

Surface area of track: $400 + 400 + 314 - 113.04 = 1{,}000.96 \text{ m}^2$

57. By the Pythagorean theorem:

$x = \sqrt{219^2 - 115^2}$

$x = \sqrt{34{,}736} \approx 186.38 \text{ m}$

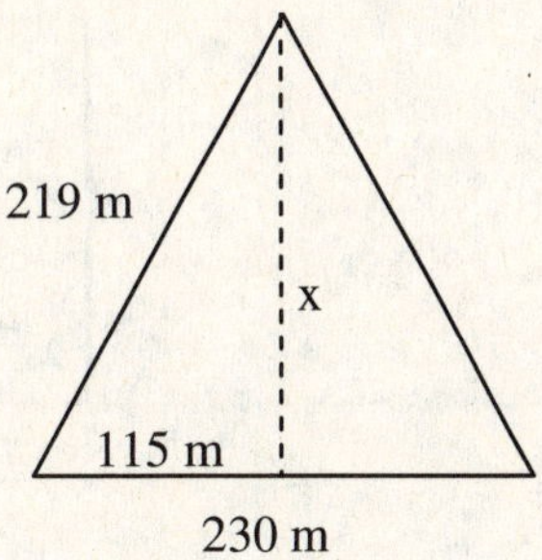

59. a) The circumference of a circle with an 18 foot radius is $C = 2 \cdot \pi r \approx 2 \cdot 3.14 \cdot 18 = 113.04$ ft. The amount of fencing we need would be $\frac{72}{360} = \frac{1}{5}$ of this or approximately 22.61 ft.

b) The area of a circle with an 18-foot radius is $A = \pi r^2 \approx 3.14 \cdot 18^2 = 3.14 \cdot 324 = 1{,}017.36 \text{ ft}^2$. One fifth of this is about 203.47 ft^2

61. rectangle, parallelogram, triangle, trapezoid

63. We had the length of the hypotenuse of ΔTMC and needed the length of another leg to find the height h.

65. No solution provided.

67. To find the area of the side of the museum, we must find the area of the trapezoid and the equilateral triangles.

The area of trapezoid is $A = \frac{1}{2} \cdot (b_1 + b_2) \cdot h = \frac{1}{2} \cdot [100 + 60] \cdot 40 = \frac{1}{2} \cdot (160) \cdot 40 = 80 \cdot 40 = 3{,}200 \text{ ft}^2$. square yards.

To find the area of each of these triangles, we will use Heron's formula.

$$s = \frac{1}{2}(a+b+c) = \frac{1}{2}(18+18+18) = \frac{54}{2} = 27$$

$$A = \sqrt{s(s-a)(s-b)(s-c)} = \sqrt{27(27-18)(27-18)(27-18)}$$

$$= \sqrt{27 \cdot 9 \cdot 9 \cdot 9} = \sqrt{19{,}683} = 81\sqrt{3} \approx 140.3 \text{ ft}^2.$$

Therefore, the area of the side of the museum is $3{,}200 - 3(140.3) = 3{,}200 - 420.9 = 2{,}779.1 \text{ ft}^2$.

69. false; it quadruples

Area of original circle: $A_0 = \pi\ r^2$

Area of circle with double radius: $A_d = \pi\ (2r)^2 = \pi \cdot 4 \cdot r^2 = 4 \cdot \pi\ r^2 = 4A_0$

71. The area of quadrilateral (parallelogram) *WXYZ* is one-half of the area of rectangle *ABCD*. Explanations may vary.

Let *b* be the base of rectangle *ABCD* and *h* the height. Although it is more common to use length and width, we will use base and height in order to mark the triangles clearly. The area of rectangle *ABCD* will be $A = h \cdot b$. The sides of the four congruent triangles will be as marked.

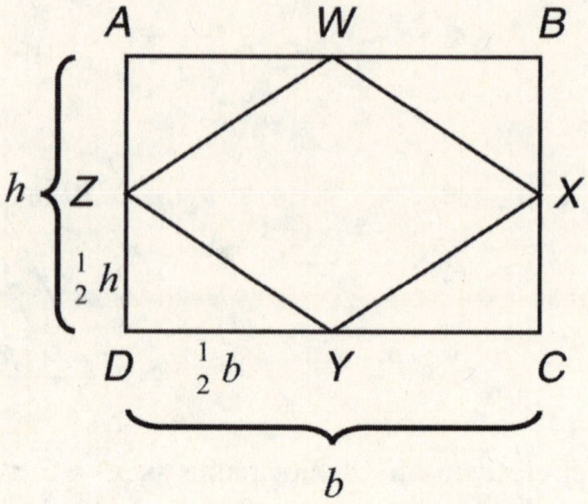

The area of each of the four congruent triangles will be $\frac{1}{2} \cdot \left(\frac{1}{2}h\right) \cdot \left(\frac{1}{2}b\right) = \frac{1}{8}hb$. The area of the four congruent triangles together will be $4 \cdot \left(\frac{1}{8}hb\right) = \frac{1}{2}hb$. Thus the area of quadrilateral (parallelogram) *WXYZ* will be $h \cdot b - \frac{1}{2}hb = \frac{1}{2}hb$ or one-half the area of rectangle *ABCD*.

73. The figure that will have the maximum area will be a square with dimensions 50 feet by 50 feet.

75. We need to first find the approximate radius of the open area.

$$A = \pi\ r^2$$
$$11{,}304 \approx 3.14 \cdot r^2$$
$$\frac{11{,}304}{3.14} \approx r^2$$

$$r \approx \sqrt{\frac{11{,}304}{3.14}} = 60 \text{ ft}$$

We are next concerned with a washer-like figure in which we need to find the area of the shaded region.

Outer Area: $A_0 = \pi\ r^2$

$$A_0 \approx 3.14 \cdot (180 + 60)^2$$
$$A_0 \approx 3.14 \cdot 240^2$$
$$A_0 \approx 3.14 \cdot 57{,}600$$
$$A_0 \approx 180{,}864 \text{ ft}^2$$

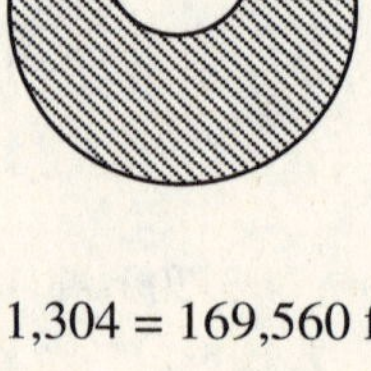

Inner Area: 11,304 ft^2

Outer Area – Inner Area = 180,864 – 11,304 = 169,560 ft^2

Each of the lots will be one-eighth of this area or about 21,195 ft^2

77. $A = (2r)^2 - \pi r^2 = 4r^2 - \pi r^2 = (4 - \pi) r^2$

Section 10.4 Volume and Surface Area

1. Both are the area of the base times the height.

3. The radius of the can that minimized its surface area is between 0.5 and 0.6 of a foot.

5. a) Surface area: $S = 2lw + 2lh + 2wh = 2\cdot6\cdot4 + 2\cdot6\cdot5 + 2\cdot4\cdot5 = 48 + 60 + 40 = 148 \text{ cm}^2$

   b) Volume: $V = lwh = 6\cdot4\cdot5 = 120 \text{ cm}^3$

7. a) Surface area: $S = \pi r\sqrt{r^2 + h^2} \approx 3.14\cdot3\cdot\sqrt{3^2 + 8^2} = 9.42\sqrt{9 + 64} = 9.42\sqrt{73} \approx 80.48 \text{ in}^2$

   b) Volume: $V = \frac{1}{3}\pi r^2 h \approx \frac{1}{3}\cdot3.14\cdot3^2\cdot8 = \frac{3.14\cdot9\cdot8}{3} = \frac{226.08}{3} = 75.36 \text{ in}^3$

9. a) Surface area: $S = 2\pi rh + 2\pi r^2 \approx 2\cdot3.14\cdot5\cdot8 + 2\cdot3.14\cdot5^2 = 251.2 + 157 = 408.2 \text{ ft}^2$

   b) Volume: $V = \pi r^2 h \approx 3.14\cdot5^2\cdot8 = 3.14\cdot25\cdot8 = 628 \text{ ft}^3$

11. a) Surface area: $S = 4\pi r^2 \approx 4\cdot3.14\cdot20^2 = 12.56\cdot400 = 5{,}024 \text{ cm}^2$

    b) Volume: $V = \frac{4}{3}\pi r^3 \approx \frac{4}{3}\cdot3.14\cdot20^3 = \frac{4\cdot3.14\cdot8{,}000}{3} = \frac{100{,}480}{3} \approx 33{,}493.33 \text{ cm}^3$

13. Volume = area of the base times height = $25\cdot3 = 75 \text{ ft}^3$

15. Area of base = $A = \frac{1}{2}\cdot(b_1 + b_2)\cdot h = \frac{1}{2}\cdot(11 + 7)\cdot4 = \frac{1}{2}\cdot(18)\cdot4 = 9\cdot4 = 36 \text{ in}^2$

    Volume = area of the base times height = $36\cdot5 = 180 \text{ in}^3$

17. $s = \frac{1}{2}(a + b + c) = \frac{1}{2}(5 + 5 + 4) = \frac{14}{2} = 7$

    Area of base applying Heron's formula is

    $A = \sqrt{s(s-a)(s-b)(s-c)} = \sqrt{7(7-5)(7-5)(7-4)} = \sqrt{7\cdot2\cdot2\cdot3} = \sqrt{84} \approx 9.17 \text{ m}^2$

    Volume = area of the base times height = $\sqrt{84}\cdot2 \approx 18.33 \text{ m}^3$

19. $12^3 = 1{,}728$

21. We need to first determine the volume of the punch bowl and the ladle.

    Punch bowl volume: $V = \frac{1}{2}\cdot\left(\frac{4}{3}\pi r^3\right) \approx \frac{4}{6}\cdot3.14\cdot9^3 = \frac{4\cdot3.14\cdot729}{6} = \frac{9{,}156.24}{6} = 1{,}526.04 \text{ in}^3$

    Ladle volume: $V = \frac{1}{2}\cdot\left(\frac{4}{3}\pi r^3\right) \approx \frac{4}{6}\cdot3.14\cdot2^3 = \frac{4\cdot3.14\cdot8}{6} = \frac{100.48}{6} \approx 16.75 \text{ in}^3$

    We then divide the volume of the punch bowl by the volume of the ladle. $\frac{1{,}526.04}{16.75} \approx 91.11$

    There are approximately 91 ladles of punch.

23. We need to first determine the volume of the cylindrical pitcher and the conical glass.

Pitcher volume: $V = \pi r^2 h \approx 3.14 \cdot (2.5)^2 \cdot 8 = 3.14 \cdot 6.25 \cdot 8 = 157 \text{ in}^3$

Glass volume: $V = \frac{1}{3}\pi r^2 h \approx \frac{1}{3} \cdot 3.14 \cdot (2)^2 \cdot 1.5 = \frac{3.14 \cdot 4 \cdot 1.5}{3} = \frac{18.84}{3} = 6.28 \text{ in}^3$

We then divide the volume of the pitcher by the volume of the glass. $\frac{157}{6.28} = 25$

There are approximately 25 glasses of punch.

25. We need to first determine the volume of the trapezoidal wheelbarrow.

Area of base = $A = \frac{1}{2} \cdot (b_1 + b_2) \cdot h = \frac{1}{2} \cdot (2+3) \cdot 1 = \frac{1}{2} \cdot (5) \cdot 1 = \frac{5}{2} \text{ ft}^2$

Volume = area of the base times height = $\frac{5}{2} \cdot \frac{5}{2} = \frac{25}{4} = 6.25 \text{ ft}^3$

The volume of the wheelbarrow is then multiplied by 16. $\frac{25}{4} \cdot 16 = 100 \text{ ft}^3$

She needs 100 ft^3 of stone and this is approximately $\frac{100}{27} \approx 3.7 \text{ yd}^3$ of stone.

27. Volume of the rectangular cake: $V = lwh = 14 \cdot 9 \cdot 3 = 378 \text{ in}^3$

Volume of the round cake: $V = \pi r^2 h \approx 3.14 \cdot 5^2 \cdot 4 = 3.14 \cdot 25 \cdot 4 = 314 \text{ in}^3$

The rectangular cake has more volume.

29. a) $V = \pi r^2 h \approx 3.14 \cdot 12^2 \cdot 5 = 3.14 \cdot 144 \cdot 5 = 2{,}260.8 \text{ in}^3$

b) $V = \pi r^2 h \approx 3.14 \cdot 10^2 \cdot 7 = 3.14 \cdot 100 \cdot 7 = 2{,}198 \text{ in}^3$

We get a larger increase by increasing the radius because in the formula for the volume of a cylinder, we square the radius. This is not the case in general, however.

31. a) $V = \frac{1}{3}\pi r^2 h \approx \frac{1}{3} \cdot 3.14 \cdot 8^2 \cdot 3 = \frac{3.14 \cdot 64 \cdot 3}{3} = \frac{602.88}{3} = 200.96 \text{ in}^3$

b) $V = \frac{1}{3}\pi r^2 h \approx \frac{1}{3} \cdot 3.14 \cdot 6^2 \cdot 5 = \frac{3.14 \cdot 36 \cdot 5}{3} = \frac{565.2}{3} = 188.4 \text{ in}^3$

We get a larger increase by increasing the radius because in the formula for the volume of a cylinder, we square the radius. This is not the case in general, however.

33. We must first find the volume of the concrete floor in cubic feet. To find the volume, we need to find the base area; that is, the area of the regular hexagon. The regular hexagon can be divided up into 6 equilateral triangles. To find the area of each of these triangles, we can either apply Heron's formula or use the Pythagorean theorem to find the height and then apply the area formula for a triangle. In this solution, we will use Heron's formula.

$s = \frac{1}{2}(a+b+c) = \frac{1}{2}(8+8+8) = \frac{24}{2} = 12$

33. (continued)

Area of base applying Heron's formula is

$$A=\sqrt{s(s-a)(s-b)(s-c)}=\sqrt{12(12-8)(12-8)(12-8)}=\sqrt{12\cdot 4\cdot 4\cdot 4}=\sqrt{768}=16\sqrt{3}\approx 27.71\text{ ft}^2$$

Since there are 6 of these triangles, we have the area of the hexagon to be $16\sqrt{3}\cdot 6=96\sqrt{3}$ $\approx 166.28\text{ ft}^2$.

Since the height of 6 inches is $\frac{1}{2}$ foot, we have

Volume = area of the base times height = $96\sqrt{3}\cdot\frac{1}{2}=48\sqrt{3}\approx 83.14\text{ ft}^3$. He must purchase 84 cubic feet of concrete.

35. The volume of the earth is approximately:

$$V=\frac{4}{3}\pi r^3\approx\frac{4}{3}\cdot 3.14\cdot\left(\frac{7{,}920}{2}\right)^3=\frac{4}{3}\cdot 3.14\cdot 3{,}960^3=\frac{4}{3}\cdot 3.14\cdot 62{,}099{,}136{,}000=259{,}988{,}382{,}720\text{ mi}^3.$$

The volume of the moon is therefore approximately $\frac{259{,}988{,}382{,}720}{49}\approx 5{,}305{,}885{,}361.63$.

$$\begin{aligned} V&=\frac{4}{3}\pi r^3\\ 5{,}305{,}885{,}361.63&\approx\frac{4}{3}\cdot 3.14\cdot r^3\\ 1{,}267{,}329{,}306.12&\approx r^3\\ r&\approx\sqrt[3]{1{,}267{,}329{,}306.12}\\ r&\approx 1{,}082.17\text{ mi}\end{aligned}$$

The diameter is therefore approximately $2\cdot 1{,}082.17\approx 2{,}164$ miles.

37. Its volume is less that the volume of a cylinder with the same height and radius, so it is some fractional part of $\pi r^2 h$, namely $\frac{1}{3}\pi r^2 h$.

39. The volume is four times as large because in the formula of the volume of a cone, the radius is squared.

$$V=\frac{1}{3}\pi r^2 h=\frac{1}{3}\cdot\pi\cdot(2r)^2\cdot h=\frac{1}{3}\pi\cdot 4\cdot r^2 h=4\cdot\left(\frac{1}{3}\pi r^2 h\right)$$

41. Let d be the diameter of the can. The height of the can is $3d$; the circumference of the can is $\pi d\approx 3.14d$, which is larger.

43. Because the volume of the tin can is to be 2 cubic feet, we set $\pi r^2 h=2$. Dividing both sides of this equation by πr^2, we get $h=\frac{2}{\pi r^2}$. Thus, the height depends on the choice of the radius. Recall the formula for finding the surface area of a tin can is $S=2\pi rh+2\pi r^2$. We can substitute $\frac{2}{\pi r^2}$ for h in this equation to get $S=2\pi r\frac{2}{\pi r^2}+2\pi r^2=\frac{4\pi r}{\pi r^2}+2\pi r^2=\frac{4}{r}+2\pi r^2$. Try various values for r.

43. (continued)

Radius: r	Surface Area: $S = \frac{4}{r} + 2\pi r^2$	
0.1	≈ 40.0628	
0.2	≈ 20.2512	decreasing
0.3	≈ 13.8985	decreasing
0.4	≈ 11.0048	decreasing
0.5	≈ 9.5700	decreasing
0.6	≈ 8.9275	decreasing
0.7	≈ 8.7915	decreasing
0.8	≈ 9.0192	increasing

The table indicates that we should start somewhere between 0.7 and 0.8. In actuality, the value should be between 0.6 and 0.7, so it would be best to start with a larger interval such as 0.6 to 0.8.

The exact value is $r = \frac{1}{\sqrt[3]{\pi}} \approx 0.6828$ ft.

45. No solution provided.

47. The smaller cubes have more surface area. For example, compare the surface area of 1 3-inch cube with the surface of 27 1-inch cubes.

49. The volume of the $4 \times 4 \times ¼$ hamburger is $4 \cdot 4 \cdot \frac{1}{4} = 4 \text{ in}^3$. The volume of the "tri-burger" will also be 4 in^3. This volume can be found from base area times height, where the height is ¼ in and the base is an equilateral triangle. Since the height is ¼ inch, the base area is 16 in^2. Since each side of an equilateral triangle has the same measure, let this measure be a. The perimeter of this triangle will be $3a$.

Applying Heron's formula we have $s = \frac{1}{2}(3a) = \frac{3a}{2}$ and

$$16 = \sqrt{\frac{3a}{2}\left(\frac{3a}{2} - a\right)\left(\frac{3a}{2} - a\right)\left(\frac{3a}{2} - a\right)}$$

$$16 = \sqrt{\frac{3a}{2} \cdot \frac{a}{2} \cdot \frac{a}{2} \cdot \frac{a}{2}}$$

$$16 = \sqrt{\frac{3a^4}{16}}$$

$$256 = \frac{3a^4}{16}$$

$$4{,}096 = 3a^4$$

$$\frac{4{,}096}{3} = a^4$$

$$a = \left(\frac{4{,}096}{3}\right)^{1/4} \approx 6.08 \text{ in.}$$

51. In this exercise, we are cutting off a cone that has a radius of $\frac{r}{2}$ and a height of $\frac{h}{2}$. The original cone has a surface area of $S = \pi r\sqrt{r^2 + h^2}$. The surface area that we are cutting off is

$$S = \pi\left(\frac{r}{2}\right)\sqrt{\left(\frac{r}{2}\right)^2 + \left(\frac{h}{2}\right)^2} = \pi\,\frac{r}{2}\sqrt{\frac{r^2}{4} + \frac{h^2}{4}} = \pi\,\frac{r}{2}\sqrt{\frac{r^2 + h^2}{4}} = \frac{1}{4}\pi r\sqrt{r^2 + h^2}.$$

The remaining surface area is therefore $\pi r\sqrt{r^2 + h^2} - \frac{1}{4}\pi r\sqrt{r^2 + h^2} = \frac{3}{4}\pi r\sqrt{r^2 + h^2}$.

53. No solution provided.

Section 10.5 The Metric System and Dimensional Analysis

1. Moving three columns to the right in the table gave us $10^3 = 1{,}000$ as many objects.

3. $\dfrac{1.6 \text{ kilometers}}{1 \text{ mile}}$

5. h; $\frac{1}{4} \text{ pound} \times \dfrac{454 \text{ grams}}{1 \text{ pound}} = 113.5 \text{ grams} = 1{,}135 \text{ dg}$

7. e; $15 \text{ feet} = 15 \text{ feet} \times \dfrac{1 \text{ yard}}{3 \text{ feet}} = 5 \text{ yards} \times \dfrac{1 \text{ meter}}{1.0936 \text{ yards}} \approx 4.572 \text{ meters} = 4{,}572 \text{ mm}$ or

$15 \text{ feet} = 15 \text{ feet} \times \dfrac{12 \text{ inches}}{1 \text{ foot}} = 180 \text{ inches} \times \dfrac{1 \text{ meter}}{39.37 \text{ inches}} \approx 4.572 \text{ meters} = 4{,}572 \text{ mm}$

9. g; $6 \text{ inches} \times \dfrac{1 \text{ meter}}{39.37 \text{ inches}} \approx 0.1524 \text{ meters} = 15.24 \text{ cm}$

11. a; $8.5 \text{ inches} \times \dfrac{1 \text{ meter}}{39.37 \text{ inches}} \approx 0.2159 \text{ meters} = 0.02159 \text{ dam}$

13. d; $6 \text{ ounces} = 6 \text{ ounces} \times \dfrac{1 \text{ quart}}{32 \text{ ounces}} = 0.1875 \text{ quarts} \times \dfrac{1 \text{ liter}}{1.0567 \text{ quarts}} \approx 0.177 \text{ liters} = 1.77 \text{ dl}$

15. 24,000 dl

17. 2,800 mm

19. 350 dL

The explanations to Exercises 21 – 25 may vary.

21. a; A juice glass is about one eighth of a quart or about one-eighth of a liter.

23. c; A nose is about 2 inches. Since 39.37 inches is about one meter or 100 centimeters, one inch is about $\dfrac{100}{39.37} \approx 2.5$ cm. Five centimeters would correspond to a two-inch nose.

25. b; The dog might be about two feet tall or about $\frac{2}{3}$ of a meter.

27. $18 \text{ meters} \times \frac{1.0936 \text{ yards}}{1 \text{ meter}} = 19.6848 \text{ yards} \times \frac{3 \text{ feet}}{1 \text{ yard}} \approx 59.05 \text{ feet}$

29. $3 \text{ kilograms} = 3000 \text{ grams} \times \frac{1 \text{ pound}}{454 \text{ grams}} \approx 6.608 \text{ pounds} \times \frac{16 \text{ ounces}}{1 \text{ pound}} \approx 105.73 \text{ ounces}$

31. $2.1 \text{ kiloliters} = 2{,}100 \text{ liters} \times \frac{1.0567 \text{ quarts}}{1 \text{ liter}} = 2{,}219.07 \text{ quarts} \times \frac{1 \text{ gallon}}{4 \text{ quarts}} \approx 554.77 \text{ gallons}$

33. $10{,}000 \text{ deciliters} = 1{,}000 \text{ liters} \times \frac{1.0567 \text{ quarts}}{1 \text{ liter}} = 1{,}056.7 \text{ quarts}$

35. $176 \text{ centimeters} = 1.76 \text{ meters} \times \frac{1.0936 \text{ yards}}{1 \text{ meter}} \approx 1.9247 \text{ yards} \times \frac{3 \text{ feet}}{1 \text{ yard}} = 5.7741 \text{ feet} \times \frac{12 \text{ inches}}{1 \text{ foot}} \approx$ 69.29 inches or $176 \text{ centimeters} = 1.76 \text{ meters} \times \frac{39.37 \text{ inches}}{1 \text{ meter}} \approx 69.29 \text{ inches}$

37. $45{,}000 \text{ kg} = 45{,}000{,}000 \text{ grams} \times \frac{1 \text{ pound}}{454 \text{ grams}} \approx 99{,}118.94 \text{ pounds} \times \frac{1 \text{ ton}}{2{,}000 \text{ pounds}} \approx 49.56 \text{ tons}$

39. $2.6 \text{ feet} \times \frac{1 \text{ yard}}{3 \text{ feet}} \approx 0.86667 \text{ yards} \times \frac{1 \text{ meter}}{1.0936 \text{ yards}} \approx 0.792 \text{ meters} = 7.92 \text{ decimeters}$ or

$2.6 \text{ feet} \times \frac{12 \text{ inches}}{1 \text{ foot}} = 31.2 \text{ inches} \times \frac{1 \text{ meter}}{39.37 \text{ inches}} \approx 0.792 \text{ meters} = 7.92 \text{ decimeters}$

41. Don't give him an inch.

43. It is first down and ten yards to go.

45. a) $V = lwh = 8 \cdot 5 \cdot 4 = 160$ cubic meters

b) $160 \text{ cubic meters} \times \frac{1{,}000 \text{ liters}}{1 \text{ cubic meter}} = 160{,}000 \text{ liters}$

c) $160{,}000 \text{ liters} \times \frac{1 \text{ kilogram}}{1 \text{ liter}} = 160{,}000 \text{ kilograms}$

47. $\frac{80\ km}{1\ hr} \times \frac{1\ mi}{1.609\ km} \approx 49.72\ mph$

49. $V = lwh = 40 \cdot 20 \cdot 6 = 4{,}800 \text{ cubic feet} \times \dfrac{1 \text{ cubic yard}}{3 \times 3 \times 3 \text{ cubic feet}} \approx 177.7778 \text{ cubic yards}$

We now convert to cubic meters.

$177.7778 \text{ cubic yards} \times \dfrac{1 \text{ cubic meter}}{1.0936 \times 1.0936 \times 1.0936 \text{ cubic yards}} \approx 135.9258 \text{ cubic meters}$

We now convert to kiloliters.

$135.9258 \text{ cubic meters} \times \dfrac{1 \text{ kiloliter}}{1 \text{ cubic meter}} \approx 135.93 \text{ kiloliters}$

51. $\dfrac{\$2.75}{1 \text{ kilogram}} \times \dfrac{1 \text{ kilogram}}{1{,}000 \text{ grams}} = \dfrac{\$2.75}{1{,}000 \text{ grams}} \times \dfrac{454 \text{ grams}}{1 \text{ pound}} = \dfrac{\$1{,}248.5}{1{,}000 \text{ pounds}} \approx \$1.25 \text{ per pound}$

53. $\dfrac{\$2.18}{1\,L} \times \dfrac{0.9464\,L}{1\,quart} \times \dfrac{4\,quart}{1\,gallon} \approx \$8.25 \text{ per gallon}$

55. Since $P = 2l + 2w$, we have $2 \cdot 62 + 2 \cdot 35 = 124 + 70 = 194$ feet of fencing. We need to convert this to meters. $194 \text{ feet} \times \dfrac{1 \text{ yard}}{3 \text{ feet}} \approx 64.6667 \text{ yards} \times \dfrac{1 \text{ meter}}{1.0936 \text{ yards}} \approx 59.13 \text{ meters}$. She will need to buy 60 meters.

57. $\dfrac{30 \text{ miles}}{1 \text{ gallon}} \times \dfrac{1 \text{ gallon}}{4 \text{ quarts}} \times \dfrac{1.0567 \text{ quarts}}{1 \text{ liter}} \times \dfrac{1.609 \text{ km}}{1 \text{ mile}} \approx \dfrac{51.0069 \text{ km}}{4 \text{ liters}} \approx 12.75 \text{ kilometers per liter}$

59. $\dfrac{\$8.00}{1 ft^2} \times \dfrac{9 ft^2}{1 yd^2} \times \dfrac{1\,yd}{0.9144\,m} \times \dfrac{1\,yd}{0.9144\,m} \approx \$86.11 \text{ per square meter}$

61. 65° C

$149 = \frac{9}{5}C + 32$

$117 = \frac{9}{5}C$

$585 = 9C$

$65 = C$

63. 140° F

$F = \frac{9}{5} \cdot 60 + 32$

$F = 9 \cdot 12 + 32$

$F = 108 + 32$

$F = 140$

65. 68° F

$F = \frac{9}{5} \cdot 20 + 32$

$F = 9 \cdot 4 + 32$

$F = 36 + 32$

$F = 68$

67. 45° C

$113 = \frac{9}{5}C + 32$

$81 = \frac{9}{5}C$

$405 = 9C$

$45 = C$

69. One hectare $= 100 \text{ m} \cdot 100 \text{ m} = 10{,}000$ square meters

71. The land that Thiep purchased is $0.75 \text{ km} \cdot 1.2 \text{ km} = 0.9 \text{ km}^2$. We know from Exercise 70 that one square kilometer is equal to 100 hectares. So, Thiep purchased 90 hectares.

73. 10

75. More decimeters because decimeters are smaller than hectometers.

77. Because dekagrams are much larger than milligrams, we would need fewer dekagrams. Therefore, we would move the decimal point to the left.

79. – 85. No solution provided.

Section 10.6 Geometric Symmetry and Tessellations

1. There are more rigid motions that can be applied to the star so that its beginning and ending position are the same.

3. The measure of an interior angle for the polygon is greater than $120°$, so if we have three or more polygons sharing a vertex, the angle sum at the vertex will exceed $360°$.

5. – 7.

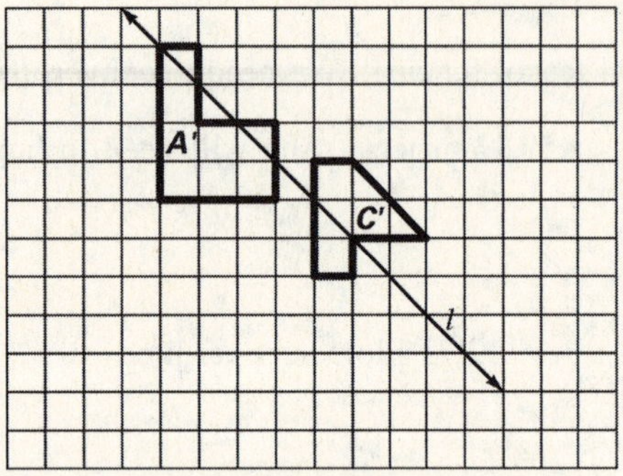

9.

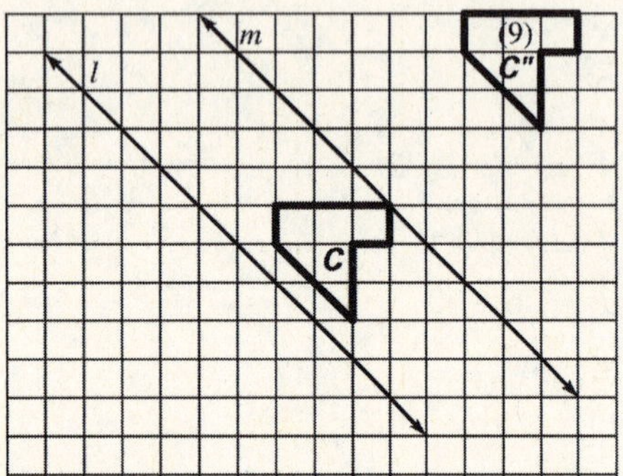

11.

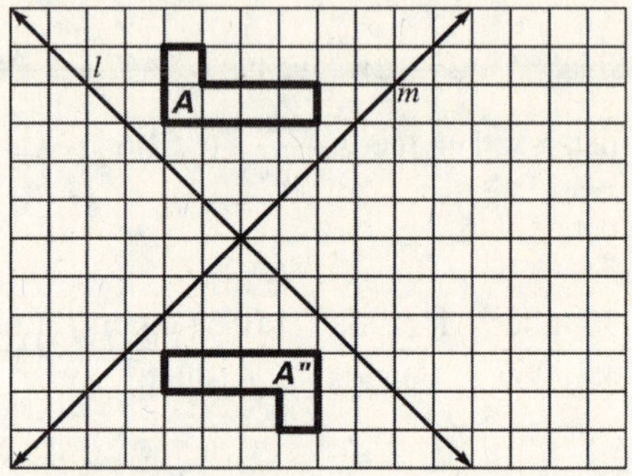

13. a) yes

b) The effect is the same as if we performed a translation.

15.

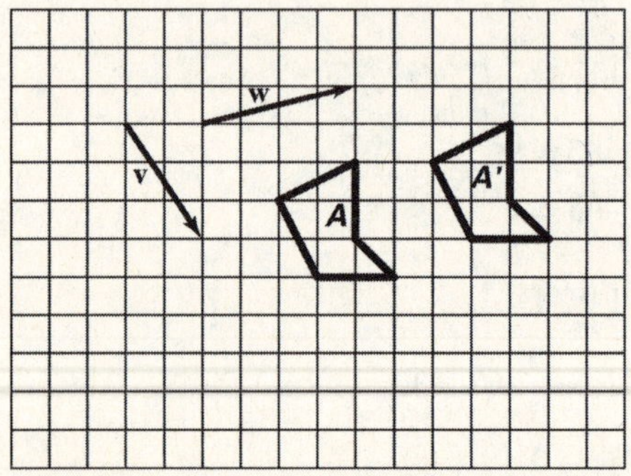

17.

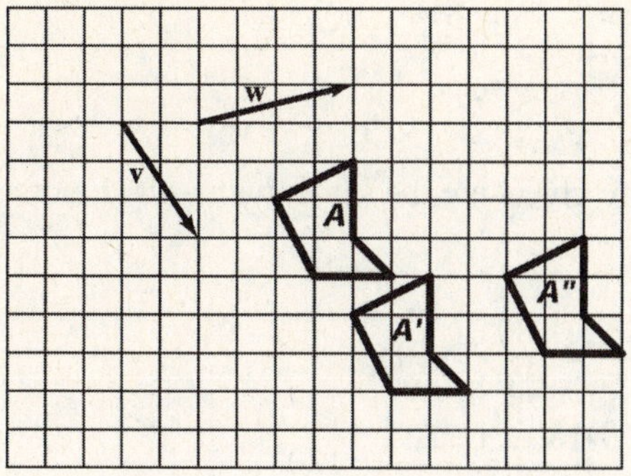

19.

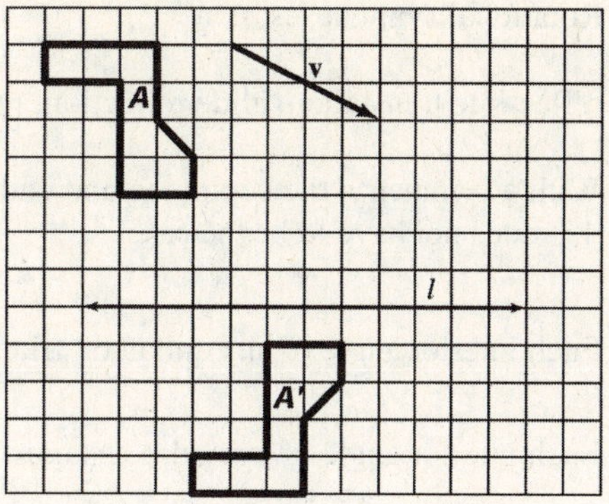

21.

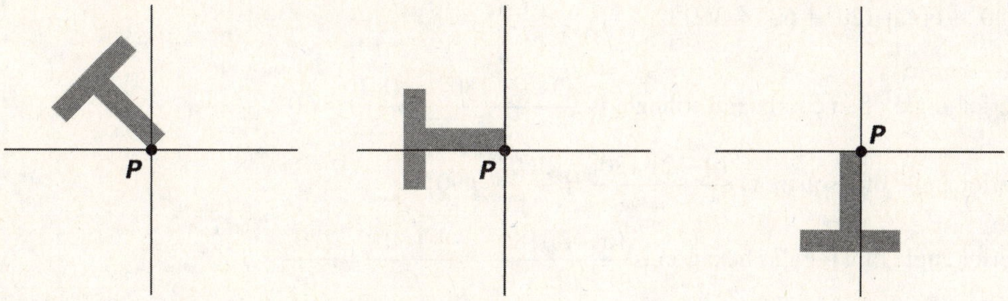

23.

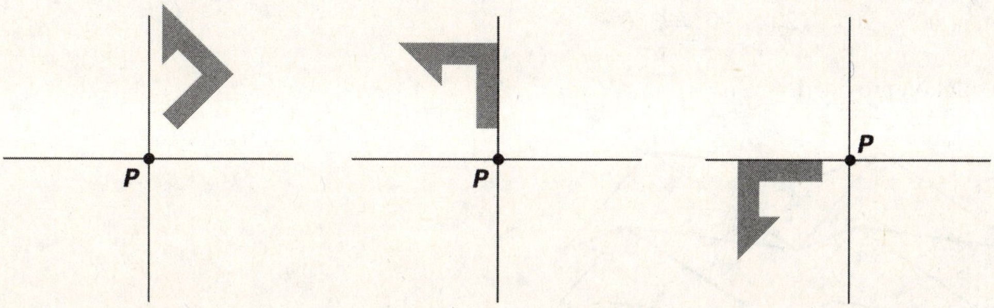

25. $(12-2)\cdot 180 = 10\cdot 180 = 1{,}800°$

$\frac{1{,}800}{12} = 150°$

27. Using the interior angles of a regular pentagon, we cannot obtain an angle sum of $360°$ around a point.

29.

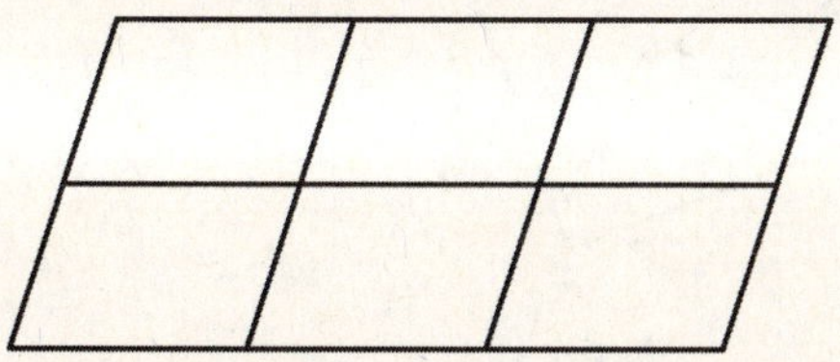

31. (c), (d)

33. (a), (e)

35. Reflectional symmetries: about a vertical line, about a horizontal line, and two diagonal lines
Rotational symmetries: $90°$, $180°$, and $270°$

37. Reflectional symmetries: about a vertical line
Rotational symmetries: none

39. Reflectional symmetries: about a vertical line
Rotational symmetries: none

41. reflection, translation, glide reflection, rotation

43. With a symmetry, the beginning and ending positions of the object are the same; with a rigid motion, this does not have to be the case.

45. Each interior angle of an equilateral triangle is $\frac{(3-2)\cdot 180°}{3}=\frac{180°}{3}=60°$.

Each interior angle of a regular hexagon is $\frac{(6-2)\cdot 180°}{6}=\frac{4\cdot 180°}{6}=\frac{720°}{6}=120°$.

At the points where the vertices of the equilateral triangles and the regular hexagons meet, the angle sum is $120°+60°+120°+60°=360°$.

47. Each interior angle of an equilateral triangle is $\frac{(3-2)\cdot 180°}{3}=\frac{180°}{3}=60°$.

Each interior angle of a square is $\frac{(4-2)\cdot 180°}{4}=\frac{360°}{4}=90°$.

Each interior angle of a regular hexagon is $\frac{(6-2)\cdot 180°}{6}=\frac{4\cdot 180°}{6}=\frac{720°}{6}=120°$.

At the points where the vertices of the equilateral triangle, squares, and the regular hexagon meet, the angle sum is $60°+90°+120°+90°=360°$.

49. – 51. No solution provided.

53.

55. In the figure, we constructed the perpendicular bisectors of segments EE' and CC'. The point where these perpendicular bisectors meet is the center of rotation.

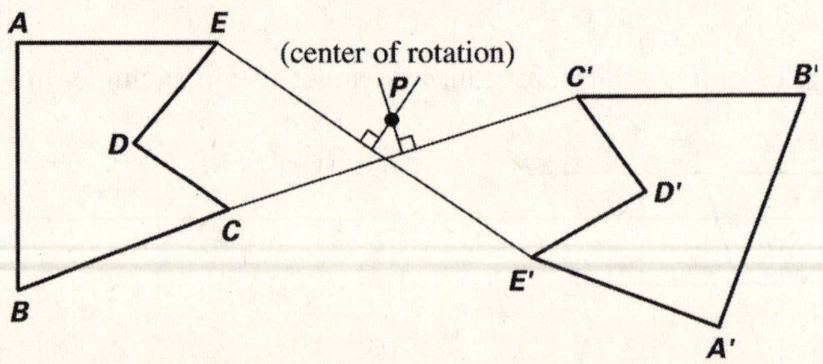

Section 10.7 Looking Deeper: Fractals

1. As the image of the cloud is magnified, the same type of pattern appears.

3. $3^D=4$; the log function

5. $4^5=1,024$

7. $\left(\frac{4}{3}\right)^{10} = \frac{1,048,576}{59,049} \approx 17.76$

9. a)

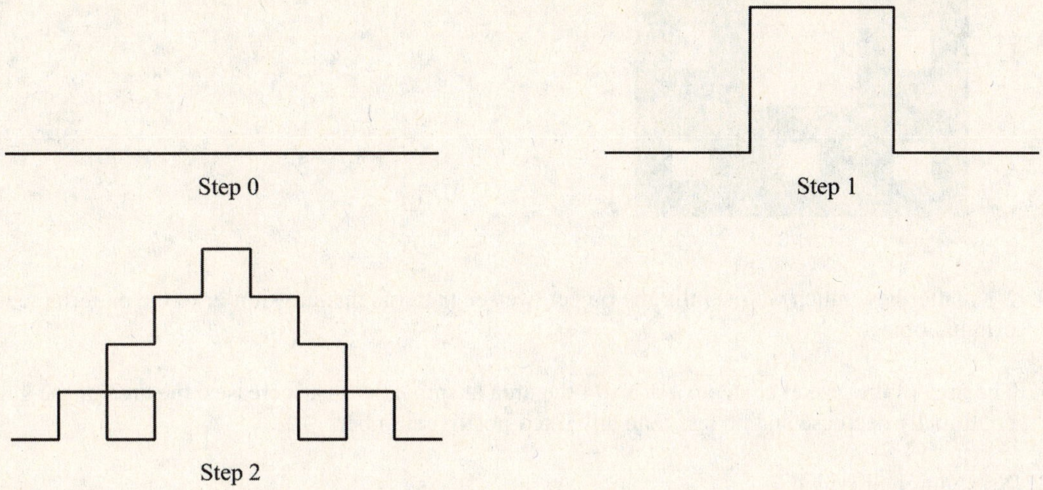

b) Length in step 5: $\left(\frac{5}{3}\right)^{5} = \frac{3,125}{243} \approx 12.86$

11.
$$\begin{aligned} 4^D &= 6 \\ \log 4^D &= \log 6 \\ D \log 4 &= \log 6 \\ D &= \frac{\log 6}{\log 4} \approx 1.29 \end{aligned}$$

13.
$$\begin{aligned} 3^D &= 5 \\ \log 3^D &= \log 5 \\ D \log 3 &= \log 5 \\ D &= \frac{\log 5}{\log 3} \approx 1.46 \end{aligned}$$

15.

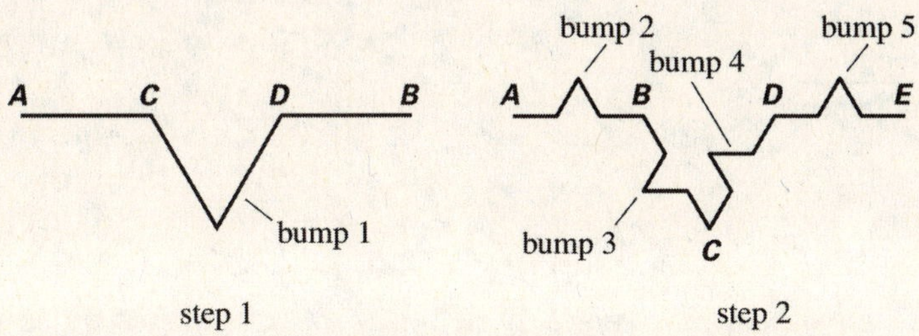

17.

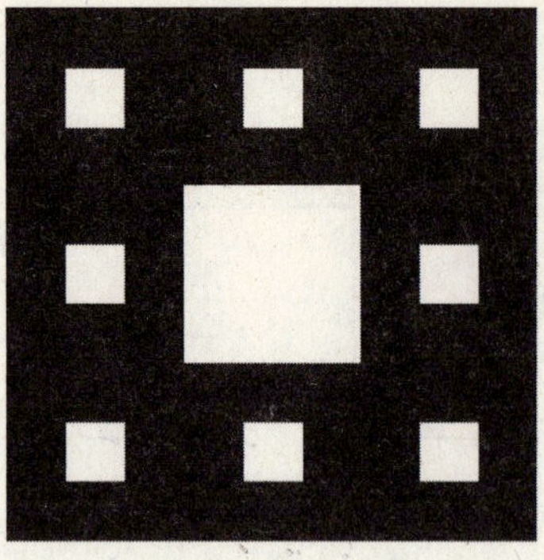

step 2

19. No matter how much we magnify the object, we see patterns that are similar to the patterns seen in the original object.

21. The area of the gasket at step $n + 1$ is 3/4 the area at step n, so as n increases, the area of the gasket will continually decrease and be less than any fixed positive number.

23. No solution provided.

25. 5″

27. At step 10: $\left(\frac{3}{4}\right)^{10} = \frac{59{,}049}{1{,}048{,}576} \approx 0.06$

At step n: $\left(\frac{3}{4}\right)^{n}$

29. No solution provided.

Chapter Review Exercises

1. a) b, g
   b) d, e

3.
$$\frac{\text{measure of central angle}}{360°} = \frac{\text{arc length}}{\text{circumference}}$$
$$\frac{m}{360°} = \frac{12-3}{24 \text{ in}}$$
$$\frac{m}{360°} = \frac{9}{24 \text{ in}}$$
$$24a = 360 \cdot 9$$
$$24a = 3{,}240$$
$$a = 135°$$

5. an arc of a great circle

7. $m\angle E = 45°$

$$\frac{EH}{8} = \frac{10}{12}$$
$$12(EH) = 10 \cdot 8$$
$$12(EH) = 80$$
$$EH = \frac{80}{12} = \frac{20}{3}$$

$$\frac{FG}{8} = \frac{6}{12}$$
$$12(FG) = 6 \cdot 8$$
$$12(FG) = 48$$
$$FG = \frac{48}{12} = 4$$

9. a) $A=\frac{1}{2}\cdot(b_1+b_2)\cdot h=\frac{1}{2}\cdot(20+12)\cdot 5=\frac{1}{2}\cdot(32)\cdot 5=16\cdot 5=80\text{ cm}^2$

b) $A=\frac{1}{2}\cdot h\cdot b=\frac{1}{2}\cdot 4\cdot 10=2\cdot 10=20\text{ in}^2$

11. a) $s=\frac{1}{2}(a+b+c)=\frac{1}{2}(10+8+4)=\frac{22}{2}=11$

$A=\sqrt{s(s-a)(s-b)(s-c)}=\sqrt{11(11-10)(11-8)(11-4)}=\sqrt{11\cdot 1\cdot 3\cdot 7}=\sqrt{231}\approx 15.2\text{ cm}^2$

b) $A=\frac{1}{2}\cdot h\cdot b$

$\sqrt{231}=\frac{1}{2}\cdot h\cdot 10$

$\sqrt{231}=5h$

$h=\frac{\sqrt{231}}{5}\approx 3.04\text{ cm}$

13. a) Area of base = $\frac{60}{360}\pi r^2\approx\frac{1}{6}\cdot 3.14\cdot 6^2=\frac{3.14\cdot 36}{6}=\frac{113.04}{6}=18.84\text{ in}^2$

Volume = area of the base times height $\approx 18.84\cdot 2=37.68\text{ in}^3$

b) Area of base = $A=\frac{1}{2}\cdot(b_1+b_2)\cdot h=\frac{1}{2}\cdot(12+10)\cdot 5=\frac{1}{2}\cdot(22)\cdot 5=11\cdot 5=55\text{ cm}^2$

Volume = area of the base times height = $55\cdot 6=330\text{ cm}^3$

15. The volume is four times as large, because in the formula of the volume of a cone, the radius is squared.

$$V=\frac{1}{3}\pi r^2h=\frac{1}{3}\cdot\pi\cdot(2r)^2\cdot h=\frac{1}{3}\pi\cdot 4\cdot r^2h=4\cdot\left(\frac{1}{3}\pi r^2h\right)$$

17. 514 decimeters $=51.4$ meters $\times\frac{1.0936\text{ yards}}{1\text{ meter}}\approx 56.21$ yards

19. $\frac{\$11.00}{1\,ft^2}\times\frac{9\,ft^2}{1\,yd^2}\times\frac{1\,yd}{0.9144\,m}\times\frac{1\,yd}{0.9144\,m}\approx\$118.40$ per square meter

21. a) c, e b) d

23.

25. zero

Chapter Test

1. a) vertical angles
   b) corresponding angles
   c) alternate exterior angles
   d) alternate interior angles

3. $m\angle c = 40°$; $m\angle b = 180° - 40° = 140°$;
   $m\angle a = 180° - (180° - 40° - 90°) = 180° - 50° = 130°$

5. a) Area of base: $s = \frac{1}{2}(a+b+c) = \frac{1}{2}(6+6+8) = \frac{20}{2} = 10$

   $A = \sqrt{s(s-a)(s-b)(s-c)} = \sqrt{10(10-6)(10-6)(10-8)} =$
   $\sqrt{10 \cdot 4 \cdot 4 \cdot 2} = \sqrt{320} \approx 17.9 \text{ cm}^2$
   Volume = area of the base times height $\approx 17.9 \cdot 4 = 71.6 \text{ cm}^3$

   b) Area of base = $A = \frac{1}{2} \cdot (b_1 + b_2) \cdot h = \frac{1}{2} \cdot (10+6) \cdot 2 = \frac{1}{2} \cdot (16) \cdot 2 = 8 \cdot 2 = 16 \text{ in}^2$
   Volume = area of the base times height = $16 \cdot 3 = 48 \text{ in}^3$

7. $$\frac{\text{measure of central angle}}{360°} = \frac{\text{arc length}}{\text{circumference}}$$
   $$\frac{180° - 150°}{360°} = \frac{a}{36 \text{ in}}$$
   $$\frac{30°}{360°} = \frac{a}{36 \text{ in}}$$
   $$360a = 30 \cdot 36$$
   $$360a = 1{,}080$$
   $$a = 3$$
   The arc length is 3 inches.

9. a) $A = \frac{1}{2} \cdot (b_1 + b_2) \cdot h = \frac{1}{2} \cdot (15+12) \cdot 2 = \frac{1}{2} \cdot (27) \cdot 2 = \frac{1}{2} \cdot 2 \cdot (27) = 27 \text{ in}^2$

   b) $A = \frac{1}{2} \cdot h \cdot b = \frac{1}{2} \cdot 4 \cdot 16 = 2 \cdot 16 = 32 \text{ cm}^2$

11. $m\angle F = 35°$
    $$\frac{FG}{12} = \frac{4}{6} \qquad \frac{GH}{10} = \frac{4}{6}$$
    $$6(FG) = 4 \cdot 12 \qquad 6(GH) = 10 \cdot 4$$
    $$6(FG) = 48 \qquad 6(GH) = 40$$
    $$FG = \frac{48}{6} = 8 \qquad GH = \frac{40}{6} = \frac{20}{3}$$

13. $4x + 5x = 90°$
    $9x = 90°$
    $x = 10°$

15. a) $s=\frac{1}{2}(a+b+c)=\frac{1}{2}(6+9+12)=\frac{27}{2}=13.5$

$$A=\sqrt{s(s-a)(s-b)(s-c)}=\sqrt{13.5(13.5-6)(13.5-9)(13.5-12)}$$
$$=\sqrt{(13.5)\cdot(7.5)\cdot(4.5)\cdot(1.5)}=\sqrt{683.4375}\approx 26.1 \text{ ft}^2$$

b) $A=\frac{1}{2}\cdot h\cdot b$

$$\sqrt{683.4375}=\frac{1}{2}\cdot h\cdot 12$$
$$\sqrt{683.4375}=6h$$
$$h=\frac{\sqrt{683.4375}}{6}\approx 4.4 \text{ ft}$$

17. a) 24 meters; b) 3,460,000 milligrams; c) 2,140 centiliters

19. $18 \text{ yards} \times \frac{1 \text{ meter}}{1.0936 \text{ yards}} \approx 16.459 \text{ meters} = 164.59 \text{ decimeters}$

21. Reflectional symmetries about lines through *AE, BF, CG, DH*; rotational symmetries through 0°, 90°, 180°, and 270°.

Chapter 11
Apportionment: How Do We Measure Fairness?

Section 11.1 Understanding Apportionment

1. The nine board members that had been previously assigned were reassigned.

3.

State	Number of customer %	Step 1: Determine exact number of seats deserved	Step 2: Assign integer parts	Examine fractional parts	Step 3: Assign 1 additional seat
California	56,000 (37.6%)	$0.376 \times 11 = 4.136$	4	0.136	4
Arizona	52,000 (34.9%)	$0.349 \times 11 = 3.839$	3	0.839	4
Nevada	41,000 (27.5%)	$0.275 \times 11 = 3.025$	3	0.025	3
Total	149,000 (100%)	11	10		11

5.

Guild	Number of members %	Step 1: Determine exact number of booths	Step 2: Assign integer parts	Examine fractional parts	Step 3: Assign 2 additional booths
Painters	87 (46.8%)	$0.468 \times 31 = 14.508$	14	0.508	14
Sculptures	46 (24.7%)	$0.247 \times 31 = 7.657$	7	0.657	8
Weavers	53 (28.5%)	$0.285 \times 31 = 8.835$	8	0.835	9
Total	186 (100%)	31	29		31

7.

State	Number of residents %	Step 1: Determine exact number of Representative	Step 2: Assign integer parts	Examine fractional parts	Step 3: Assign 1 additional Representative
Alabama	4,041 (37.3%)	$0.373 \times 19 = 7.087$	7	0.087	7
Mississippi	2,573 (23.7%)	$0.237 \times 19 = 4.503$	4	0.503	5
Louisiana	4,220 (39.0%)	$0.390 \times 19 = 7.410$	7	0.410	7
Total	10,834 (100%)	19	18		19

9. $$\text{average constituency} = \frac{\text{population}}{\text{number of representatives}} = \frac{177{,}408}{3} = 59{,}136$$

11. A is more poorly represented.

$$\text{average constituency of A} = \frac{\text{population}}{\text{number of representatives}} = \frac{27{,}600}{16} = 1{,}725$$

$$\text{average constituency of B} = \frac{\text{population}}{\text{number of representatives}} = \frac{23{,}100}{14} = 1{,}650$$

The absolute unfairness is $1{,}725 - 1{,}650 = 75$

$$\text{relative unfairness} = \frac{\text{absolute unfairness}}{\text{smaller average consituency}} = \frac{75}{1{,}650} \approx 0.045$$

13. $\text{average constituency of Naxxon} = \dfrac{\text{stockholders}}{\text{number of members}} = \dfrac{4,700}{5} = 940$

$\text{average constituency of Aroco} = \dfrac{\text{stockholders}}{\text{number of members}} = \dfrac{3,700}{4} = 925$

The absolute unfairness is 940 – 925 = 15.

$\text{relative unfairness} = \dfrac{\text{absolute unfairness}}{\text{smaller average consituency}} = \dfrac{15}{925} \approx 0.016$

15. $\text{average constituency of Colorado} = \dfrac{\text{population}}{\text{number of representatives}} = \dfrac{2,200,000}{5} = 440,000$

$\text{average constituency of Delaware} = \dfrac{\text{population}}{\text{number of representatives}} = \dfrac{600,000}{1} = 600,000$

The absolute unfairness is 600,000 – 440,000 = 160,000

$\text{relative unfairness} = \dfrac{\text{absolute unfairness}}{\text{smaller average consituency}} = \dfrac{160,000}{440,000} \approx 0.364$

17. a)

Division	Number of employees %	Step 1: Determine exact number of council members	Step 2: Assign integer parts	Examine fractional parts	Step 3: Assign 2 additional members
(D)igital	140 (54.9%)	$0.549 \times 12 = 6.588$	6	0.588	7
(C)omputers	85 (33.3%)	$0.333 \times 12 = 3.996$	3	0.996	4
(B)usiness Products	30 (11.8%)	$0.118 \times 12 = 1.416$	1	0.416	1
Total	255 (100%)	12	10		12

b)

Division	Number of employees %	Step 1: Determine exact number of council members	Step 2: Assign integer parts	Examine fractional parts	Step 3: Assign 1 additional member
(D)igital	140 (54.9%)	$0.549 \times 13 = 7.137$	7	0.137	7
(C)omputers	85 (33.3%)	$0.333 \times 13 = 4.329$	4	0.329	4
(B)usiness Products	30 (11.8%)	$0.118 \times 13 = 1.534$	1	0.534	2
Total	255 (100%)	13	12		13

Division	Number of employees %	Step 1: Determine exact number of council members	Step 2: Assign integer parts	Examine fractional parts	Step 3: Assign 2 additional members
(D)igital	140 (54.9%)	$0.549 \times 14 = 7.686$	7	0.686	8
(C)omputers	85 (33.3%)	$0.333 \times 14 = 4.662$	4	0.662	5
(B)usiness Products	30 (11.8%)	$0.118 \times 14 = 1.652$	1	0.652	1
Total	255 (100%)	14	12		14

c) Business Products loses a seat when the council size is increased from 13 to 14.

19. a)

Division	Number of patients %	Step 1: Determine exact number of doctors	Step 2: Assign integer parts	Examine fractional parts	Step 3: Assign 1 additional doctor
Center City	213 (37%)	$0.37\times 23 = 8.51$	8	0.51	8
South Street	65 (11%)	$0.11\times 23 = 2.53$	2	0.53	3
West Side	300 (52%)	$0.52\times 23 = 11.96$	11	0.96	12
Total	100 (100%)	23	21		23

b) In going from 31 to 32, South Street loses a physician.

Division	Number of patients %	Step 1: Determine exact number of doctors	Step 2: Assign integer parts	Examine fractional parts	Step 3: Assign 1 additional doctor
Center City	213 (37%)	$0.37\times 31 = 11.42$	11	0.42	11
South Street	65 (11%)	$0.11\times 31 = 3.49$	3	0.49	4
West Side	300 (52%)	$0.52\times 31 = 16.09$	16	0.09	16
Total	100 (100%)	31	30		31

Division	Number of patients %	Step 1: Determine exact number of doctors	Step 2: Assign integer parts	Examine fractional parts	Step 3: Assign 1 additional doctor
Center City	213 (37%)	$0.37\times 32 = 11.79$	11	0.79	12
South Street	65 (11%)	$0.11\times 32 = 3.60$	3	0.60	3
West Side	300 (52%)	$0.52\times 32 = 16.61$	16	0.61	17
Total	100 (100%)	32	30		32

21. Once representatives have been assigned, do not reassign them at a later time.

23. Absolute unfairness is the difference in the average constituencies; it does not take into account the size of the constituencies as relative unfairness does.

25. – 27. No solution provided.

29. a) $\text{average constituency of Naxxon} = \dfrac{\text{stockholders}}{\text{number of members}} = \dfrac{4,700}{7} \approx 671$

$$\text{average constituency of Eurobile} = \frac{\text{stockholders}}{\text{number of members}} = \frac{1,600}{2} = 800$$

The absolute unfairness is $800 - 671 = 129$.

$$\text{relative unfairness} = \frac{\text{absolute unfairness}}{\text{smaller average consituency}} = \frac{129}{671} \approx 0.192$$

29. (continued)

b) $\text{average constituency of Naxxon} = \dfrac{\text{stockholders}}{\text{number of members}} = \dfrac{4,700}{6} \approx 783$

$\text{average constituency of Eurobile} = \dfrac{\text{stockholders}}{\text{number of members}} = \dfrac{1,600}{3} \approx 533$

The absolute unfairness is 783 – 533 = 250.

$\text{relative unfairness} = \dfrac{\text{absolute unfairness}}{\text{smaller average consituency}} = \dfrac{250}{533} \approx 0.469$

c) Giving the extra representative to Naxxon will result in a smaller unfairness.

Section 11.2 The Huntington–Hill Apportionment Principle

1. A smaller relative unfairness occurs if the next representative is given to B.

3. It was assigned to the company with the highest Huntington–Hill number.

5. a) $\text{average constituency of Musicians' Guild} = \dfrac{\text{members}}{\text{number of representatives}} = \dfrac{908}{5} = 181.6$

$\text{average constituency of Artists' Alliance} = \dfrac{\text{members}}{\text{number of representatives}} = \dfrac{633}{3} = 211$

$$\text{relative unfairness} = \dfrac{\text{larger average consituency} - \text{smaller average consituency}}{\text{smaller average consituency}}$$
$$= \dfrac{211 - 181.6}{181.6} \approx 0.162$$

Note: If you round the average constituency of the Musicians' Guild to the nearest whole person, it would be approximately 182. This implies the relative unfairness would be approximately 0.159.

b) $\text{average constituency of Musicians' Guild} = \dfrac{\text{members}}{\text{number of representatives}} = \dfrac{908}{4} = 227$

$\text{average constituency of Artists' Alliance} = \dfrac{\text{members}}{\text{number of representatives}} = \dfrac{633}{4} \approx 158$

$$\text{relative unfairness} = \dfrac{\text{larger average consituency} - \text{smaller average consituency}}{\text{smaller average consituency}}$$
$$= \dfrac{227 - 158}{158} \approx 0.437$$

Note: If you don't round the average constituency of the Artists' Alliance to the nearest whole person, it would be 156.25. The relative unfairness would be approximately 0.434.

c) Giving the additional representative to Musicians' Guild will result in a smaller unfairness.

7. a) $\text{average number of passengers on red line} = \dfrac{\text{passengers}}{\text{number of cars}} = \dfrac{405}{10} = 40.5$

$\text{average number of passengers on blue line} = \dfrac{\text{passengers}}{\text{number of cars}} = \dfrac{287}{7} = 41$

$\text{relative unfairness} = \dfrac{\text{larger average} - \text{smaller average}}{\text{smaller average}} = \dfrac{41 - 40.5}{40.5} \approx 0.012$

b) $\text{average number of passengers on red line} = \dfrac{\text{passengers}}{\text{number of cars}} = \dfrac{405}{9} = 45$

$\text{average number of passengers on blue line} = \dfrac{\text{passengers}}{\text{number of cars}} = \dfrac{287}{8} = 35.875$

$\text{relative unfairness} = \dfrac{\text{larger average} - \text{smaller average}}{\text{smaller average}} = \dfrac{45 - 35.875}{35.875} \approx 0.254$

c) Giving the additional car to red line will result in a smaller unfairness.

9. If Alabama receives the additional representative

$\text{average constituency of Alabama} = \dfrac{\text{population}}{\text{number of representatives}} = \dfrac{4,461,130}{8} \approx 557,641$

$\text{average constituency of New York} = \dfrac{\text{population}}{\text{number of representatives}} = \dfrac{19,004,973}{29} \approx 655,344$

$$\text{relative unfairness} = \dfrac{\text{larger average consituency} - \text{smaller average consituency}}{\text{smaller average consituency}}$$

$$= \dfrac{655,344 - 557,641}{557,641} \approx 0.175$$

If New York receives the additional representative

$\text{average constituency of Alabama} = \dfrac{\text{population}}{\text{number of representatives}} = \dfrac{4,461,130}{7} = 637,304$

$\text{average constituency of New York} = \dfrac{\text{population}}{\text{number of representatives}} = \dfrac{19,004,973}{30} \approx 633,499$

$$\text{relative unfairness} = \dfrac{\text{larger average consituency} - \text{smaller average consituency}}{\text{smaller average consituency}}$$

$$= \dfrac{637,304 - 633,499}{633,499} \approx 0.006$$

Giving the additional representative to New York will result in a smaller unfairness.

11. If Colorado receives the additional representative

$$\text{average constituency of Colorado} = \frac{\text{population}}{\text{number of representatives}} = \frac{4,311,882}{8} \approx 538,985$$

$$\text{average constituency of Delaware} = \frac{\text{population}}{\text{number of representatives}} = \frac{758,068}{1} = 666,168$$

$$\text{relative unfairness} = \frac{\text{larger average consituency} - \text{smaller average consituency}}{\text{smaller average consituency}}$$

$$= \frac{666,168 - 538,985}{538,985} \approx 0.236$$

If Delaware receives the additional representative

$$\text{average constituency of Colorado} = \frac{\text{population}}{\text{number of representatives}} = \frac{4,311,882}{7} \approx 615,983$$

$$\text{average constituency of Delaware} = \frac{\text{population}}{\text{number of representatives}} = \frac{758,068}{2} = 379,034$$

$$\text{relative unfairness} = \frac{\text{larger average consituency} - \text{smaller average consituency}}{\text{smaller average consituency}}$$

$$= \frac{615,983 - 379,034}{379,034} \approx 0.625$$

Giving the additional representative to Colorado will result in a smaller unfairness.

13. $$\text{Huntington-Hill number} = \frac{(\text{population } X)^2}{x(x+1)} = \frac{908^2}{4 \cdot 5} = 41,223.2$$

15. $$\text{Huntington-Hill number} = \frac{(\text{population } X)^2}{x(x+1)} = \frac{287^2}{7 \cdot 8} \approx 1,470.9$$

17. $$\text{Huntington-Hill number} = \frac{(\text{population } X)^2}{x(x+1)} = \frac{4.5^2}{7 \cdot 8} \approx 0.362$$

19. $$\text{Huntington-Hill number} = \frac{(\text{population } X)^2}{x(x+1)} = \frac{4.3^2}{7 \cdot 6} \approx 0.330$$

In Exercises 21 – 23, the state populations will be reported in millions.

21. $$\text{Huntington-Hill number for Indiana} = \frac{(\text{population } X)^2}{x(x+1)} = \frac{6.09^2}{9 \cdot 10} \approx 0.412$$

$$\text{Huntington-Hill number for Illinois} = \frac{(\text{population } X)^2}{x(x+1)} = \frac{12.44^2}{19 \cdot 20} \approx 0.407$$

Since $0.407 < 0.412$, Indiana should receive the additional representative.

23. Huntington-Hill number for Alaska $= \dfrac{(\text{population } X)^2}{x(x+1)} = \dfrac{0.63^2}{1 \cdot 2} \approx 0.198$

Huntington-Hill number for New Hampshire $= \dfrac{(\text{population } X)^2}{x(x+1)} = \dfrac{1.24^2}{2 \cdot 3} \approx 0.256$

Huntington-Hill number for Wyoming $= \dfrac{(\text{population } X)^2}{x(x+1)} = \dfrac{0.51^2}{1 \cdot 2} \approx 0.13$

Since $0.13 < 0.198 < 0.256$, New Hampshire should receive the additional representative.

25. Huntington-Hill number for Musicians' Guild $= \dfrac{(\text{population } X)^2}{x(x+1)} = \dfrac{908^2}{4 \cdot 5} \approx 41,223$

Huntington-Hill number for Artists' Alliance $= \dfrac{(\text{population } X)^2}{x(x+1)} = \dfrac{633^2}{3 \cdot 4} \approx 33,391$

Huntington-Hill number for Actors' Coalition $= \dfrac{(\text{population } X)^2}{x(x+1)} = \dfrac{420^2}{2 \cdot 3} = 29,400$

Since $29400 < 33391 < 41223$, the Musicians' Guild should receive the additional representative.

27. Huntington-Hill number for red line $= \dfrac{(\text{population } X)^2}{x(x+1)} = \dfrac{405^2}{9 \cdot 10} = 1,822.5$

Huntington-Hill number for blue line $= \dfrac{(\text{population } X)^2}{x(x+1)} = \dfrac{287^2}{7 \cdot 8} = 1,470.875$

Huntington-Hill number for yellow line $= \dfrac{(\text{population } X)^2}{x(x+1)} = \dfrac{156^2}{3 \cdot 4} = 2,028$

Since 1,470.875 < 1,822.5 < 2,028 the yellow line should receive the additional car.

29. Huntington-Hill number for Musicians' Guild $= \dfrac{(\text{population } X)^2}{x(x+1)} = \dfrac{908^2}{5 \cdot 6} \approx 27,482$

Huntington-Hill number for Artists' Alliance $= \dfrac{(\text{population } X)^2}{x(x+1)} = \dfrac{633^2}{3 \cdot 4} \approx 33,391$

Huntington-Hill number for Actors' Coalition $= \dfrac{(\text{population } X)^2}{x(x+1)} = \dfrac{420^2}{2 \cdot 3} = 29,400$

Since 27,482 < 29,400 < 33,391 the Artists' Alliance should receive the second additional representative.

31. According to the assignment of the 11 representatives in Exercise 29, the Musicians' Guild has 5; the Artists' Alliance has 4; and the Actors Coalition has 2. This is consistent with the Hamilton apportionment method

Groups	Number of members %	Step 1: Determine exact number of representatives	Step 2: Assign integer parts	Examine fractional parts	Step 3: Assign additional representative
Musicians' Guild	908 (46.30%)	$0.4630 \times 11 = 5.0930$	5	0.0930	5
Artists' Alliance	633 (32.28%)	$0.3228 \times 11 = 3.5508$	3	0.5508	4
Actors' Coalition	420 (21.42%)	$0.2142 \times 11 = 2.3562$	2	0.3562	2
Total	1961 (100%)	11	10		11

33. Assign representatives so as to give the smallest relative unfairness.

35. After assigning one representative to each state, the rest of the representatives are assigned to the states according to the size of their Huntington–Hill numbers, in decreasing order, starting with the largest.

37. This would be similar to the situation that Michigan has 15 representatives and Wisconsin has 9 and we want to assign one more representative.

$$\text{Huntington-Hill number for Michigan} = \frac{(\text{population } X)^2}{x(x+1)} = \frac{9.3^2}{15 \cdot 16} \approx 0.360$$

$$\text{Huntington-Hill number for Wisconsin} = \frac{(\text{population } X)^2}{x(x+1)} = \frac{X^2}{9 \cdot 10} = \frac{X^2}{90}$$

$0.360 < \frac{X^2}{90} \Rightarrow 32.4 < X^2 \Rightarrow 5.6921 < X$ Therefore, Wisconsin's population needs to be approximately 5.7 million in order to take a seat away from Michigan.

39. No solution provided.

Section 11.3 Applications of the Apportionment Principle

1. Naxxon had the largest Huntington–Hill number.

3. We use the entries in row 1. Since the carpenters have the largest number, they get the fourth seat.

5. We look at the entries that read 104.2, 162.0, and 160.2. Since the plumbers have the largest number, they get the fifth seat.

7. We use the entries in row 1. Since the Carney building has the largest number, they get the fourth seat.

9. We look at the entries that read 264.5, 181.5, and 308.2. Since the Carney building has the largest number, they get the sixth seat.

11. Theater, Music, and Dance each get one representative in no particular order. The remaining seven assignments in order would be MTMTMTD.

13. a)

Boroughs	Number of residents %	Step 1: Determine exact number of seats	Step 2: Assign integer parts	Examine fractional parts	Step 3: Assign 2 additional seats
Bronx	1.4 (26.9%)	$0.269 \times 10 = 2.69$	2	0.69	3
Manhattan	1.6 (30.8%)	$0.308 \times 10 = 3.08$	3	0.08	3
Queens	2.2 (42.3%)	$0.423 \times 10 = 4.23$	4	0.23	4
Total	5.2 (100%)	10	9		10

b) Make the following table.

Current Representation	(B)ronx	(M)anhattan	(Q)ueens
1	$\frac{1.4^2}{1 \cdot 2} = 0.98$	$\frac{1.6^2}{1 \cdot 2} = 1.28$	$\frac{2.2^2}{1 \cdot 2} = 2.42$
2	$\frac{1.4^2}{2 \cdot 3} \approx 0.33$	$\frac{1.6^2}{2 \cdot 3} \approx 0.43$	$\frac{2.2^2}{2 \cdot 3} \approx 0.81$
3	$\frac{1.4^2}{3 \cdot 4} \approx 0.16$	$\frac{1.6^2}{3 \cdot 4} \approx 0.21$	$\frac{2.2^2}{3 \cdot 4} \approx 0.40$
4	$\frac{1.4^2}{4 \cdot 5} \approx 0.10$	$\frac{1.6^2}{4 \cdot 5} = 0.13$	$\frac{2.2^2}{4 \cdot 5} \approx 0.24$

Bronx, Manhattan, and Queens each get one representative in no particular order. The remaining seven assignments in order would be QMBQMQB. So Bronx should have 3 representatives, Manhattan should have 3, and Queens should have 4, according to the Huntington–Hill apportionment principle.

15. Make the following table.

Current Representation	Utah	Idaho	Oregon
1	$\frac{2.4^2}{1 \cdot 2} = 2.880$	$\frac{1.5^2}{1 \cdot 2} = 1.125$	$\frac{3.6^2}{1 \cdot 2} = 6.480$
2	$\frac{2.4^2}{2 \cdot 3} \approx 0.960$	$\frac{1.5^2}{2 \cdot 3} = 0.375$	$\frac{3.6^2}{2 \cdot 3} \approx 2.160$
3	$\frac{2.4^2}{3 \cdot 4} \approx 0.480$	$\frac{1.5^2}{3 \cdot 4} \approx 0.188$	$\frac{3.6^2}{3 \cdot 4} \approx 1.08$
4	$\frac{2.4^2}{4 \cdot 5} = 0.288$	$\frac{1.5^2}{4 \cdot 5} \approx 0.113$	$\frac{3.6^2}{4 \cdot 5} = 0.648$

Utah, Idaho, and Oregon each get one representative in no particular order. If we assume "UT" represents Utah, "ID" represents Idaho, and "OR" represents Oregon then the remaining seven assignments in order would be OR, UT, OR, ID, OR, UT, OR. So Utah would have 3 representatives, Idaho would have 2, and Oregon would have 5 according to the Huntington–Hill method.

17. Make the following table.

Current Representation	Arkansas	Kansas	Nebraska
1	$\frac{2.75^2}{1\cdot 2} \approx 3.781$	$\frac{2.76^2}{1\cdot 2} \approx 3.809$	$\frac{1.8^2}{1\cdot 2} = 1.620$
2	$\frac{2.75^2}{2\cdot 3} \approx 1.260$	$\frac{2.76^2}{2\cdot 3} \approx 1.270$	$\frac{1.8^2}{2\cdot 3} = 0.540$
3	$\frac{2.75^2}{3\cdot 4} \approx 0.630$	$\frac{2.76^2}{3\cdot 4} \approx 0.635$	$\frac{1.8^2}{3\cdot 4} = 0.270$
4	$\frac{2.75^2}{4\cdot 5} \approx 0.378$	$\frac{2.76^2}{4\cdot 5} \approx 0.381$	$\frac{1.8^2}{4\cdot 5} = 0.162$
5	$\frac{2.75^2}{5\cdot 6} \approx 0.252$	$\frac{2.76^2}{5\cdot 6} \approx 0.254$	$\frac{1.8^2}{5\cdot 6} = 0.108$

Arkansas, Kansas, and Nebraska each get one representative in no particular order. If we assume "AR" represents Arkansas, "KS" represents Kansas, and "NE" represents Nebraska then the remaining eight assignments in order would be KS, AR, NE, KS, AR, KS, AR, NE. So Arkansas would have 4 representatives, Kansas would have 4, and Nebraska would have 3 according to the Huntington–Hill method.

19. BB; We need to find one more entry for row 6 for Business. That entry is $\frac{40^2}{6\cdot 7} \approx 38.1$. By comparing 45.0, 33.3, and 38.1 we can see that the twelfth representative should be assigned to H. Humanities would have 4 representatives, Science would have 2, and Business would have 5.

21. Make the following table.

Current Representation	Region 1	Region 2	Region 3
1	$\frac{123^2}{1\cdot 2} \approx 7565$	$\frac{44^2}{1\cdot 2} = 968$	$\frac{79^2}{1\cdot 2} \approx 3121$
2	$\frac{123^2}{2\cdot 3} \approx 2522$	$\frac{44^2}{2\cdot 3} \approx 323$	$\frac{79^2}{2\cdot 3} \approx 1040$
3	$\frac{123^2}{3\cdot 4} \approx 1261$	$\frac{44^2}{3\cdot 4} \approx 161$	$\frac{79^2}{3\cdot 4} \approx 520$
4	$\frac{123^2}{4\cdot 5} \approx 756$	$\frac{44^2}{4\cdot 5} \approx 97$	$\frac{79^2}{4\cdot 5} \approx 312$

Region 1, region 2, and region 3 each get one representative in no particular order. If we assume "A" represents region 1, "B" represents region 2, and "C" represents region 3, then the remaining four assignments in order would be ACAA. So region 1 would have 4 representatives, region 2 would have 1, and region 3 would have 2, according to the Huntington–Hill method.

23. Calculate a table of Huntington–Hill numbers.

25. No solution provided.

27. 8 representatives are needed. In order to do this, A would have 2 representatives and C would have 3.

29. No solution provided.

Section 11.4 Other Paradoxes and Apportionment Methods

1. The standard divisor is the total population divided by the number of representatives; the standard quota is the state's population divided by the standard divisor.

3. by trial and error

5. a) $\text{Standard Divisor} = \dfrac{\text{Total Population}}{\text{Number of Representatives Allocated}} = \dfrac{512,000}{16} = 32,000$

$$\text{Standard Quota} = \frac{\text{State's Population}}{\text{Standard Divisor}} = \frac{84,160}{32,000} = 2.63$$

b) 2.63 rounded down is 2.

c) 2.63 rounded up is 3.

d) 2.63 rounded as usual is 3.

7. a) $\text{Standard Divisor} = \dfrac{\text{Total Population}}{\text{Number of Representatives Allocated}} = \dfrac{102,000}{11} = 9,272.73$

$$\text{Standard Quota} = \frac{\text{State's Population}}{\text{Standard Divisor}} = \frac{19,975}{9,272.73} = 2.15$$

b) 2.15 rounded down is 2.

c) 2.15 rounded up is 3.

d) 2.15 rounded as usual is 2.

9. The total population considered is $56,000 + 52,000 + 41,000 = 149,000$.

$\text{Standard Divisor} = \dfrac{149,000}{11} \approx 13,545.45$; California's Standard Quota $= \dfrac{56,000}{13,545.45} \approx 4.134$;

Nevada's Standard Quota $= \dfrac{41,000}{13,545.45} \approx 3.027$; Arizona's Standard Quota $= \dfrac{52,000}{13,545.45} \approx 3.839$

11. With the standard divisor of 13,545.45, a modified divisor is 14,000.

	California	Nevada	Arizona
Number of Customers	56,000	41,000	52,000
Standard Quota	4.134	3.027	3.839
Modified Quota	$\dfrac{56,000}{14,000} = 4$	$\dfrac{41,000}{14,000} \approx 2.93$	$\dfrac{52,000}{14,000} \approx 3.71$
Round Modified Quota Up	**4**	**3**	**4**
			Total = 11

13. The total number of workers considered is $213+273+178=664$.

Standard Divisor $= \dfrac{664}{20} = 33.2$; Performers' Standard Quota $= \dfrac{213}{33.2} \approx 6.416$;

Food Workers' Standard Quota $= \dfrac{273}{33.2} \approx 8.223$;

Maintenance Workers' Standard Quota $= \dfrac{178}{33.2} \approx 5.361$

15. With the standard divisor of 33.2, a modified divisor is 35.

	Performers	Food Workers	Maintenance
Number of Workers	213	273	178
Standard Quota	6.416	8.223	5.361
Modified Quota	$\frac{213}{35} \approx 6.09$	$\frac{273}{35} = 7.8$	$\frac{178}{35} \approx 5.09$
Round Modified Quota Up	7	8	6
			Total = 21

This total was too high, so try a modified divisor of 36.

	Performers	Food Workers	Maintenance
Number of Workers	213	273	178
Modified Quota	$\frac{213}{36} \approx 5.92$	$\frac{273}{36} \approx 7.58$	$\frac{178}{36} \approx 4.94$
Round Modified Quota Up	6	8	5
			Total = 19

This total was too low, so try a modified divisor of 35.5.

	Performers	Food Workers	Maintenance
Number of Workers	213	273	178
Modified Quota	$\frac{213}{35.5} = 6$	$\frac{273}{35.5} \approx 7.69$	$\frac{178}{35.5} \approx 5.01$
Round Modified Quota Up	**6**	**8**	**6**
			Total = 20

17. The total number of members considered is $25+18+29+31=103$.

Standard Divisor $= \dfrac{103}{20} = 5.15$; Electricians Union's Standard Quota $= \dfrac{25}{5.15} \approx 4.854$;

Plumber Union's Standard Quota $= \dfrac{18}{5.15} \approx 3.495$; Painters Union's Standard Quota $= \dfrac{29}{5.15} \approx 5.631$;

Carpenters Union's Standard Quota $= \dfrac{31}{5.15} \approx 6.019$

19. With the standard divisor of 5.15, a modified divisor is 5.5.

	Electricians	Plumbers	Painters	Carpenters
Number of Members	25	18	29	31
Standard Quota	4.85	3.50	5.63	6.02
Modified Quota	$\frac{25}{5.5} \approx 4.55$	$\frac{18}{5.5} \approx 3.27$	$\frac{29}{5.5} \approx 5.27$	$\frac{31}{5.5} \approx 5.64$
Round Modified Quota Up	5	4	6	6
				Total = 21

This total was too high, so try a modified divisor of 6.

	Electricians	Plumbers	Painters	Carpenters
Number of Members	25	18	29	31
Modified Quota	$\frac{25}{6} \approx 4.17$	$\frac{18}{6} = 3$	$\frac{29}{6} \approx 4.83$	$\frac{31}{6} \approx 5.17$
Round Modified Quota Up	5	3	5	6
				Total = 19

This total was too low, so try a modified divisor of 5.8.

	Electricians	Plumbers	Painters	Carpenters
Number of Members	25	18	29	31
Modified Quota	$\frac{25}{5.8} \approx 4.31$	$\frac{18}{5.8} \approx 3.10$	$\frac{29}{5.8} = 5$	$\frac{31}{5.8} \approx 5.34$
Round Modified Quota Up	**5**	**4**	**5**	**6**
				Total = 20

21. The total number of graduate students considered is $30+20+17+14=81$.

Standard Divisor $= \frac{81}{19} \approx 4.263$; Fiction Majors' Standard Quota $= \frac{30}{4.263} \approx 7.037$;

Poetry Majors' Standard Quota $= \frac{20}{4.263} \approx 4.692$;

Technical Writing Majors' Standard Quota $= \frac{17}{4.263} \approx 3.988$;

Media Writers Majors' Standard Quota $= \frac{14}{4.26} \approx 3.284$

23. With the standard divisor of 4.263 a modified divisor is 4.3.

	Fiction	Poetry	Technical	Media
Number of Students	30	20	17	14
Standard Quota	7.037	4.692	3.988	3.284
Modified Quota	$\frac{30}{4.3} \approx 6.98$	$\frac{20}{4.3} \approx 4.65$	$\frac{17}{4.3} \approx 3.95$	$\frac{14}{4.3} \approx 3.26$
Round Modified Quota Up	7	5	4	4
				Total = 20

23. (continued)

This total was too high, so try a modified divisor of 5.

	Fiction	Poetry	Technical	Media
Number of Students	30	20	17	14
Modified Quota	$\frac{30}{5}=6$	$\frac{20}{5}=4$	$\frac{17}{5}=3.4$	$\frac{14}{5}=2.8$
Round Modified Quota Up	6	4	4	3
				Total = 17

This total was too low, so try a modified divisor of 4.8.

	Fiction	Poetry	Technical	Media
Number of Students	30	20	17	14
Modified Quota	$\frac{30}{4.8}=6.25$	$\frac{20}{4.8}\approx 4.17$	$\frac{17}{4.8}\approx 3.54$	$\frac{14}{4.8}\approx 2.92$
Round Modified Quota Up	**7**	**5**	**4**	**3**
				Total = 19

25. The total number of students considered is $56+29+11+4=100$.

Standard Divisor $= \frac{100}{6}\approx 16.667$; Pilates' Standard Quota $= \frac{56}{16.667}\approx 3.36$;

Kick Boxing's Standard Quota $= \frac{29}{16.667}\approx 1.74$;

Yoga's Standard Quota $= \frac{11}{16.667}\approx 0.66$;

Spinning's Standard Quota $= \frac{4}{16.667}\approx 0.24$

27. With the standard divisor of 16.67 a modified divisor of 27.5.

	Pilates	kick boxing	yoga	spinning
Number of Students	56	29	11	4
Standard Quota	3.36	1.74	0.66	0.24
Modified Quota	$\frac{56}{27.5}\approx 2.04$	$\frac{29}{27.5}\approx 1.05$	$\frac{11}{27.5}=0.4$	$\frac{4}{27.5}\approx 0.15$
Round Modified Quota Up	3	2	1	1
				Total = 7

This total was too high, so try a modified divisor of 29.

	Pilates	kick boxing	yoga	spinning
Number of Students	56	29	11	4
Modified Quota	$\frac{56}{29}\approx 1.93$	$\frac{29}{29}=1$	$\frac{11}{29}\approx 0.38$	$\frac{4}{29}\approx 0.14$
Round Modified Quota Up	2	1	1	1
				Total = 5

27. (continued)

This total was too low, so try a modified divisor of 28

	Pilates	kick boxing	yoga	spinning
Number of Students	56	29	11	4
Modified Quota	$\frac{56}{28}=2$	$\frac{29}{28}\approx 1.04$	$\frac{11}{28}\approx 0.39$	$\frac{4}{28}\approx 0.14$
Round Modified Quota Up	**2**	**2**	**1**	**1**
				Total = 6

29. We must first find how the ten representatives were originally apportioned according to the Hamilton method. The standard divisor $=\frac{10,000}{10}=1,000$.

State	Number of residents	Standard Quota	Integer parts	Fractional Parts	Assign 2 additional representatives
A	570	$\frac{570}{1,000}=0.570$	0	0.570	1
B	2,557	$\frac{2,557}{1,000}=2.557$	2	0.557	2
C	6,873	$\frac{6,873}{1,000}=6.873$	6	0.873	7
Total	10,000		8		10

Determine the apportionment ten years later. The standard divisor $=\frac{10,080}{10}=1,008$.

State	Number of residents	Standard Quota	Integer parts	Fractional Parts	Assign 2 additional representatives
A	590	$\frac{590}{1,008}\approx 0.585$	0	0.585	0
B	2,617	$\frac{2,617}{1,008}\approx 2.596$	2	0.596	3
C	6,873	$\frac{6,873}{1,008}\approx 6.818$	6	0.818	7
Total	10,080		8		10

Rate of increase in A's population $=\frac{20}{570}\approx 3.5\%$;

Rate of increase in B's population $=\frac{60}{2,557}\approx 2.3\%$

Yes, because A's population has increased faster than B's, but A loses a representative to B in the reapportionment.

31. We must first find how the 150 security personnel were originally apportioned according to the Hamilton method. The standard divisor $= \dfrac{150,000}{150} = 1,000$.

Airport	Number of passengers	Standard Quota	Integer parts	Fractional Parts	Assign 2 security personnel
Allenport	120,920	$\dfrac{120,920}{1,000} = 120.920$	120	0.920	121
Bakerstown	5,550	$\dfrac{5,550}{1,000} = 5.550$	5	0.550	6
Columbia City	23,530	$\dfrac{23,530}{1,000} = 23.530$	23	0.530	23
Total	150,000		148		150

Determine the apportionment one year later. The standard divisor $= \dfrac{151,000}{150} \approx 1006.67$.

Airport	Number of passengers	Standard Quota	Integer parts	Fractional Parts	Assign 2 security personnel
Allenport	121,420	$\dfrac{121,420}{1,006.67} \approx 120.615$	120	0.615	121
Bakerstown	5,650	$\dfrac{5,650}{1,006.67} \approx 5.613$	5	0.613	5
Columbia City	23,930	$\dfrac{23,930}{1,006.67} \approx 23.771$	23	0.771	24
Total	151,000		148		150

Rate of increase at Columbia City $= \dfrac{23,930 - 23,530}{23,530} = \dfrac{400}{23,530} \approx 1.7\%$

Rate of increase at Bakerstown $= \dfrac{5,650 - 5,550}{5,550} = \dfrac{100}{5,550} \approx 1.8\%$

Yes, because the number of passengers per week at Bakerstown has increased faster than at Columbia City, but Bakerstown loses a security guard to Columbia City.

33. Consider the apportionment among only states A and B. The standard divisor $= \frac{9,770}{100} = 97.7$.

State	Number of residents	Standard Quota	Integer parts	Fractional Parts	Assign 1 additional computer labs
A	2,000	$\frac{2,000}{97.7} \approx 20.471$	20	0.471	20
B	7,770	$\frac{7,770}{97.7} \approx 79.529$	79	0.529	**80**
Total	9,770		99		100

Now consider approximately the number of representatives C would get with the standard divisor of 97.7. C would get $\frac{600}{97.7} \approx 6.141$ representatives.

Determine the apportionment with 6 additional committee members.

The standard divisor $= \frac{10,370}{106} \approx 97.83$.

State	Number of residents	Standard Quota	Integer parts	Fractional Parts	Assign 1 additional computer labs
A	2,000	$\frac{2,000}{97.83} \approx 20.444$	20	0.444	21
B	7,770	$\frac{7,770}{97.83} \approx 79.423$	79	0.423	**79**
C	600	$\frac{600}{97.83} \approx 6.133$	6	0.133	6
Total	10,370		105		106

Yes, when C's representatives are added to the commission, B loses a representative to A.

35. Consider the apportionment among only states A, B, and C. The standard divisor $= \frac{16,200}{56} \approx 289.29$.

State	Number of residents	Standard Quota	Integer parts	Fractional Parts	Assign 1 additional computer labs
A	8,700	$\frac{8,700}{289.29} \approx 30.074$	30	0.074	30
B	4,300	$\frac{4,300}{289.29} \approx 14.864$	14	0.864	15
C	3,200	$\frac{3,200}{289.29} \approx 11.062$	11	0.062	11
Total	16,200		55		56

Now consider approximately the number of representatives D would get with the standard divisor of 289.29. D would get $\frac{6,500}{289.29} \approx 22.469$ representatives.

Determine the apportionment with 22 additional committee members.

35. (continued)

The standard divisor $= \frac{22,700}{78} \approx 291.03$.

State	Number of residents	Standard Quota	Integer parts	Fractional Parts	Assign 3 additional computer labs
A	8,700	$\frac{8,700}{291.03} \approx 29.894$	29	0.894	30
B	4,300	$\frac{4,300}{291.03} \approx 14.775$	14	0.775	15
C	3,200	$\frac{3,200}{291.03} \approx 10.995$	10	0.995	11
D	6,500	$\frac{6,500}{291.03} \approx 22.334$	22	0.334	22
Total	22,700		75		78

No, when D's representatives are added to the commission, there is no change in A's, B's, or C's apportionment.

37. Total Population $= 700 + 1,500 + 820 + 4,530 + 2,200 + 550 = 10,300$

Standard Divisor $= \frac{\text{Total Population}}{\text{Number of Representatives Allocated}} = \frac{10,300}{200} = 51.5$

State	Number of residents	Standard Quota	Lower Quota	Upper Quota
A	700	$\frac{700}{51.5} \approx 13.592$	13	14
B	1,500	$\frac{1,500}{51.5} \approx 29.126$	29	30
C	820	$\frac{820}{51.5} \approx 15.922$	15	16
D	4,530	$\frac{4,530}{51.5} \approx 87.961$	87	**88**
E	2,200	$\frac{2,200}{51.5} \approx 42.718$	42	43
F	550	$\frac{550}{51.5} \approx 10.680$	10	11

Jefferson apportionment method: With the standard divisor of 51.5 a modified divisor is 50.5.

State	Population	Modified Quota	Round Modified Quota Down
A	700	$\frac{700}{50.5} \approx 13.861$	13
B	1,500	$\frac{1,500}{50.5} \approx 29.703$	29
C	820	$\frac{820}{50.5} \approx 16.238$	16
D	4,530	$\frac{4,530}{50.5} \approx 89.703$	**89**
E	2,200	$\frac{2,200}{50.5} \approx 43.564$	43
F	550	$\frac{550}{50.5} \approx 10.891$	10
Total	10,300		200

37. (continued)

Adams apportionment method: With the standard divisor of 51.5, a modified divisor is 52.1.

State	Population	Modified Quota	Round Modified Quota Up
A	700	$\frac{700}{52.1} \approx 13.436$	14
B	1,500	$\frac{1,500}{52.1} \approx 28.791$	29
C	820	$\frac{820}{52.1} \approx 15.739$	16
D	4,530	$\frac{4,530}{52.1} \approx 86.948$	87
E	2,200	$\frac{2,200}{52.1} \approx 42.226$	43
F	550	$\frac{550}{52.1} \approx 10.557$	11
Total	10,300		200

Using the Jefferson method, D would receive 89 representatives, which is greater than its upper quota.

39. Total Population $= 1,400 + 2,000 + 1,500 + 9,000 + 4,300 + 500 = 18,700$

$$\text{Standard Divisor} = \frac{\text{Total Population}}{\text{Number of Representatives Allocated}} = \frac{18,700}{500} = 37.4$$

State	Number of residents	Standard Quota	Lower Quota	Upper Quota
A	1,400	$\frac{1,400}{37.4} \approx 37.433$	37	38
B	2,000	$\frac{2,000}{37.4} \approx 53.476$	53	54
C	1,500	$\frac{1,500}{37.4} \approx 40.107$	40	41
D	9,000	$\frac{9,000}{37.4} \approx 240.642$	**240**	**241**
E	4,300	$\frac{4,300}{37.4} \approx 114.973$	114	115
F	500	$\frac{500}{37.4} \approx 13.369$	13	14

39. (continued)

Jefferson apportionment method: With the standard divisor of 37.4 a modified divisor is 37.1.

State	Population	Modified Quota	Round Modified Quota Down
A	1,400	$\frac{1,400}{37.1} \approx 37.736$	37
B	2,000	$\frac{2,000}{37.1} \approx 53.908$	53
C	1,500	$\frac{1,500}{37.1} \approx 40.431$	40
D	9,000	$\frac{9,000}{37.1} \approx 242.588$	**242**
E	4,300	$\frac{4,300}{37.1} \approx 115.903$	115
F	500	$\frac{500}{37.1} \approx 13.477$	13
Total	18,700		500

Adams apportionment method: With the standard divisor of 37.4 a modified divisor is 37.7.

State	Population	Modified Quota	Round Modified Quota Up
A	1,400	$\frac{1,400}{37.7} \approx 37.135$	38
B	2,000	$\frac{2,000}{37.7} \approx 53.050$	54
C	1,500	$\frac{1,500}{37.7} \approx 39.788$	40
D	9,000	$\frac{9,000}{37.7} \approx 238.727$	**239**
E	4,300	$\frac{4,300}{37.7} \approx 114.058$	115
F	500	$\frac{500}{37.7} \approx 13.263$	14
Total	18,700		500

Using the Jefferson method, D would receive 242 representatives, which is greater than its upper quota. Using the Adams method, D would receive 239 representatives, which is less than its lower quota.

41. The standard divisor is the average number of people that each representative represents. The standard quota is the number of representatives that a state deserves.

43. Because the Jefferson method rounds the modified quotas down, we often need larger modified quotas and hence smaller modified divisors.

45. She needs to try a larger modified divisor.

47. No solution provided.

49. smallest 926; largest 940

51. smallest 1,058; largest 1,066

53. No solution provided.

Section 11.5 Looking Deeper: Fair Division

1. In discrete fair division, the objects cannot be divided like in continuous fair division.

3. He had the highest bids.

5. a) discrete
b) continuous
c) discrete

7.

	Darnell $\left(\frac{1}{2}\right)$	Joy $\left(\frac{1}{2}\right)$
Bid on car	\$110,000	\$85,000
Fair share of estate	\$110,000/2 = \$55,000	\$85,000/2 = \$42,500
Item obtained with the highest bid	Car	

9.

	Dennis (40%)	Zadie (60%)
Bid on copyright	\$20,000	\$28,000
Fair share of copyright	$\$20{,}000 \times 0.40 = \$8{,}000$	$\$28{,}000 \times 0.60 = \$16{,}800$
Item obtained with the highest bid		Copyright

11.

	Ed $\left(\frac{1}{3}\right)$	Al $\left(\frac{1}{3}\right)$	Jerry $\left(\frac{1}{3}\right)$
Bid on painting	\$13,000	\$ 8,000	\$ 9,000
Bid on statue	\$14,000	\$13,000	\$15,000
Total value	\$27,000	\$21,000	\$24,000
Fair share of estate	$\frac{1}{3} \cdot 27000 = \$9{,}000$	$\frac{1}{3} \cdot 21000 = \$7{,}000$	$\frac{1}{3} \cdot 24000 = \$8{,}000$

13.

	Darnell $\left(\frac{1}{2}\right)$	Joy $\left(\frac{1}{2}\right)$
Pays to estate $(+)$ or receives from estate $(-)$	\$55,000 $(+)$	\$42,500 $(-)$
Division of estate balance (\$12,500)	\$6,250 $(-)$	\$6,250 $(-)$
Summary of cash	Pays \$48,750	Receives \$48,750

15.

	Dennis (40%)	Zadie (60%)
Pays to pool $(+)$ or receives from pool $(-)$	\$8,000 $(-)$	\$11,200 $(+)$
Division of pool balance (\$3,200)	\$1,280 $(-)$	\$ 1,920 $(-)$
Summary of cash	Receives \$9,280	Pays \$9,280

17.

	Ed $\left(\frac{1}{3}\right)$	Al $\left(\frac{1}{3}\right)$	Jerry $\left(\frac{1}{3}\right)$
Items obtained with highest bid	Painting		Statue
Pays to estate $(+)$ or receives from estate $(-)$	$13{,}000 - 9{,}000$ \$4,000.00 $(+)$	\$7,000.00 $(-)$	$15{,}000 - 8{,}000$ \$7,000.00 $(+)$
Division of estate balance (\$4,000)	\$1,333.33 $(-)$	\$1,333.33 $(-)$	\$1,333.33 $(-)$
Summary of cash	Pays \$2,666.67	Receives \$8,333.33	Pays \$5,666.67

19.

	Betty $\left(\frac{1}{2}\right)$	Dennis $\left(\frac{1}{2}\right)$
Bid on ring	\$16,000	\$18,000
Bid on desk	\$ 4,500	\$ 5,000
Bid on books	\$ 4,000	\$ 3,000
Total value	\$24,500	\$26,000
Fair share of estate	\$12,250	\$13,000

	Betty $\left(\frac{1}{2}\right)$	Dennis $\left(\frac{1}{2}\right)$
Item with highest bid	Books	Ring and Desk
Pays to pool $(+)$ or receives from pool $(-)$	\$8,250 $(-)$	\$10,000 $(+)$
Division of estate balance (\$1,750)	\$ 875 $(-)$	\$ 875 $(-)$
Summary of Cash	Receives \$9,125	Pays \$9,125

Betty receives the books, Dennis receives the ring and desk; Dennis pays Betty \$9,125.

21. Each person has a different estimate of the value of the estate and does not know the others' estimates.

23. Those who think that items in the estate are worth more receive the items and then must contribute cash to the estate that is then (partially) distributed to those who made low estimates of the items in the estate.

25. – 33. No solution provided.

Chapter Review Exercises

1. When adding a member to the legislature and without changing populations, a state loses a representative.

3. No solution provided.

5. a) Huntington-Hill number for Florida $= \dfrac{(\text{population } X)^2}{x(x+1)} = \dfrac{16.03^2}{25 \cdot 26} \approx 0.395$

   Huntington-Hill number for Texas $= \dfrac{(\text{population } X)^2}{x(x+1)} = \dfrac{21.49^2}{32 \cdot 33} \approx 0.437$

   b) Since $0.395 < 0.437$, Texas should receive the additional representative.

7. Qinhuangdao, Beijing, Qingdao, and Tianjin routes each get one bus in no particular order. The remaining nine buses in order would be given to HBQHBHQBT. So the Beijing route should have 4 buses, the Qingdao route should have 3, the Tianjin route should have 2, and the Qinhuandao route should have 4 according to the Huntington–Hill apportionment principle.

9. Total Responses $= 8+64+11+31=114$

$$\text{Standard Divisor} = \frac{\text{Total Responses}}{\text{Number of Classes}} = \frac{114}{8} = 14.25$$

Jefferson apportionment method: With the standard divisor of 14.25, a modified divisor is 11.

Class	Responses	Modified Quota	Round Modified Quota Down
Pilates	8	$\frac{8}{11} \approx 0.727$	0
Kick boxing	64	$\frac{64}{11} \approx 5.818$	5
Yoga	11	$\frac{11}{11} = 1$	1
Spinning	31	$\frac{31}{11} \approx 2.818$	2
Total	114		8

11. Total Responses $= 8+64+11+31=114$

$$\text{Standard Divisor} = \frac{\text{Total Responses}}{\text{Number of Classes}} = \frac{114}{8} = 14.25$$

Webster apportionment method: With the standard divisor of 14.25, a modified divisor is 15.

Class	Responses	Modified Quota	Round Modified Quota
Pilates	8	$\frac{8}{15} \approx 0.533$	1
Kick boxing	64	$\frac{64}{15} \approx 4.267$	4
Yoga	11	$\frac{11}{15} \approx 0.733$	1
Spinning	31	$\frac{31}{15} \approx 2.067$	2
Total	114		8

13. The new states paradox occurs when a new state is added, and its share of seats is added to the legislature causing a change in the allocation of seats previously given to another state.

15.

	Tito $\left(\frac{1}{3}\right)$	Omarosa $\left(\frac{1}{3}\right)$	Piers $\left(\frac{1}{3}\right)$
Bid on rifle	\$12,000	\$17,000	\$10,000
Bid on sword	\$15,000	\$13,000	\$11,000
Total value	\$27,000	\$30,000	\$21,000
Fair share of estate	$\frac{1}{3} \cdot 27000 = \$9{,}000$	$\frac{1}{3} \cdot 30000 = \$10{,}000$	$\frac{1}{3} \cdot 21000 = \$7{,}000$

Chapter Test

1. When adding a member to the legislature and without changing populations, a state loses a representative.

3.

Division	Number of members %	Step 1:Determine exact number of seats	Step 2: Assign integer parts	Examine fractional parts	Step 3: Assign additional seats
Art	47 (23.86%)	$0.2386 \times 8 = 1.9088$	1	0.9088	2
Music	111 (56.34%)	$0.5634 \times 8 = 4.5072$	4	0.5072	4
Theater	39 (19.80%)	$0.1980 \times 8 = 1.5840$	1	0.5840	2
Total	197 (100%)	8	6		8

5. a) $\text{Huntington} - \text{Hill number for Oregon} = \dfrac{(\text{population } X)^2}{x(x+1)} = \dfrac{3.61^2}{5 \cdot 6} \approx 0.434$

$\text{Huntington} - \text{Hill number for Arizona} = \dfrac{(\text{population } X)^2}{x(x+1)} = \dfrac{5.23^2}{8 \cdot 9} \approx 0.380$

b) Since $0.380 < 0.434$, Oregon should receive the additional representative.

7. Mystery Mountain, City Sidewalks, Jungle Village, and Great Frontier routes each get one bus in no particular order. The remaining ten buses in order would be given to MCJMGCJMGC. So the Mystery Mountain route should have 4 buses, the City Sidewalks route should have 4, the Jungle Village route should have 3, and the Great Frontier route should have 3 according to the Huntington–Hill apportionment principle.

9. Consider the apportionment among only airports A, B, and C.

The standard divisor $= \dfrac{180{,}600}{150} = 1204.$

Town	Number of customers	Standard Quota	Integer parts	Fractional Parts	Assign 1 additional security officer
Allenwood	80,500	$\dfrac{80{,}500}{1204} \approx 66.86$	66	0.86	**67**
Black Hills	6,800	$\dfrac{6{,}800}{1204} \approx 5.65$	5	0.65	**6**
Colombia	93,300	$\dfrac{93{,}300}{1204} \approx 77.49$	77	0.49	77
Total	180,600		148		150

Now consider approximately the number of security personnel Devon would get with the standard divisor of 1204. Devon would get $\dfrac{41{,}400}{1204} \approx 34.39$ representatives.

Determine the apportionment with 34 additional committee members.

The standard divisor $= \dfrac{222{,}000}{184} \approx 1206.5.$

9. (continued)

Town	Number of customers	Standard Quota	Integer parts	Fractional Parts	Assign 1 additional security officer
Allenwood	80,500	$\frac{80,500}{1206.5} \approx 66.72$	66	0.72	**67**
Black Hills	6,800	$\frac{6,800}{1206.5} \approx 5.64$	5	0.64	**6**
Colombia	93,300	$\frac{93,300}{1206.5} \approx 77.33$	77	0.33	77
Devon	41,400	$\frac{41,400}{1206.5} \approx 34.31$	34	0.31	34
Total	222,000		182		184

No, because when Devon is added, no mall loses any security personnel.

11. Total Responses $= 47+32+41+21=141$

$$\text{Standard Divisor} = \frac{\text{Total Responses}}{\text{Number of Classes}} = \frac{141}{9} = 15.67$$

Webster apportionment method: With the standard divisor of 15.67, a modified divisor is 15.

Class	Responses	Modified Quota	Round Modified Quota
massage	47	$\frac{47}{15} \approx 3.133$	3
aromatherapy	32	$\frac{32}{15} \approx 2.133$	2
yoga	41	$\frac{41}{15} \approx 2.733$	3
meditation	21	$\frac{21}{15} = 1.4$	1
Total	141		9

13. Total Responses $= 47+32+41+21=141$

$$\text{Standard Divisor} = \frac{\text{Total Responses}}{\text{Number of Classes}} = \frac{141}{9} = 15.67$$

Adams apportionment method: With the standard divisor of 15.67, a modified divisor is 20.8.

Class	Responses	Modified Quota	Round Modified Quota Up
massage	47	$\frac{47}{20.8} \approx 2.260$	3
aromatherapy	32	$\frac{32}{20.8} \approx 1.538$	2
yoga	41	$\frac{41}{20.8} \approx 1.971$	2
meditation	21	$\frac{21}{20.8} \approx 1.010$	2
Total	141		9

15.

	Larry $\left(\frac{1}{3}\right)$	Moe $\left(\frac{1}{3}\right)$	Curly $\left(\frac{1}{3}\right)$
Bid on Branch A	\$13 million	\$9 million	\$10 million
Bid on Branch B	\$17 million	\$18 million	\$14 million
Total value	\$30 million	\$27 million	\$24 million
Fair share of estate	$\frac{1}{3}\cdot 30 = \$10$ million	$\frac{1}{3}\cdot 27 = \$9$ million	$\frac{1}{3}\cdot 24 = \$8$ million

	Larry $\left(\frac{1}{3}\right)$	Moe $\left(\frac{1}{3}\right)$	Curly $\left(\frac{1}{3}\right)$
Items obtained with highest bid	Branch A	Branch B	
Pays to estate $(+)$ or receives from estate $(-)$	$13-10$ \$3 million $(+)$	$18-9$ \$9 million $(+)$	\$8 million $(-)$
Division of estate balance (\$4 million)	\$1,333,333.33 $(-)$	\$1,333,333.33 $(-)$	\$1,333,333.33 $(-)$
Summary of cash	Pays \$1,666,666.67	Pays \$7,666,666.67	Receives \$9,333,333.33

Chapter 12
Voting: Using Mathematics to Make Choices

Section 12.1 Voting Methods

1. Her eighteen second-place votes gave her more points than Carim's one extra first-place vote.

3. Each received ½ point because the number that preferred T over B was the same as the number that preferred B over T.

5. a) No. The total number of votes is 2,156 + 1,462 + 986 + 428 = 5,032; $2156 \div 5032 \approx 0.428$, which is less than 50%

   b) Edelson

7. Athletic facilities with 33 votes

9. Since Student Union Building had the least votes, it is removed. The new table would be

Preference	15	30	18	17	10	2
1st	A	D	A	C	C	A
2nd	C	C	D	D	A	C
3rd	D	A	C	A	D	D

By combining identical columns, we have

Preference	17	30	18	17	10
1st	A	D	A	C	C
2nd	C	C	D	D	A
3rd	D	A	C	A	D

Since Campus Security (C) has the fewest first-place votes, it is then eliminated

Preference	17	30	18	17	10
1st	A	D	A	D	A
2nd	D	A	D	A	D

By combining identical columns, we have

Preference	45	47
1st	A	D
2nd	D	A

Since Dining Facilities (D) now has the most first-place votes, it wins the election.

11. Comedy with 23 votes

13. Since Greek Tragedy had the least votes, it is removed. The new table would be

Preference	10	15	13	12	5	7
1st	C	D	C	M	M	C
2nd	M	M	D	D	C	M
3rd	D	C	M	C	D	D

13. (continued)

By combining identical columns, we have

Preference	17	15	13	12	5
1st	C	D	C	M	M
2nd	M	M	D	D	C
3rd	D	C	M	C	D

Since Drama (D) has the fewest first-place votes, it is then eliminated.

Preference	17	15	13	12	5
1st	C	M	C	M	M
2nd	M	C	M	C	C

By combining identical columns, we have

Preference	30	32
1st	C	M
2nd	M	C

Since Mystery (M) now has the most first-place votes, it wins the election.

15. Technology with 15 votes

17. Since Families (F) had the least votes, it is removed. The new table would be

Preference	15	7	13	5	2
1st	T	E	G	P	E
2nd	P	P	E	G	G
3rd	E	G	P	E	P
4th	G	T	T	T	T

Since Poverty in Third World Countries (P) has the fewest first-place votes, it is then eliminated.

Preference	15	7	13	5	2
1st	T	E	G	G	E
2nd	E	G	E	E	G
3rd	G	T	T	T	T

By combining identical columns we have

Preference	15	9	18
1st	T	E	G
2nd	E	G	E
3rd	G	T	T

Since Environment (E) has the fewest first-place votes, it is then eliminated.

Preference	15	9	18
1st	T	G	G
2nd	G	T	T

By combining identical columns, we have

Preference	15	27
1st	T	G
2nd	G	T

Since Gender Roles (G) now has the most first-place votes, it wins the election.

19. Londram with 17 votes

21. Since E-Mall (E) had the least votes, it is removed. The new table would be

Preference	15	11	9	10	2
1st	L	F	S	T	L
2nd	F	S	L	F	S
3rd	S	L	F	L	F
4th	T	T	T	S	T

Since Securenet (S) has the fewest first-place votes, it is then eliminated.

Preference	15	11	9	10	2
1st	L	F	L	T	L
2nd	F	L	F	F	F
3rd	T	T	T	L	T

By combining identical columns, we have

Preference	26	11	10
1st	L	F	T
2nd	F	L	F
3rd	T	T	L

Since Techenium (T) has the fewest first-place votes, it is then eliminated.

Preference	26	11	10
1st	L	F	F
2nd	F	L	L

By combining identical columns, we have

Preference	26	21
1st	L	F
2nd	F	L

Since Londram (L) now has the most first-place votes, it wins the election.

23. The Borda count is being used. Three points for 1st, two points for 2nd, and one point for 3rd.

Player, Team	1st	2nd	3rd	Total	
Tim Tebow, FL	462	**229**	113	1,957	$(1957-462\times3-113\times1)\div2=229$
Darren McFadden, AR	**291**	355	120	1,703	$(1703-255\times2-120\times1)\div3=291$
Colt Brennan, HI	54	114	**242**	632	$(632-54\times3-114\times2)\div1=242$
Chase Daniel, MO	25	84	182	425	$98\times3+112\times2+117\times1=635$

25. Answers may vary.
The first one eliminated should be in last place. The second one eliminated should be in the second-to-last place. This process continues until the winner is determined.

27. 1st : Athletic Facilities, 2nd :Dining Facilities, 3rd :Campus Security, and 4th : Student Union Building

29. 1st :Dining Facilities, 2nd :Athletic Facilities, 3rd :Campus Security, and 4th : Student Union Building

31. 1st :Technology, 2nd :Gender Roles, 3rd :Environment, 4th : Poverty in Third World Countries, and 5th :Families

33. 1st :Gender Roles, 2nd:Technology, 3rd :Environment, 4th : Poverty in Third World Countries, and 5th :Families

35. The possible points awarded are 4, 3, 2, or 1. $(4+3+2+1)\cdot 20=10\cdot 20=200$

37. If the candidate is in first place in all selections, then $5 \cdot 22 = 110$ is the Borda count.

39. Assume the four candidates are A, B, C, and D. The comparisons you need to make are A to B, A to C, A to D, B to C, B to D, and C to D. Since 1 point is awarded to each of the winners of these comparisons, there would be 6 points total.

41. Assume the five candidates are A, B, C, D and E. The comparisons you need to make are A to B, A to C, A to D, A to E, B to C, B to D, B to E, C to D, C to E, and D to E. Since 1 point is awarded to each of the winners of these comparisons and any individual candidate appears in a comparison only 4 times, then the maximum that could be earned is 4 points.

43. Voters can state their second, third, and fourth preferences.

45. A candidate who wins an election should be able to beat the other candidates head to head.

47. Answers may vary.
The candidate that was eliminated was the one with the fewest (fewer) votes. Thus, the candidate that wins was the candidate with the majority of the votes.

49. No solution provided.

51. S, with 92 votes.

Candidate	Votes
A	15 + 0 + 18 + 10 + 2 = 45
C	15 + 30 + 0 + 17 + 10 + 2 = 74
D	0 + 30 + 18 + 17 + 0 + 0 = 65
S	**15 + 30 + 18 + 17 + 10 + 2 = 92**

53. E, with 37 votes.

Candidate	Votes
E	15 + 7 + 13 + 0 + 2 = 37
F	0 + 0 + 13 + 5 + 2 = 20
G	0 + 7 + 13 + 5 + 2 = 27
P	15 + 7 + 0 + 5 + 0 = 27
T	15 + 0 + 0 + 0 + 0 = 15

Section 12.2 Defects in Voting Methods

1. Example 1; the Borda method violates the majority criterion.

3. the plurality and pairwise comparison methods

5. Option B with a total of 149 points

Preference	1st place votes × 4 (points)	2nd place votes × 3 (points)	3rd place votes × 2 (points)	4th place votes × 1 (points)	Total Points
A	25×4	0×3	15×2	9×1	139
B	15×4	21×3	13×2	0×1	149
C	9×4	15×3	12×2	13×1	118
D	0×4	13×3	9×2	27×1	84

7. No solution provided.

9.

Preference		1st place votes × 4 (points)	2nd place votes × 3 (points)	3rd place votes × 2 (points)	4th place votes × 1 (points)	Total Points
Lower the age to 18	(A)	3×4	18×3	10×2	14×1	100
Lower the age to 19	(B)	10×4	14×3	21×2	0×1	124
Lower the age to 20	(C)	22×4	10×3	0×2	13×1	131
Keep the age at 21	(D)	10×4	3×3	14×2	18×1	95

Option C is the winner by the Borda count. The Condorcet's criterion is not satisfied in this example because option B can beat all of the other options in a head-to-head competition, but C is the winner by the Borda count. 24 out of 45 prefer B to A, 23 out of 45 prefer B to C, and 32 out of 45 prefer B to D.

11.

Preference		1st place votes × 4 (points)	2nd place votes × 3 (points)	3rd place votes × 2 (points)	4th place votes × 1 (points)	Total Points
Atlanta	(A)	2×4	16×3	29×2	0×1	114
Boston	(B)	18×4	19×3	2×2	8×1	141
Chicago	(C)	23×4	2×3	8×2	14×1	128
Denver	(D)	4×4	10×3	8×2	25×1	87

Option B is the winner by the Borda count. The independence-of-irrelevant-alternatives criterion is not satisfied in this example because if option A were taken out, then the preference table would be

Preference	15	4	8	10	8	2
1st	C	D	C	B	B	C
2nd	B	B	D	D	C	B
3rd	D	C	B	C	D	D

According to the Borda count, option C would be declared the winner.

Preference		1st place votes × 3 (points)	2nd place votes × 2 (points)	3rd place votes × 1 (points)	Total Points
Chicago	(C)	25×3	8×2	14×1	105
Boston	(B)	18×3	21×2	8×1	104
Denver	(D)	4×3	18×2	25×1	73

13. Since option B (reduce expenditures on art and music programs) had the least votes, it is removed.

The new table would be

Preference	9	12	4	5	4	1
1st	C	D	C	A	A	C
2nd	A	A	D	D	C	A
3rd	D	C	A	C	D	D

By combining identical columns, we have

Preference	10	12	4	5	4
1st	C	D	C	A	A
2nd	A	A	D	D	C
3rd	D	C	A	C	D

13. (continued)

Since option A (reduce sports programs) has the fewest first-place votes, it is then eliminated.

Preference	10	12	4	5	4
1st	C	D	C	D	C
2nd	D	C	D	C	D

By combining identical columns, we have

Preference	18	17
1st	C	D
2nd	D	C

Since option C (increase class size) now has the most first-place votes, it wins the election. The Condorcet's criterion is not satisfied in this example because option A (reduce sport programs) can beat all of the other options in a head-to-head competition, but C (increase class size) is the winner by the plurality-with-elimination method. 30 out of 35 prefer A to B, 21 out of 35 prefer A to C, and 19 out of 35 prefer A to D.

15. We must compare a) A with B, b) A with C, and c) B with C.

a) In comparing A with B, we ignore C. Thus, 13 prefer A over B, and 10+5=15 prefer B over A. Thus, we award 1 point to option B.

d) In comparing A with C, we ignore B. Thus, 13 prefer A over C, and 10+5=15 prefer C over A. Thus, we award 1 point to option C.

e) In comparing B with C, we ignore A. Thus, 13+10=23 prefer B over C, and 5 prefer C over B. Thus, we award 1 point to option B.

Since B has 2 points and C has 1, option B is declared the winner.

This example doesn't violate the majority criterion because the criterion states that <u>if</u> a majority of the voters rank a candidate as their first choice, <u>then</u> that candidate should win the election. The hypothesis of this conditional criterion was not satisfied. The majority (over half) did not vote for any one candidate.

17. Answers may vary.

Preference	**5**	**4**
1st	B	A
2nd	A	C
3rd	C	B

Preference	1st place votes × 3 (points)	2nd place votes × 2 (points)	3rd place votes × 1 (points)	Total Points
A	4×3	5×2	0×1	22
B	5×3	0×2	4×1	19
C	0×3	4×2	5×1	13

Option A is the winner by the Borda count. Condorcet's criterion is not satisfied in this example because option B can beat all of the other options in a head-to-head competition, but A is the winner by Borda count. 5 out of 9 prefer B to A and 5 out of 9 prefer B to C.

19. Answers may vary.

Preference									
1st	A	A	A	A	B	B	B	C	C
2nd	B	B	B	B	A	A	A	B	B
3rd	C	C	C	C	C	C	C	A	A

Option A is the winner by the plurality method. In head-to-head competition, B defeats both A and C, so Condorcet's criterion is not satisfied.

21. Answers may vary.

Preference	20	**30**	8	8	12
1st	C	D	A	A	B
2nd	A	A	D	B	C
3rd	B	B	B	C	A
4th	D	C	C	D	D

Since option B had the least votes, it is removed. The new table would be

Preference	20	30	8	8	12
1st	C	D	A	A	C
2nd	A	A	D	C	A
3rd	D	C	C	D	D

By combining identical columns, we have

Preference	32	30	8	8
1st	C	D	A	A
2nd	A	A	D	C
3rd	D	C	C	D

Since option A had the least votes, it is removed. The new table would be

Preference	32	30	8	8
1st	C	D	D	C
2nd	D	C	C	D

By combining identical columns, we have

Preference	40	38
1st	C	D
2nd	D	C

Since option C now has the most first-place votes, it wins the election. Condorcet's criterion is not satisfied in this example because option A can beat all of the other options in a head-to-head competition, but option C is the winner by the plurality-with-elimination method. 66 out of 78 prefer A to B, 46 out of 78 prefer A to C, and 48 out of 78 prefer A to D

23. Answers may vary.
In the plurality-with-elimination method, if an option has the majority (over 50%) of the votes, then that option will not be eliminated during this process. If anything, the option will gain strength as other options are eliminated and remain the winner of the election.

25. majority, Condorcet's, independence-of-irrelevant-alternatives, monotonicity

27. In any election involving more than two candidates, no voting method will satisfy all four fairness criteria.

29. Answers may vary.
Essentially the Condorcet's Criterion is automatically satisfied in the pairwise comparison method. An option will achieve the best possible outcome, winning every pairwise competition.

31. – 37. No solution provided.

39. Answers may vary.
Yes.

Preference	5	4
1st	A	B
2nd	B	C
3rd	C	A

Preference	1st place votes × 3 (points)	2nd place votes × 2 (points)	3rd place votes × 1 (points)	Total Points
A	5×3	0×2	4×1	19
B	4×3	5×2	0×1	22
C	0×3	4×2	5×1	13

Option B is the winner by the Borda count even though option A has the majority (approximately 56% of the vote).

41. No solution provided.

Section 12.3 Weighted Voting Systems

1. 51 was the quota; 26, 26, 12, 12, 12, and 12 were the weights of the six voters.

3. They were critical members in all 11 winning coalitions.

5. a) The quota is 5.
   b) There are 5 voters. Each voter has 1 vote.
   c) There are no dictators.
   d) The sum of the votes is the same as the quota. Each voter has veto power.

7. a) The quota is 11.
   b) There are 4 voters. Voter A has 10 votes, voter B has 3, voter C has 4, and voter D has 5.
   c) There are no dictators.
   d) No voter has veto power.

9. a) The quota is 15.
   b) There are 5 voters. Voter A has 1 vote, voter B has 2, two voters (C and D) have 3, and voter E has 4. Notice that the sum of the votes is less than the quota. No resolutions can be passed.
   c) There are no dictators.
   d) No voter has veto power, because no resolutions can be passed.

11. a) The quota is 12.
    b) There are 4 voters. Voter A has 1 vote, voter B has 3, voter C has 5, and voter D has 7.
    c) There are no dictators.
    d) Two voters have veto power. Voters C and D each have veto power.

13. a) The quota is 25.
    b) There are 5 voters. Two voters (A and B) have 4 votes, voter C has 6, voter D has 7, and voter E has 9.
    c) There are no dictators.
    d) Three voters have veto power. Voters C, D, and E each have veto power.

15. a) The quota is 51.
    b) There are 5 voters. Three voters (A, B, and C) have 20 votes and two (D and E) have 10.
    c) There are no dictators.
    d) No voter has veto power.

17.

Coalition	Weight	
{A}	1	
{B}	3	
{C}	5	
{D}	7	
{A,B}	4	
{A,C}	6	
{A,D}	8	
{B,C}	8	
{B,D}	10	
{C,D}	12	Winning
{A,B,C}	9	
{A,B,D}	11	
{A,C,D}	13	Winning
{B,C,D}	15	Winning
{A,B,C,D}	16	Winning

19.

Coalition	Weight	
{A}	4	
{B}	4	
{C}	6	
{D}	7	
{E}	9	
{A,B}	8	
{A,C}	10	
{A,D}	11	
{A,E}	13	
{B,C}	10	
{B,D}	11	
{B,E}	13	
{C,D}	13	
{C,E}	15	
{D,E}	16	
{A,B,C}	14	
{A,B,D}	15	
{A,B,E}	17	
{A,C,D}	17	
{A,C,E}	19	
{A,D,E}	20	
{B,C,D}	17	
{B,C,E}	19	
{B,D,E}	20	
{C,D,E}	22	
{A,B,C,D}	21	
{A,B,C,E}	23	
{A,B,D,E}	24	
{A,C,D,E}	26	Winning
{B,C,D,E}	26	Winning
{A,B,C,D,E}	30	Winning

21.

Coalition	Weight		Critical Voters
{C,D}	12	Winning	C,D
{A,C,D}	13	Winning	C,D
{B,C,D}	15	Winning	C,D
{A,B,C,D}	16	Winning	C,D

23.

Coalition	Weight		Critical Voters
{A,C,D,E}	26	Winning	A,C,D,E
{B,C,D,E}	26	Winning	B,C,D,E
{A,B,C,D,E}	30	Winning	C,D,E
{A,B,C,D,E}	31	Winning	E

25. A simple majority is 6, since there are 11 total voters.

Coalition	Weight	
{P}	5	
{T}	4	
{S}	2	
{P,T}	9	Winning
{P,S}	7	Winning
{T,S}	6	Winning
{P,S,T}	11	Winning

27.

Coalition	Weight	
{A}	3	
{C}	4	
{T}	3	
{N}	2	
{A,C}	7	
{A,T}	6	
{A,N}	5	
{C,T}	7	
{C,N}	6	
{T,N}	5	
{A,C,T}	10	Winning
{A,C,N}	9	Winning
{A,T,N}	8	Winning
{C,T,N}	9	Winning
{A,C,T,N}	12	Winning

29.

Coalition	Weight		Critical Voters
{P,T}	9	Winning	P,T
{P,S}	7	Winning	P,S
{T,S}	6	Winning	T,S
{P,S,T}	11	Winning	None

31.

Coalition	Weight		Critical Voters
{A,C,T}	10	Winning	A,C,T
{A,C,N}	9	Winning	A,C,N
{A,T,N}	8	Winning	A,T,N
{C,T,N}	9	Winning	C,T,N
{A,C,T,N}	12	Winning	None

33. The only winning coalition in Exercise 5 is when all five members vote for the issue. Each voter would therefore have a Banzhaf power index of $\frac{1}{5}$.

35.

Coalition	Weight		Critical Voters
{A}	4		
{B}	4		
{C}	6		
{D}	7		
{E}	9		
{A,B}	8		
{A,C}	10		
{A,D}	11		
{A,E}	13		
{B,C}	10		
{B,D}	11		
{B,E}	13		
{C,D}	13		
{C,E}	15		
{D,E}	16		
{A,B,C}	14		
{A,B,D}	15		
{A,B,E}	17		
{A,C,D}	17		
{A,C,E}	19		
{A,D,E}	20		
{B,C,D}	17		
{B,C,E}	19		
{B,D,E}	20		
{C,D,E}	22		
{A,B,C,D}	21		
{A,B,C,E}	23		
{A,B,D,E}	24		
{A,C,D,E}	26	Winning	A,C,D,E
{B,C,D,E}	26	Winning	B,C,D,E
{A,B,C,D,E}	30	Winning	C,D,E

$$\frac{\text{The number of times A is critical in winning coalitions}}{\text{The total number of times voters are critical in winning coalitions}} = \frac{1}{11}$$

$$\frac{\text{The number of times B is critical in winning coalitions}}{\text{The total number of times voters are critical in winning coalitions}} = \frac{1}{11}$$

$$\frac{\text{The number of times C is critical in winning coalitions}}{\text{The total number of times voters are critical in winning coalitions}} = \frac{3}{11}$$

$$\frac{\text{The number of times D is critical in winning coalitions}}{\text{The total number of times voters are critical in winning coalitions}} = \frac{3}{11}$$

$$\frac{\text{The number of times E is critical in winning coalitions}}{\text{The total number of times voters are critical in winning coalitions}} = \frac{3}{11}$$

37.

Coalition	Weight		Critical Voters
{C,D}	12	Winning	C,D
{A,C,D}	13	Winning	C,D
{B,C,D}	15	Winning	C,D
{A,B,C,D}	16	Winning	C,D

$$\frac{\text{The number of times A is critical in winning coalitions}}{\text{The total number of times voters are critical in winning coalitions}} = \frac{0}{8} = 0$$

$$\frac{\text{The number of times B is critical in winning coalitions}}{\text{The total number of times voters are critical in winning coalitions}} = \frac{0}{8} = 0$$

$$\frac{\text{The number of times C is critical in winning coalitions}}{\text{The total number of times voters are critical in winning coalitions}} = \frac{4}{8} = \frac{1}{2}$$

$$\frac{\text{The number of times D is critical in winning coalitions}}{\text{The total number of times voters are critical in winning coalitions}} = \frac{4}{8} = \frac{1}{2}$$

39. a) No solution provided.

b) Let the members of the firm be represented by {K,C,V,W,X,Y,Z}. The winning coalitions are given in the following table.

Coalition	Critical Voters
{K,C,V,W}	K,C,V,W
{K,C,V,X}	K,C,V,X
{K,C,V,Y}	K,C,V,Y
{K,C,V,Z}	K,C,V,Z
{K,C,W,X}	K,C,W,X
{K,C,W,Y}	K,C,W,Y
{K,C,W,Z}	K,C,W,Z
{K,C,X,Y}	K,C,X,Y
{K,C,X,Z}	K,C,X,Z
{K,C,Y,Z}	K,C,Y,Z
{K,C,V,W,X}	K,C
{K,C,V,W,Y}	K,C
{K,C,V,W,Z}	K,C
{K,C,V,X,Y}	K,C
{K,C,V,X,Z}	K,C
{K,C,V,Y,Z}	K,C
{K,C,W,X,Y}	K,C
{K,C,W,X,Z}	K,C
{K,C,W,Y,Z}	K,C
{K,C,X,Y,Z}	K,C
{K,C,V,W,X,Y}	K,C
{K,C,V,W,X,Z}	K,C
{K,C,V,W,Y,Z}	K,C
{K,C,V,X,Y,Z}	K,C
{K,C,W,X,Y,Z}	K,C
{K,C,V,W,X,Y,Z}	K,C

39. (continued)

$$\frac{\text{The number of times K is critical in winning coalitions}}{\text{The total number of times voters are critical in winning coalitions}} = \frac{26}{72} = \frac{13}{36}$$

$$\frac{\text{The number of times C is critical in winning coalitions}}{\text{The total number of times voters are critical in winning coalitions}} = \frac{26}{72} = \frac{13}{36}$$

$$\frac{\text{The number of times V is critical in winning coalitions}}{\text{The total number of times voters are critical in winning coalitions}} = \frac{4}{72} = \frac{1}{18}$$

$$\frac{\text{The number of times W is critical in winning coalitions}}{\text{The total number of times voters are critical in winning coalitions}} = \frac{4}{72} = \frac{1}{18}$$

$$\frac{\text{The number of times X is critical in winning coalitions}}{\text{The total number of times voters are critical in winning coalitions}} = \frac{4}{72} = \frac{1}{18}$$

$$\frac{\text{The number of times Y is critical in winning coalitions}}{\text{The total number of times voters are critical in winning coalitions}} = \frac{4}{72} = \frac{1}{18}$$

$$\frac{\text{The number of times Z is critical in winning coalitions}}{\text{The total number of times voters are critical in winning coalitions}} = \frac{4}{72} = \frac{1}{18}$$

41. a) No solution provided.

b) Let the members of the firm be represented by {K,C,H,X,Y,Z}. The winning coalitions are given in the following table.

Coalition	Critical Voters
{K,C,X,Y}	K,C,X,Y
{K,C,X,Z}	K,C,X,Z
{K,C,Y,Z}	K,C,Y,Z
{K,H,X,Y}	K,H,X,Y
{K,H,X,Z}	K,H,X,Z
{K,H,Y,Z}	K,H,Y,Z
{C,H,X,Y}	C,H,X,Y
{C,H,X,Z}	C,H,X,Z
{C,H,Y,Z}	C,H,Y,Z
{K,C,H,X,Y}	X,Y
{K,C,H,X,Z}	X,Z
{K,C,H,Y,Z}	Y,Z
{K,C,X,Y,Z}	K,C
{K,H,X,Y,Z}	K,H
{C,H,X,Y,Z}	C,H
{K,C,H,X,Y,Z}	None

$$\frac{\text{The number of times K is critical in winning coalitions}}{\text{The total number of times voters are critical in winning coalitions}} = \frac{8}{48} = \frac{1}{6}$$

$$\frac{\text{The number of times H is critical in winning coalitions}}{\text{The total number of times voters are critical in winning coalitions}} = \frac{8}{48} = \frac{1}{6}$$

$$\frac{\text{The number of times C is critical in winning coalitions}}{\text{The total number of times voters are critical in winning coalitions}} = \frac{8}{48} = \frac{1}{6}$$

$$\frac{\text{The number of times X is critical in winning coalitions}}{\text{The total number of times voters are critical in winning coalitions}} = \frac{8}{48} = \frac{1}{6}$$

41. (continued)

$$\frac{\text{The number of times Y is critical in winning coalitions}}{\text{The total number of times voters are critical in winning coalitions}} = \frac{8}{48} = \frac{1}{6}$$

$$\frac{\text{The number of times Z is critical in winning coalitions}}{\text{The total number of times voters are critical in winning coalitions}} = \frac{8}{48} = \frac{1}{6}$$

43. Without that voter's support, a resolution cannot pass.

45. Consider the votes to be labels A, B, C, D, E. There are $2^5 - 1 = 31$ different coalitions. Many of them have the same weight, however. The sixth line of Pascal's triangle (recall from earlier chapters) is

1 5 10 10 5 1.

This means there are 1 coalition with none (not counted), 5 with 2 (not a winning coalition because the quota is 3), 10 with 3, 10 with 4, 5 with 5, and 1 with 6. Of the 25 winning coalitions, only 10 of them will have critical voters.

Coalition	Critical Voters
{A,B,C}	A,B,C
{A,B,D}	A,B,D
{A,B,E}	A,B,E
{A,C,D}	A,C,D
{A,C,E}	A,C,E
{A,D,E}	A,D,E
{B,C,D}	B,C,D
{B,C,E}	B,C,E
{B,D,E}	B,D,E
{C,D,E}	C,D,E

Each voter would have the same Banzhaf power index of

$$\frac{\text{The number of times A is critical in winning coalitions}}{\text{The total number of times voters are critical in winning coalitions}} = \frac{6}{30} = \frac{1}{5}$$

47. a) The only winning coalitions are those that have the dictator included. Let voter A be the voter with 15 votes, voter B be the voter with 2 votes, voters C and D each with 3 votes, and voter E with 5 votes. The winning coalitions are

Coalition	Critical Voters
{A}	A
{A,B}	A
{A,C}	A
{A,D}	A
{A,E}	A
{A,B,C}	A
{A,B,D}	A
{A,B,E}	A
{A,C,D}	A
{A,C,E}	A
{A,D,E}	A
{A,B,C,D}	A
{A,B,C,E}	A
{A,B,D,E}	A
{A,C,D,E}	A
{A,B,C,D,E}	A

$$\frac{\text{The number of times A is critical in winning coalitions}}{\text{The total number of times voters are critical in winning coalitions}} = \frac{16}{16} = 1$$

The Banzhaf power index of A is 1, the index of the other voters is 0.

b) No solution provided.

49. – 55. No solution provided.

Section 12.4: Looking Deeper: The Shapley–Shubik Index

1. With the notation $\{A, B, C\}$, the order of the voters does not matter; with the notation (A, B, C), the order of the voters does matter.

3. We divided the number of times that A was pivotal in some permutation of the voters by total number of permutations of the voters.

5.

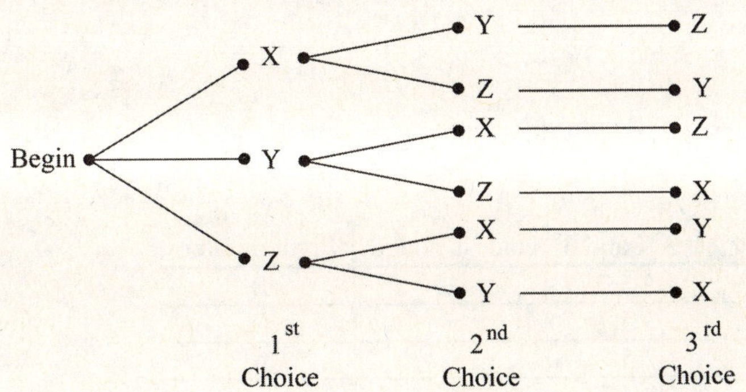

The permutations are (X,Y,Z), (X,Z,Y), (Y,X,Z), (Y,Z,X), (Z,X,Y), and (Z,Y,X).

7.

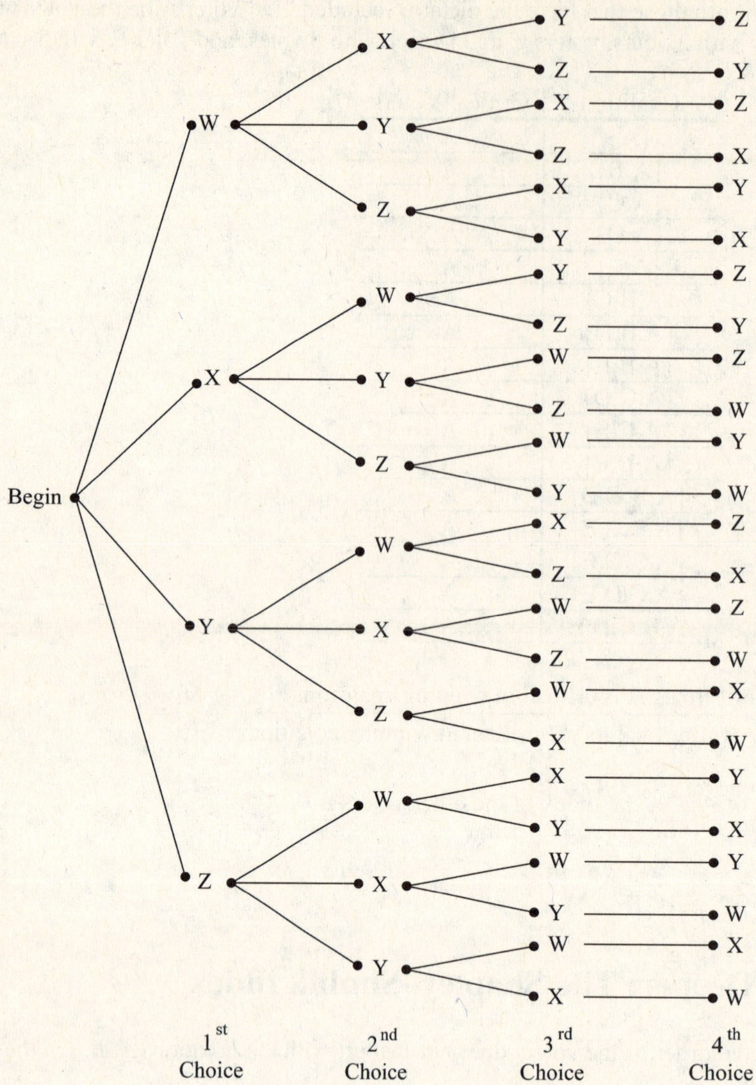

The permutations are (W,X,Y,Z), (W,X,Z,Y), (W,Y,X,Z), (W,Y,Z,X), (W,Z,X,Y), (W,Z,Y,X), (X,W,Y,Z), (X,W,Z,Y), (X,Y,W,Z), (X,Y,Z,W), (X,Z,W,Y), (X,Z,Y,W), (Y,W,X,Z), (Y,W,Z,X), (Y,X,W,Z), (Y,X,Z,W), (Y,Z,W,X), (Y,Z,X,W), (Z,W,X,Y), (Z,W,Y,X), (Z,X,W,Y), (Z,X,Y,W), (Z,Y,W,X), and (Z,Y,X,W).

9. $6! = 6\times5\times4\times3\times2\times1 = 720$

11. $12! = 12\times11\times10\times9\times8\times7\times6\times5\times4\times3\times2\times1 = 479,001,600$

13.

Coalition	Weight after 1st, 2nd, and 3rd voter is added	Pivotal voter
a) (A,B,C)	3,5,7	B
b) (A,C,B)	3,5,7	C
c) (B,A,C)	2,5,7	A
d) (B,C,A)	2,4,7	A
e) (C,A,B)	2,5,7	A
f) (C,B,A)	2,4,7	A

15. Let the voters be labeled A, B, and C.

Coalition	Weight after 1st, 2nd, and 3rd voter is added	Pivotal voter
(A,B,C)	3,6,8	B
(A,C,B)	3,5,8	B
(B,A,C)	3,6,8	A
(B,C,A)	3,5,8	A
(C,A,B)	2,5,8	B
(C,B,A)	2,5,8	A

$$\frac{\text{The number of times A is pivotal in some permutation of voters}}{\text{The total number of permutations of voters}} = \frac{3}{6} = \frac{1}{2}$$

$$\frac{\text{The number of times B is pivotal in some permutation of voters}}{\text{The total number of permutations of voters}} = \frac{3}{6} = \frac{1}{2}$$

$$\frac{\text{The number of times C is pivotal in some permutation of voters}}{\text{The total number of permutations of voters}} = \frac{0}{6} = 0$$

17.

Coalition	Weight after 1st, 2nd, 3rd and 4th voter is added	Pivotal voter
(A,B,C,D)	3,6,8,10	C
(A,B,D,C)	3,6,8,10	D
(A,C,B,D)	3,5,8,10	B
(A,C,D,B)	3,5,7,10	B
(A,D,B,C)	3,5,8,10	B
(A,D,C,B)	3,5,7,10	B
(B,A,C,D)	3,6,8,10	C
(B,A,D,C)	3,6,8,10	D
(B,C,A,D)	3,5,8,10	A
(B,C,D,A)	3,5,7,10	A
(B,D,A,C)	3,5,8,10	A
(B,D,C,A)	3,5,7,10	A
(C,A,B,D)	2,5,8,10	B
(C,A,D,B)	2,5,7,10	B
(C,B,A,D)	2,5,8,10	A
(C,B,D,A)	2,5,7,10	A
(C,D,A,B)	2,4,7,10	B
(C,D,B,A)	2,4,7,10	A
(D,A,B,C)	2,5,8,10	B
(D,A,C,B)	2,5,7,10	B
(D,B,A,C)	2,5,8,10	A
(D,B,C,A)	2,5,7,10	A
(D,C,A,B)	2,4,7,10	B
(D,C,B,A)	2,4,7,10	A

$$\frac{\text{The number of times A is pivotal in some permutation of voters}}{\text{The total number of permutations of voters}} = \frac{10}{24} = \frac{5}{12}$$

$$\frac{\text{The number of times B is pivotal in some permutation of voters}}{\text{The total number of permutations of voters}} = \frac{10}{24} = \frac{5}{12}$$

$$\frac{\text{The number of times C is pivotal in some permutation of voters}}{\text{The total number of permutations of voters}} = \frac{2}{24} = \frac{1}{12}$$

$$\frac{\text{The number of times D is pivotal in some permutation of voters}}{\text{The total number of permutations of voters}} = \frac{2}{24} = \frac{1}{12}$$

19.

Coalition	Weight after 1st, 2nd, 3rd and 4th voter is added	Pivotal voter
(A,B,C,D)	2,3,4,5	C
(A,B,D,C)	2,3,4,5	D
(A,C,B,D)	2,3,4,5	B
(A,C,D,B)	2,3,4,5	D
(A,D,B,C)	2,3,4,5	B
(A,D,C,B)	2,3,4,5	C
(B,A,C,D)	1,3,4,5	C
(B,A,D,C)	1,3,4,5	D
(B,C,A,D)	1,2,4,5	A
(B,C,D,A)	1,2,3,5	A
(B,D,A,C)	1,2,4,5	A
(B,D,C,A)	1,2,3,5	A
(C,A,B,D)	1,3,4,5	B
(C,A,D,B)	1,3,4,5	D
(C,B,A,D)	1,2,4,5	A
(C,B,D,A)	1,2,3,5	A
(C,D,A,B)	1,2,4,5	A
(C,D,B,A)	1,2,3,5	A
(D,A,B,C)	1,3,4,5	B
(D,A,C,B)	1,3,4,5	C
(D,B,A,C)	1,2,4,5	A
(D,B,C,A)	1,2,3,5	A
(D,C,A,B)	1,2,4,5	A
(D,C,B,A)	1,2,3,5	A

$$\frac{\text{The number of times A is pivotal in some permutation of voters}}{\text{The total number of permutations of voters}} = \frac{12}{24} = \frac{1}{2}$$

$$\frac{\text{The number of times B is pivotal in some permutation of voters}}{\text{The total number of permutations of voters}} = \frac{4}{24} = \frac{1}{6}$$

$$\frac{\text{The number of times C is pivotal in some permutation of voters}}{\text{The total number of permutations of voters}} = \frac{4}{24} = \frac{1}{6}$$

$$\frac{\text{The number of times D is pivotal in some permutation of voters}}{\text{The total number of permutations of voters}} = \frac{4}{24} = \frac{1}{6}$$

21. a) The Shapley–Shubik index for each person will be $\frac{1}{5}$.

b) No solution provided.

c) No solution provided.

23. a) The Shapley–Shubik index for A is 1. It will be 0 for each of the others.

b) No solution provided.

c) No solution provided.

25.

Coalition	Pivotal voter
(A,B,C,D)	B
(A,B,D,C)	B
(A,C,B,D)	C
(A,C,D,B)	C
(A,D,B,C)	D
(A,D,C,B)	D
(B,A,C,D)	A
(B,A,D,C)	A
(B,C,A,D)	A
(B,C,D,A)	A
(B,D,A,C)	A
(B,D,C,A)	A
(C,A,B,D)	A
(C,A,D,B)	A
(C,B,A,D)	A
(C,B,D,A)	A
(C,D,A,B)	A
(C,D,B,A)	A
(D,A,B,C)	A
(D,A,C,B)	A
(D,B,A,C)	A
(D,B,C,A)	A
(D,C,A,B)	A
(D,C,B,A)	A

$$\frac{\text{The number of times A is pivotal in some permutation of voters}}{\text{The total number of permutations of voters}} = \frac{18}{24} = \frac{3}{4}$$

$$\frac{\text{The number of times B is pivotal in some permutation of voters}}{\text{The total number of permutations of voters}} = \frac{2}{24} = \frac{1}{12}$$

$$\frac{\text{The number of times C is pivotal in some permutation of voters}}{\text{The total number of permutations of voters}} = \frac{2}{24} = \frac{1}{12}$$

$$\frac{\text{The number of times D is pivotal in some permutation of voters}}{\text{The total number of permutations of voters}} = \frac{2}{24} = \frac{1}{12}$$

27. a) (A,C,T,N),(A,C,N,T),(A,T,C,N),(A,T,N,C),(A,N,C,T),(A,N,T,C),(C,A,T,N),(C,A,N,T),(C,T,A,N), (C,T,N,A),(C,N,T,A),(C,N,A,T),(T,A,C,N),(T,A,N,C),(T,C,A,N),(T,C,N,A),(T,N,A,C),(T,N,C,A), (N,A,B,T),(N,A,C,C),(N,C,A,T),(N,C,T,A),(N,T,A,C), and (N,T,C,A)

27. (continued)

b)

Coalition	Weight after 1st, 2nd, 3rd and 4th voter is added	Pivotal group
(A,C,T,N)	3,7,10,12	T
(A,C,N,T)	3,7,9,12	N
(A,T,C,N)	3,6,10,12	C
(A,T,N,C)	3,6,8,12	N
(A,N,C,T)	3,5,9,12	C
(A,N,T,C)	3,5,8,12	T
(C,A,T,N)	4,7,10,12	T
(C,A,N,T)	4,7,9,12	N
(C,T,A,N)	4,7,10,12	A
(C,T,N,A)	4,7,9,12	N
(C,N,A,T)	4,6,9,12	A
(C,N,T,A)	4,6,9,12	T
(T,A,C,N)	3,6,10,12	C
(T,A,N,C)	3,6,8,12	N
(T,C,A,N)	3,7,10,12	A
(T,C,N,A)	3,7,9,12	N
(T,N,A,C)	3,5,8,12	A
(T,N,C,A)	3,5,9,12	C
(N,A,C,T)	2,5,9,12	C
(N,A,T,C)	2,5,8,12	T
(N,C,A,T)	2,6,9,12	A
(N,C,T,A)	2,6,9,12	T
(N,T,A,C)	2,5,8,12	A
(N,T,C,A)	2,5,9,12	C

c) $\dfrac{\text{The number of times A is pivotal in some permutation of voters}}{\text{The total number of permutations of voters}} = \dfrac{6}{24} = \dfrac{1}{4}$

$\dfrac{\text{The number of times F is pivotal in some permutation of voters}}{\text{The total number of permutations of voters}} = \dfrac{6}{24} = \dfrac{1}{4}$

$\dfrac{\text{The number of times T is pivotal in some permutation of voters}}{\text{The total number of permutations of voters}} = \dfrac{6}{24} = \dfrac{1}{4}$

$\dfrac{\text{The number of times N is pivotal in some permutation of voters}}{\text{The total number of permutations of voters}} = \dfrac{6}{24} = \dfrac{1}{4}$

29. A pivotal voter for a coalition is the first player, as voters are added, who makes the coalition a winning coalition

31. Before B can be pivotal, the managing editor and one other member of the board must already have been chosen.

33. No solution provided.

Chapter Review Exercises

1. a) No. The total number of votes is 2,156 + 1,462 + 986 + 428 = 5,032.
   $2156 \div 5032 \approx 0.428$, which is less than 50%.

   b) Myers

3. Since education in the future (E) didn't have any first-place votes, it is removed. The new table would be

Preference	1,531	1,102	906	442	375
1st	G	R	S	S	G
2nd	R	S	G	R	S
3rd	S	G	R	G	R

Since role of government in a free society ® has the fewest first-place votes, it is then eliminated

Preference	1,906	1,102	906	442
1st	G	S	S	S
2nd	S	G	G	G

By combining identical columns, we have

Preference	1,906	2,450
1st	G	S
2nd	S	G

Since social justice (S) now has the most first-place votes, it wins the election.

5.

Preference	1st place votes × 4 (points)	2nd place votes × 3 (points)	3rd place votes × 2 (points)	4th place votes × 1 (points)	Total Points
R	2×4	0×3	0×2	1×1	9
D	1×4	2×3	0×2	0×1	10
P	0×4	1×3	1×2	1×1	6
Q	0×4	0×3	2×2	1×1	5

Option D is the winner by the Borda method. No, this election does not satisfy the majority criterion. Option R had the majority of the 1st place votes.

7.

Preference	1st place votes × 4 (points)	2nd place votes × 3 (points)	3rd place votes × 2 (points)	4th place votes × 1 (points)	Total Points
A	3×4	10×3	23×2	0×1	88
B	13×4	14×3	3×2	6×1	106
C	17×4	3×3	4×2	12×1	97
D	3×4	9×3	6×2	18×1	69

Option B is the winner by the Borda count. The independence-of-irrelevant-alternatives criterion is not satisfied in this example because if option A were taken out, then the preference table would be

Preference	11	3	6	9	4	3
1st	C	D	C	B	B	C
2nd	B	B	D	D	C	B
3rd	D	C	B	C	D	D

7. (continued)

According to the Borda count, C would now be declared the winner.

Preference	1st place votes $\times$ 3 (points)	2nd place votes $\times$ 2 (points)	3rd place votes $\times$ 1 (points)	Total Points
C	20×3	4×2	12×1	80
B	13×3	17×2	6×1	79
D	3×3	15×2	18×1	57

9. The quota is 17; There are 4 voters. Voter A has 1 vote, voter B has 5, voter C has 7, and voter D has 8; There are no dictators; Voters B, C, and D each have veto power.

11. Let voter A be the voter with 2 votes, voter B with 3, voter C with 5, and voter D with 7.

Coalition	Weight	Critical Voters
{C,D}	12	C,D
{A,B,D}	12	A,B,D
{A,C,D}	14	C,D
{B,C,D}	15	C,D
{A,B,C,D}	17	D

$$\frac{\text{The number of times A is critical in winning coalitions}}{\text{The total number of times voters are critical in winning coalitions}} = \frac{1}{10}$$

$$\frac{\text{The number of times B is critical in winning coalitions}}{\text{The total number of times voters are critical in winning coalitions}} = \frac{1}{10}$$

$$\frac{\text{The number of times C is critical in winning coalitions}}{\text{The total number of times voters are critical in winning coalitions}} = \frac{3}{10}$$

$$\frac{\text{The number of times D is critical in winning coalitions}}{\text{The total number of times voters are critical in winning coalitions}} = \frac{5}{10} = \frac{1}{2}$$

13.

Coalition	Weight	Critical Voters
{A,B}	12	A,B
{A,C}	14	A,C
{A,D}	14	A,D
{A,B,C}	15	A
{A,B,D}	15	A
{A,C,D}	17	A
{A,B,C,D}	18	A

Voter A has a Banzhaf power index of $\frac{7}{10}$

$$\frac{\text{The number of times A is critical in winning coalitions}}{\text{The total number of times voters are critical in winning coalitions}} = \frac{7}{10}$$

while the other four voter have a Banzhaf power index of $\frac{1}{10}$

$$\frac{\text{The number of times B,C,or D is critical in winning coalitions}}{\text{The total number of times voters are critical in winning coalitions}} = \frac{1}{10}$$

15.

Sum of Weights of Coalition Members	Pivotal voter
(A, B, C) 4 + 3	B
(A, C, B) 4 + 2	C
(B, A, C) 3 + 4	A
(B, C, A) 3 + 2 + 4	A
(C, A, B) 2 + 4	A
(C, B, A) 2 + 3 + 4	A

17. The Shapley–Shubik index for each person will be $\frac{1}{6}$.

Chapter Test

1. a) No. The total number of votes is 2,543 + 1,532 + 892 + 473 = 5,440. $2{,}543 \div 5{,}440 \approx 0.467$, which is less than 50%.

   b) Molina

3.

Preference	1st place votes × 4 (points)	2nd place votes × 3 (points)	3rd place votes × 2 (points)	4th place votes × 1 (points)	Total Points
A	$1{,}612 \times 4$	$1{,}754 \times 3$	849×2	0×1	13,409
B	$1{,}754 \times 4$	$1{,}612 \times 3$	0×2	849×1	12,701
C	849×4	0×3	$2{,}457 \times 2$	909×1	9,219
D	0×4	849×3	909×2	$2{,}457 \times 1$	6,822

Option A is the winner by the Borda count. The Condorcet's criterion is satisfied in this example, because option A can beat all of the other options in a head-to-head competition and is the winner by Borda count. 2,461 out of 4,215 prefer A to B, 3,366 out of 4,215 prefer A to C, and 3,366 out of 4,215 prefer A to D.

5. $6! = 720$

7.

Preference	1st place votes × 4 (points)	2nd place votes × 3 (points)	3rd place votes × 2 (points)	4th place votes × 1 (points)	Total Points
A	2×4	0×3	0×2	1×1	9
B	1×4	2×3	0×2	0×1	10
C	0×4	0×3	3×2	0×1	6
D	0×4	1×3	0×2	2×1	5

Option B is the winner by the Borda method. No, this election does not satisfy the majority criterion. Option A had the majority of the 1st place votes.

9. Remove option D, since it had the least number of first-place votes. The new table would be

Preference	35	71	36	14
1st	A	B	C	A
2nd	B	A	A	C
3rd	C	C	B	B

Then remove option C, since it had the least number of first-place votes. The new table would be

Preference	35	71	36	14
1st	A	B	A	A
2nd	B	A	B	B

By combining identical columns, we have

Preference	85	71
1st	A	B
2nd	B	A

Since option A now has the most first-place votes, it wins the election.

Now if we remove option C, the new table would be

Preference	35	71	36	14
1st	A	B	D	D
2nd	B	A	A	A
3rd	D	D	B	B

By combining identical columns we have

Preference	35	71	50
1st	A	B	D
2nd	B	A	A
3rd	D	D	B

Remove option A, since it had the least number of first-place votes. The new table would be

Preference	35	71	50
1st	B	B	D
2nd	D	D	B

By combining identical columns we have

Preference	104	50
1st	B	D
2nd	D	B

Since option B now has the most first-place votes, it wins the election instead of A. The independence-of-irrelevant-alternatives criterion is not satisfied.

11.

Sum of Weights of Coalition Members	Pivotal voter
(A, B, C) 5 + 3	B
(A, C, B) 5 + 3	C
(B, A, C) 3 + 5	A
(B, C, A) 3 + 3 + 5	A
(C, A, B) 3 + 5	A
(C, B, A) 3 + 3 + 5	A

13.

Preference	1st place votes × 4 (points)	2nd place votes × 3 (points)	3rd place votes × 2 (points)	4th place votes × 1 (points)	Total Points
A	19×4	21×3	6×2	8×1	159
B	21×4	7×3	20×2	6×1	151
C	6×4	8×3	23×2	17×1	111
D	8×4	18×3	5×2	23×1	119

Option A is the winner by the Borda count. The independence-of-irrelevant-alternatives criterion is not satisfied in this example because if option D were taken out, then the preference table would be

Preference	7	5	6	12	16	8
1st	A	B	C	A	B	C
2nd	B	A	A	B	A	B
3rd	C	C	B	C	C	A

According to the Borda count, B would now be declared the winner.

Preference	1st place votes × 3 (points)	2nd place votes × 2 (points)	3rd place votes × 1 (points)	Total Points
A	19×3	27×2	8×1	119
B	21×3	27×2	6×1	123
C	14×3	0×2	40×1	96

15.

Sum of Weights of Coalition Members	Pivotal voter
(A, B, C) 4 + 4	B
(A, C, B) 4 + 2 + 4	B
(B, A, C) 4 + 4	A
(B, C, A) 4 + 2 + 4	A
(C, A, B) 2 + 4 + 4	B
(C, B, A) 2 + 4 + 4	A

15. (continued)

$$\frac{\text{The number of times A is pivotal in some permutation of voters}}{\text{The total number of permutations of voters}} = \frac{3}{6} = \frac{1}{2}$$

$$\frac{\text{The number of times B is pivotal in some permutation of voters}}{\text{The total number of permutations of voters}} = \frac{3}{6} = \frac{1}{2}$$

$$\frac{\text{The number of times C is pivotal in some permutation of voters}}{\text{The total number of permutations of voters}} = \frac{0}{6} = 0$$

Chapter 13
Counting: Just How Many Are There?

Section 13.1 Introduction to Counting Methods

1. We first listed pairs that began with C, then all pairs that began with O, then all pairs that began with F, and finally all pairs that began with T.

3. A mathematician who applies mathematics to reveal the organization of terrorist cells.

5. JT, JC, JF, JR, TC, TF, TR, CF, CR, FR

7. JT, JC, JF, JR, TJ, TC, TF, TR, CJ, CT, CF, CR, FJ, FT, FC, FR, RJ, RT, RC, RF

9. The answer would be larger if repetition is allowed.

Use the diagram on the next page for the solutions to Exercises 11 – 13.

11. 4, the ways are HTTT, THTT, TTHT, and TTTH

13. 6, the ways are HHTT, HTHT, HTTH, THHT, THTH, and TTHH

15. Without drawing the tree, you can imagine that the tree begins and there are six branches on the first number choice. Each of those branches split into five branches for the second number choice for a total of thirty branches. There would therefore be **thirty** two-digit numbers that could be formed.

17. Without drawing the tree, you can imagine that the tree begins and there are six branches on the first number choice. Each of those branches split into five branches for the second number choice for a total of thirty branches. Each of those branches split into four branches for the third number choice for a total of 120 branches. There would therefore be **120** three-digit numbers that could be formed.

Use the diagram on page 279 for the solutions to Exercises 19 – 23.

19. 4 ways; (1,4), (2,3), (3,2), (4,1)

21. 6 ways; (1,1), (2,2), (3,3), (4,4), (5,5), (6,6)

23. 10 ways; (1,1), (1,2), (1,3), (1,4), (2,1), (2,2), (2,3), (3,1), (3,2), (4,1)

25. Without drawing the tree, you can imagine that the tree begins and there are six branches on the first top choice. Each of those branches split into five branches for the second choice of pants for a total of thirty branches. Each of those branches split into four branches for the choice of a jacket for a total of 120 branches. There would therefore be **120** outfits that could be formed.

27. Assume that the ordering is that the first person buys beverages, the second arranges for food and the third sends out invitations. Let the first letter of the name indicate the name of Susan's friend, except for H representing Phyllis . The possible arrangements are:

    PHJ, PHD, PJD, PDJ, PDH, PJH
    HPJ, HPD, HJD, HDJ, HDP, HJP
    JDP, JDH, JPH, JHP, JHD, JPD
    DPH, DPJ, DHJ, DJH, DJP, DHP for a total of 24 ways.

Use the following diagram for the solutions to Exercises 11 – 13.

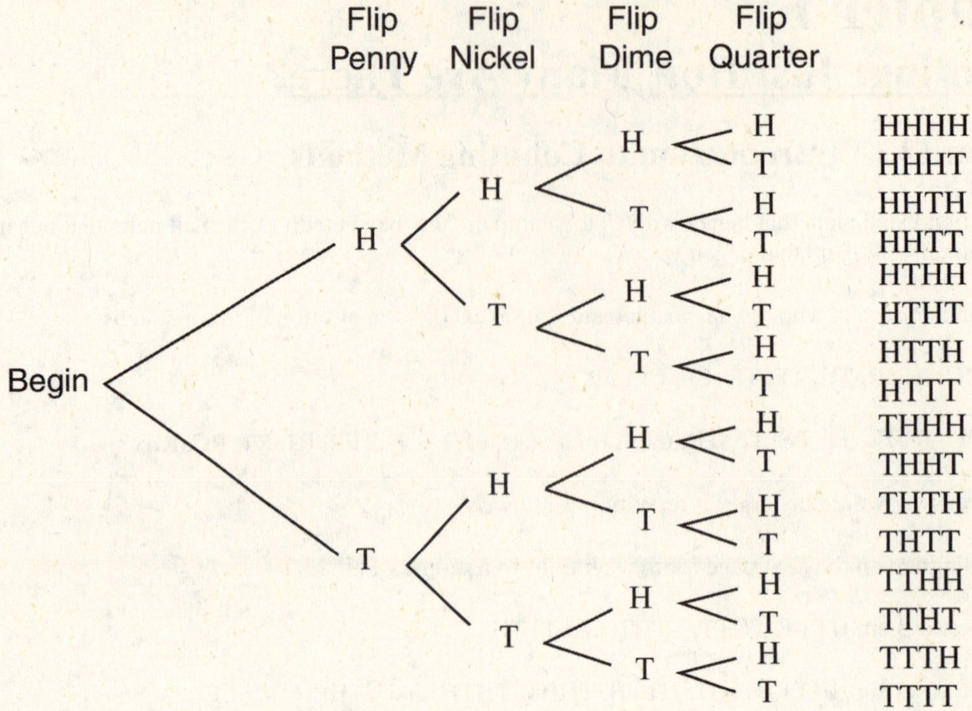

29. Without drawing the tree, you can imagine that the tree begins and there are two branches on the first flip. Each of those branches split into two branches on the second flip for a total of four branches. Each of those branches split into two branches on the third flip for a total of eight branches. Each of those branches split into two branches on the fourth flip for a total of sixteen branches. Finally, each of those branches split into two branches on the fifth flip for a total of thirty-two branches. There would therefore be **thirty-two** ways five coins could be flipped.

31. Without drawing the tree, you can imagine that the tree begins and there are four branches on the first roll. Each of those branches split into four branches on the second roll for a total of sixteen branches. There would therefore be **sixteen** ways two tetrahedral dice could be rolled.

33. 16

35. 9

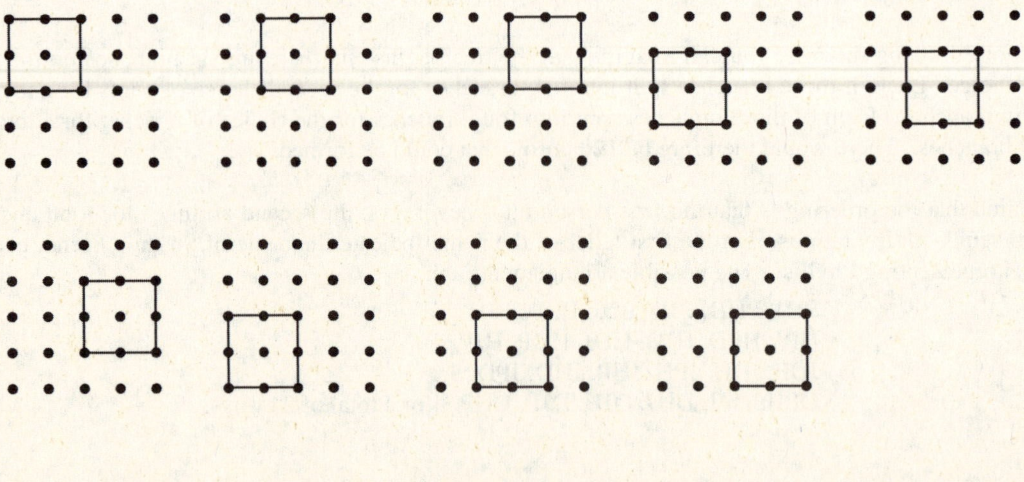

Use the following diagram for the solutions to Exercises 19 – 23.

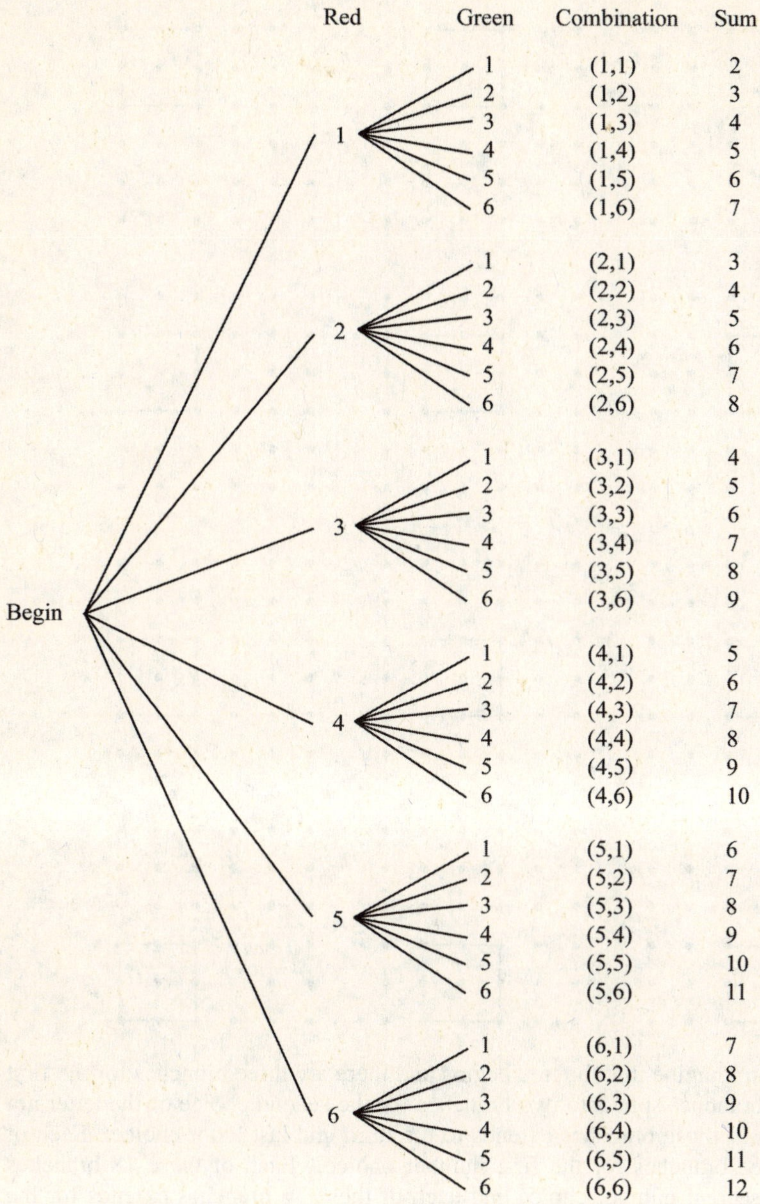

37. 16

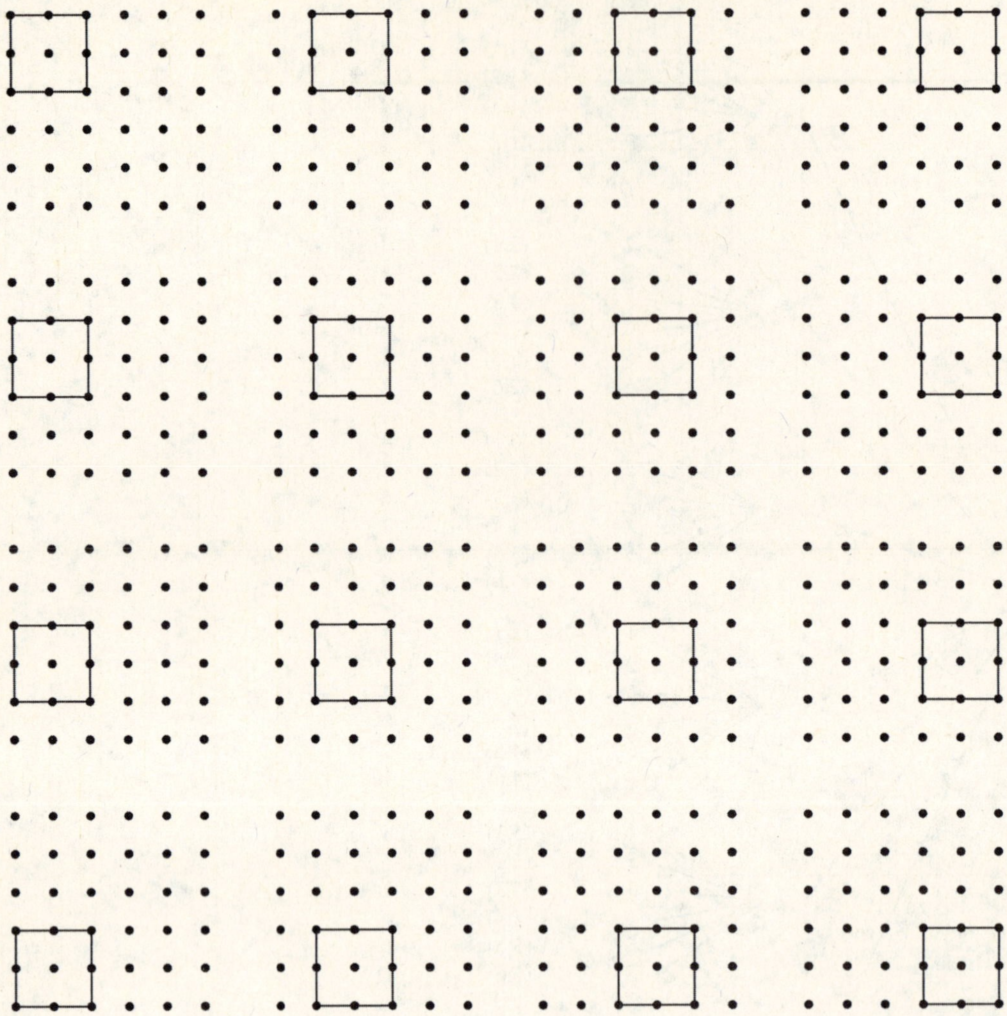

39. Without drawing the tree, you can imagine that the tree begins and there are three branches for the first choice of letters. Each of those branches split into two branches for the second choice of the letter not chosen on the first choice. Each of those branches extends to the third and last letter choice. Each of these six branches split into three branches for the first number choice. Each of these 18 branches splits into two branches for the second number choice and each of these 36 branches extends for the third and final number choice. There would therefore be **thirty-six** license plates to be investigated.

41. a) $2^5 = 32$
    b) 1; 5; 10
    c) 16 out of 32

43. 24

Seat 1	Seat 2	Seat 3	Seat 4
Jay-Z	Beyonce	Brad	Angelina
Jay-Z	Beyonce	Angelina	Brad
Jay-Z	Brad	Beyonce	Angelina
Jay-Z	Brad	Angelina	Beyonce
Jay-Z	Angelina	Beyonce	Brad
Jay-Z	Angelina	Brad	Beyonce
Beyonce	Jay-Z	Brad	Angelina
Beyonce	Jay-Z	Angelina	Brad
Beyonce	Brad	Jay-Z	Angelina
Beyonce	Brad	Angelina	Jay-Z
Beyonce	Angelina	Jay-Z	Brad
Beyonce	Angelina	Brad	Jay-Z

Seat 1	Seat 2	Seat 3	Seat 4
Brad	Jay-Z	Beyonce	Angelina
Brad	Jay-Z	Angelina	Beyonce
Brad	Beyonce	Jay-Z	Angelina
Brad	Beyonce	Angelina	Jay-Z
Brad	Angelina	Jay-Z	Beyonce
Brad	Angelina	Beyonce	Jay-Z
Angelina	Jay-Z	Brad	Beyonce
Angelina	Jay-Z	Beyonce	Brad
Angelina	Beyonce	Jay-Z	Brad
Angelina	Beyonce	Brad	Jay-Z
Angelina	Brad	Jay-Z	Beyonce
Angelina	Brad	Beyonce	Jay-Z

45. 12

Seat 1	Seat 2	Seat 3	Seat 4
Beyonce	Jay-Z	Angelina	Brad
Angelina	Jay-Z	Beyonce	Brad
Beyonce	Brad	Angelina	Jay-Z
Angelina	Brad	Beyonce	Jay-Z
Beyonce	Jay-Z	Brad	Angelina
Angelina	Jay-Z	Brad	Beyonce
Beyonce	Brad	Jay-Z	Angelina
Angelina	Brad	Jay-Z	Beyonce
Jay-Z	Beyonce	Brad	Angelina
Jay-Z	Angelina	Brad	Beyonce
Brad	Beyonce	Jay-Z	Angelina
Brad	Angelina	Jay-Z	Beyonce

47. In the bottom row, there are $5 \times 5 = 25$ oranges. In the next row there will be $4 \times 4 = 16$ oranges. This is followed by oranges, then $2 \times 2 = 4$ oranges, and finally one orange on top. There is a total of $25 + 16 + 9 + 4 + 1 = 55$ oranges.

Use the diagram on the next page for the solutions to Exercises 49 – 51.

49. 27

51. 18

Vanilla – Vanilla – Strawberry, Vanilla – Vanilla – Chocolate, Vanilla – Strawberry – Vanilla, Vanilla – Strawberry – Strawberry, Vanilla – Chocolate – Vanilla, Vanilla-Chocolate – Chocolate, Strawberry – Vanilla – Vanilla, Strawberry – Vanilla – Strawberry, Strawberry – Strawberry – Vanilla, Strawberry – Strawberry – Chocolate, Strawberry – Chocolate – Strawberry, Strawberry – Chocolate – Chocolate, Chocolate – Vanilla – Vanilla, Chocolate – Vanilla – Chocolate, Chocolate – Strawberry – Strawberry, Chocolate – Strawberry – Chocolate, Chocolate – Chocolate – Vanilla, and Chocolate -Chocolate – Strawberry

53. We thought of flipping the coins one at a time. We also thought of rolling the dice one at a time.

55. No solution provided.

Use this diagram for the solutions to Exercises 49 – 51.

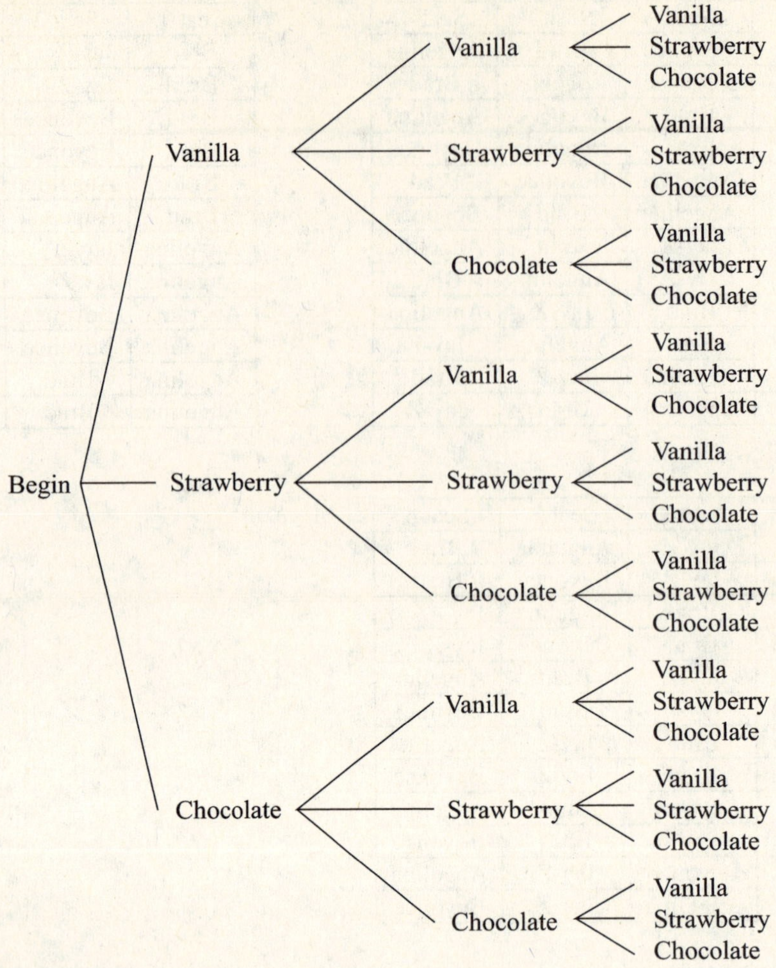

57. Follow the branches in the tree below to make the schedules.

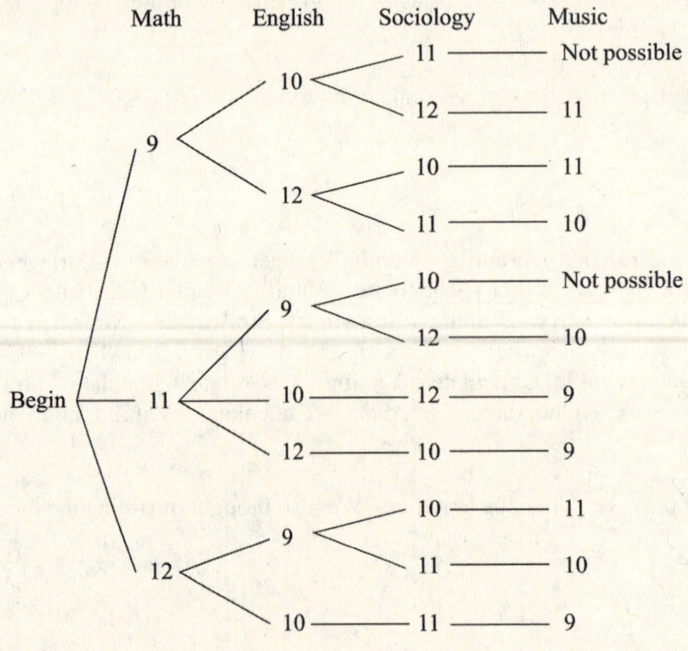

57. (continued)

His nine possible schedules are:

9:00	10:00	11:00	12:00
Math	English	Music	Sociology
Math	Sociology	Music	English
Math	Music	Sociology	English
English	Music	Math	Sociology
Music	English	Math	Sociology
Music	Sociology	Math	English
English	Sociology	Music	Math
English	Music	Sociology	Math
Music	English	Sociology	Math

Use the following diagram for the solutions to Exercises 59 – 61.

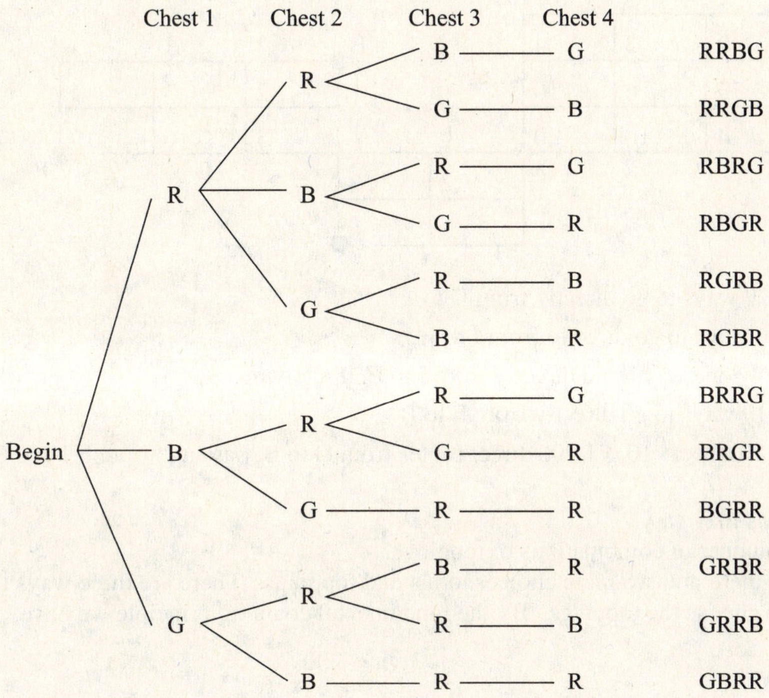

59. RRBG, RBRG, BRRG

61. RBGR, RGBR, BRGR, BGRR, GRBR, GBRR

Section 13.2 The Fundamental Counting Principle

1. In (a), we thought of flipping the four coins one at a time instead of flipping them all at once. In (b), we thought of rolling the red die, then the blue die, and then the green die.

3. A skilled player can win at blackjack.

5. $7 \times 6 = 42$

7. $8 \times 7 \times 6 = 336$

9. $4 \times 8 \times 3 \times 5 = 480$

11. $5 \times 6 \times 13 \times 4 = 1{,}560$

13. $8 \times 8 = 64$

15. $8\times12=96$

17. $9\times10\times10\times10=9{,}000$

19. If the number must be greater than 5,000 and odd, the first digit must be 5 or greater and the last digit must be odd, so $5\times10\times10\times5=2{,}500$.

21. $1\times26\times26\times26=17{,}576$

23. $12\times11\times10\times9\times8\times7\times6\times5\times4\times3\times2\times1=479{,}001{,}600$

25. $3\times3\times3\times3\times3=243$

27. To choose direct routes, you must always be traveling down and/or to the right. The numbers in the diagrams below indicate how many ways there are to get to the intersection below and to the right of the number between each stop.

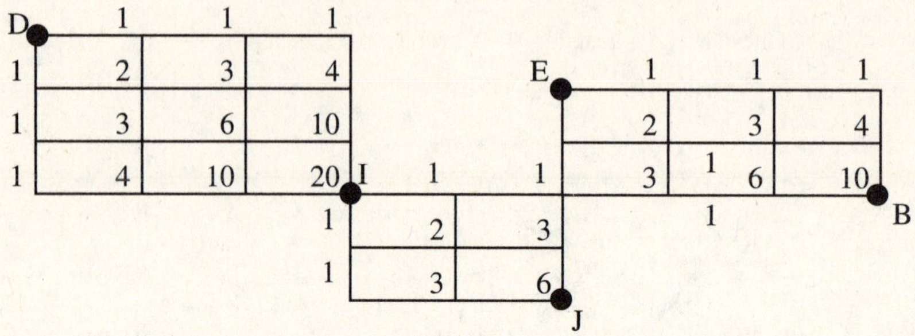

a) There are 20 ways to go directly from D to I.

b) There are 6 ways to go directly from I to J.

c) There is only one way to go directly from J to E, directly up.

d) There are 10 ways to go directly from E to B.

e) There are $20\times6\times1\times10=1{,}200$ direct routes from D to B, passing through J and E.

29. 6; Explanations may vary.
Let n be the number of combinations of toppings.
Consider that there are two main choices, crust and toppings. There are three ways to choose the crust and n ways to choose the toppings. By the fundamental counting principle we have

$$3\times n<200$$
$$n<66\tfrac{2}{3}$$

From Chapter 1, we have that there will be 2^t combinations, where t is the number of toppings. Since $2^6=64=n$, there must be 6 toppings for $3\cdot64=192$ combinations.

31. Louise and her tutor can sit in the front in 7 different ways. Since the tutor must sit to the right of Louise, we would have $7\times1\times10\times9\times8\times7\times6\times5=1{,}058{,}400$ ways to arrange the front row.

33. $6\times5\times4\times3\times2\times1=720$

35. $6\times2\times1\times3\times2\times1=72$

37. In example 5, we considered the condition that Louise and her tutor must sit next to each other first.

39. Answers may vary.
We can use a tree diagram to visualize the total number of possibilities that we count when we use the Fundamental Counting Principle.

41. – 47. No solution provided.

Section 13.3 Permutations and Combinations

1. Objects cannot be repeated in a permutation.

3. One combination of three people corresponds to 3! = 6 permutations.

5. $4! = 4\cdot 3\cdot 2\cdot 1 = 24$

7. $(8-5)! = 3! = 3\cdot 2\cdot 1 = 6$

9. $\frac{10!}{7!} = \frac{10\cdot 9\cdot 8\cdot 7\cdot 6\cdot 5\cdot 4\cdot 3\cdot 2\cdot 1}{7\cdot 6\cdot 5\cdot 4\cdot 3\cdot 2\cdot 1} = 10\cdot 9\cdot 8 = 720$

11. $\frac{10!}{7!3!} = \frac{10\cdot 9\cdot 8\cdot 7\cdot 6\cdot 5\cdot 4\cdot 3\cdot 2\cdot 1}{7\cdot 6\cdot 5\cdot 4\cdot 3\cdot 2\cdot 1\cdot 3\cdot 2\cdot 1} = \frac{10\cdot 9\cdot 8}{3\cdot 2\cdot 1} = 10\cdot 3\cdot 4 = 120$

13. $P(6,2) = \frac{6!}{(6-2)!} = \frac{6!}{4!} = \frac{6\cdot 5\cdot 4\cdot 3\cdot 2\cdot 1}{4\cdot 3\cdot 2\cdot 1} = 6\cdot 5 = 30$

15. $C(10,3) = \frac{10!}{3!(10-3)!} = \frac{10!}{3!7!} = \frac{10\cdot 9\cdot 8\cdot 7\cdot 6\cdot 5\cdot 4\cdot 3\cdot 2\cdot 1}{3\cdot 2\cdot 1\cdot 7\cdot 6\cdot 5\cdot 4\cdot 3\cdot 2\cdot 1} = \frac{10\cdot 9\cdot 8}{3\cdot 2\cdot 1} = 10\cdot 3\cdot 4 = 120$

17. $P(8,3) = \frac{8!}{(8-3)!} = \frac{8!}{5!} = \frac{8\cdot 7\cdot 6\cdot 5\cdot 4\cdot 3\cdot 2\cdot 1}{5\cdot 4\cdot 3\cdot 2\cdot 1} = 8\cdot 7\cdot 6 = 336$

19. $P(10,8) = \frac{10!}{(10-8)!} = \frac{10!}{2!} = \frac{10\cdot 9\cdot 8\cdot 7\cdot 6\cdot 5\cdot 4\cdot 3\cdot 2\cdot 1}{2\cdot 1} = 10\cdot 9\cdot 8\cdot 7\cdot 6\cdot 5\cdot 4\cdot 3 = 1{,}814{,}400$

21. $C(8,3) = \frac{8!}{3!(8-3)!} = \frac{8!}{3!5!} = \frac{8\cdot 7\cdot 6\cdot 5\cdot 4\cdot 3\cdot 2\cdot 1}{3\cdot 2\cdot 1\cdot 5\cdot 4\cdot 3\cdot 2\cdot 1} = \frac{8\cdot 7\cdot 6}{3\cdot 2\cdot 1} = 8\cdot 7 = 56$

23. $C(10,8) = \frac{10!}{8!(10-8)!} = \frac{10!}{8!2!} = \frac{10\cdot 9\cdot 8\cdot 7\cdot 6\cdot 5\cdot 4\cdot 3\cdot 2\cdot 1}{8\cdot 7\cdot 6\cdot 5\cdot 4\cdot 3\cdot 2\cdot 1\cdot 2\cdot 1} = \frac{10\cdot 9}{2\cdot 1} = 5\cdot 9 = 45$

25. The number of permutations of three selections that can be made from 10 objects.

27.

1
1 1
1 2 1
1 3 3 1
1 4 6 4 1
1 5 10 10 5 1
1 6 15 20 15 6 1
1 7 21 35 35 21 7 1
1 8 28 56 70 56 28 8 1

29. Seventh Row: 1 7 **21** 35 35 21 7 1

31. The second entry of the 18th row

33. The sixth entry of the 20th row

35. $P(8,8)$

37. $C(17,3)$

39. $P(5,5)$

41. $C(10,3)$

43. $C(9,5)$

45. $C(17,8)$

47. $C(25,6)=\dfrac{25!}{6!(25-6)!}=\dfrac{25!}{6!19!}$

$$=\frac{25\cdot 24\cdot 23\cdot 22\cdot 21\cdot 20\cdot 19\cdot 18\cdot 17\cdot 16\cdot 15\cdot 14\cdot 13\cdot 12\cdot 11\cdot 10\cdot 9\cdot 8\cdot 7\cdot 6\cdot 5\cdot 4\cdot 3\cdot 2\cdot 1}{6\cdot 5\cdot 4\cdot 3\cdot 2\cdot 1\cdot 19\cdot 18\cdot 17\cdot 16\cdot 15\cdot 14\cdot 13\cdot 12\cdot 11\cdot 10\cdot 9\cdot 8\cdot 7\cdot 6\cdot 5\cdot 4\cdot 3\cdot 2\cdot 1}$$

$$=\frac{25\cdot 24\cdot 23\cdot 22\cdot 21\cdot 20}{6\cdot 5\cdot 4\cdot 3\cdot 2\cdot 1}=5\cdot 23\cdot 11\cdot 7\cdot 20=177{,}100$$

49. $P(17,8)=\dfrac{17!}{(17-8)!}=\dfrac{17!}{9!}=\dfrac{17\cdot 16\cdot 15\cdot 14\cdot 13\cdot 12\cdot 11\cdot 10\cdot 9\cdot 8\cdot 7\cdot 6\cdot 5\cdot 4\cdot 3\cdot 2\cdot 1}{9\cdot 8\cdot 7\cdot 6\cdot 5\cdot 4\cdot 3\cdot 2\cdot 1}$

$$=17\cdot 16\cdot 15\cdot 14\cdot 13\cdot 12\cdot 11\cdot 10=980{,}179{,}200 \text{ minutes}$$

There are $60\cdot 24\cdot 365=525{,}600$ minutes per year, so it would take $\dfrac{980{,}179{,}200}{525{,}600}\approx 1{,}864.9$ years or about 1,865 years.

51. $P(15,5)=\dfrac{15!}{(15-5)!}=\dfrac{15!}{10!}=\dfrac{15\cdot 14\cdot 13\cdot 12\cdot 11\cdot 10\cdot 9\cdot 8\cdot 7\cdot 6\cdot 5\cdot 4\cdot 3\cdot 2\cdot 1}{10\cdot 9\cdot 8\cdot 7\cdot 6\cdot 5\cdot 4\cdot 3\cdot 2\cdot 1}=15\cdot 14\cdot 13\cdot 12\cdot 11=360{,}360$

53. $P(15,5)\times P(15,5)\times P(15,4)\times P(15,5)\times P(15,5)=552{,}446{,}474{,}061{,}128{,}648{,}601{,}600{,}000$ (a VERY large number)

55. $C(6,2)\cdot C(8,3)=\dfrac{6!}{2!(6-2)!}\cdot\dfrac{8!}{3!(8-3)!}=\dfrac{6!}{2!4!}\cdot\dfrac{8!}{3!5!}=\dfrac{6\cdot 5\cdot 4\cdot 3\cdot 2\cdot 1}{2\cdot 1\cdot 4\cdot 3\cdot 2\cdot 1}\cdot\dfrac{8\cdot 7\cdot 6\cdot 5\cdot 4\cdot 3\cdot 2\cdot 1}{3\cdot 2\cdot 1\cdot 5\cdot 4\cdot 3\cdot 2\cdot 1}$

$$=\frac{6\cdot 5\cdot}{2\cdot 1}\cdot\frac{8\cdot 7\cdot 6}{3\cdot 2\cdot 1}=(3\cdot 5)\cdot(8\cdot 7)=15\cdot 56=840$$

57. $C(7,3)\cdot C(5,2)\cdot C(11,3)=\dfrac{7!}{3!(7-3)!}\cdot\dfrac{5!}{2!(5-2)!}\cdot\dfrac{11!}{3!(11-3)!}=\dfrac{7!}{3!4!}\cdot\dfrac{5!}{2!3!}\cdot\dfrac{11!}{3!8!}=$

$$\frac{7\cdot 6\cdot 5\cdot 4\cdot 3\cdot 2\cdot 1}{3\cdot 2\cdot 1\cdot 4\cdot 3\cdot 2\cdot 1}\cdot\frac{5\cdot 4\cdot 3\cdot 2\cdot 1}{2\cdot 1\cdot 3\cdot 2\cdot 1}\cdot\frac{11\cdot 10\cdot 9\cdot 8\cdot 7\cdot 6\cdot 5\cdot 4\cdot 3\cdot 2\cdot 1}{3\cdot 2\cdot 1\cdot 8\cdot 7\cdot 6\cdot 5\cdot 4\cdot 3\cdot 2\cdot 1}=\frac{7\cdot 6\cdot 5}{3\cdot 2\cdot 1}\cdot\frac{5\cdot 4}{2\cdot 1}\cdot\frac{11\cdot 10\cdot 9}{3\cdot 2\cdot 1}=$$

$$(7\cdot 5)\cdot(5\cdot 2)\cdot(11\cdot 5\cdot 3)=35\cdot 10\cdot 165=57{,}750$$

59. $C(5,2)\cdot C(4,2)=\dfrac{5!}{2!(5-2)!}\cdot\dfrac{4!}{2!(4-2)!}=\dfrac{5!}{2!3!}\cdot\dfrac{4!}{2!2!}=\dfrac{5\cdot 4\cdot 3\cdot 2\cdot 1}{2\cdot 1\cdot 3\cdot 2\cdot 1}\cdot\dfrac{4\cdot 3\cdot 2\cdot 1}{2\cdot 1\cdot 2\cdot 1}=\dfrac{5\cdot 4}{2\cdot 1}\cdot\dfrac{4\cdot 3}{2\cdot 1}$

$$=(5\cdot 2)\cdot(2\cdot 3)=10\cdot 6=60$$

61. $P(12,2)\cdot C(10,2)=\dfrac{12!}{(12-2)!}\cdot\dfrac{10!}{2!(10-2)!}=\dfrac{12!}{10!}\cdot\dfrac{10!}{2!8!}$

$$=\frac{12\cdot 11\cdot 10\cdot 9\cdot 8\cdot 7\cdot 6\cdot 5\cdot 4\cdot 3\cdot 2\cdot 1}{10\cdot 9\cdot 8\cdot 7\cdot 6\cdot 5\cdot 4\cdot 3\cdot 2\cdot 1}\cdot\frac{10\cdot 9\cdot 8\cdot 7\cdot 6\cdot 5\cdot 4\cdot 3\cdot 2\cdot 1}{2\cdot 1\cdot 8\cdot 7\cdot 6\cdot 5\cdot 4\cdot 3\cdot 2\cdot 1}=12\cdot 11\cdot\frac{10\cdot 9}{2\cdot 1}$$

$$=132\cdot(5\cdot 9)=132\cdot 45=5{,}940$$

63. $P(5,2)\cdot C(11,3)=\dfrac{5!}{(5-2)!}\cdot\dfrac{11!}{3!(11-3)!}=\dfrac{5!}{3!}\cdot\dfrac{11!}{3!8!}$

$$=\frac{5\cdot 4\cdot 3\cdot 2\cdot 1}{3\cdot 2\cdot 1}\cdot\frac{11\cdot 10\cdot 9\cdot 8\cdot 7\cdot 6\cdot 5\cdot 4\cdot 3\cdot 2\cdot 1}{3\cdot 2\cdot 1\cdot 8\cdot 7\cdot 6\cdot 5\cdot 4\cdot 3\cdot 2\cdot 1}=5\cdot 4\cdot\frac{11\cdot 10\cdot 9}{3\cdot 2\cdot 1}$$

$$=20\cdot(11\cdot 5\cdot 3)=20\cdot 165=3{,}300$$

65. As you look at the intersections, you'll see that there is a pattern as to how many routes can be created as you leave the mall on the way to the bank. To choose direct routes, you must always be traveling down and/or to the right. The numbers indicate how many ways there are to get to the intersection below and to the right of the number. The total length of the route is 8 blocks. There are 70 possible routes.

Mall	1	1	1	1
1	2	3	4	5
1	3	6	10	15
1	4	10	20	35
1	5	15	35	70

Bank

Also, since there are 8 blocks to be traveled and you could turn right at any of these intersections four times. So there are $C(8,4)=\dfrac{8!}{4!(8-4)!}=\dfrac{8!}{4!4!}=\dfrac{8\cdot 7\cdot 6\cdot 5\cdot 4\cdot 3\cdot 2\cdot 1}{4\cdot 3\cdot 2\cdot 1\cdot 4\cdot 3\cdot 2\cdot 1}=\dfrac{8\cdot 7\cdot 6\cdot 5}{4\cdot 3\cdot 2\cdot 1}=7\cdot 2\cdot 5=70$ paths.

Also, since there are 8 blocks to be traveled and you could go down at any of these intersections four times. So there are $C(8,4)=\dfrac{8!}{4!(8-4)!}=\dfrac{8!}{4!4!}=\dfrac{8\cdot 7\cdot 6\cdot 5\cdot 4\cdot 3\cdot 2\cdot 1}{4\cdot 3\cdot 2\cdot 1\cdot 4\cdot 3\cdot 2\cdot 1}=\dfrac{8\cdot 7\cdot 6\cdot 5}{4\cdot 3\cdot 2\cdot 1}=7\cdot 2\cdot 5=70$ paths.

67. In permutation problems, the order of the arrangement matters; in combination problems, the order of arrangement does not matter.

69. the rth entry in the nth row

71. Because of the special conditions on the numbers, we cannot use all digits at each position in the number.

73. The number of ways to do something must be an integer.

75. – 77. No solution provided.

79. $C(13,2)\cdot C(13,3)=\dfrac{13!}{2!(13-2)!}\cdot\dfrac{13!}{3!(13-3)!}=\dfrac{13!}{2!11!}\cdot\dfrac{13!}{3!10!}=$

$$\frac{13\cdot12\cdot11\cdot10\cdot9\cdot8\cdot7\cdot6\cdot5\cdot4\cdot3\cdot2\cdot1}{2\cdot1\cdot11\cdot10\cdot9\cdot8\cdot7\cdot6\cdot5\cdot4\cdot3\cdot2\cdot1}\cdot\frac{13\cdot12\cdot11\cdot10\cdot9\cdot8\cdot7\cdot6\cdot5\cdot4\cdot3\cdot2\cdot1}{3\cdot2\cdot1\cdot10\cdot9\cdot8\cdot7\cdot6\cdot5\cdot4\cdot3\cdot2\cdot1}=\frac{13\cdot12}{2\cdot1}\cdot\frac{13\cdot12\cdot11}{3\cdot2\cdot1}=$$

$(13\cdot6)\cdot(13\cdot2\cdot11)=78\cdot286=22{,}308$

81. 

83. 6; for six lines

85. $C(n-1,r-1)$

87. $C(n-1,r-1)+C(n-1,r)=C(n,r)$

89. The number of ways that we can choose an *r*-element set from *n* elements is the same as the number of ways that we can choose $n-r$ elements from *n* elements; for example, the number of ways that we can choose two elements from five elements is the same as the number of ways that we can choose three elements from five elements.

Section 13.4 Looking Deeper: Counting and Gambling

1. The number of ways each of the three wheels could stop were multiplied together.

3. $2\times5\times12=120$

5. $5\times2\times2=20$

7. $1\times5\times2=10$

9. 4, since there are four suits

11. a) 4

b) $C(13,5)=\dfrac{13!}{5!(13-5)!}=\dfrac{13!}{5!8!}=\dfrac{13\cdot12\cdot11\cdot10\cdot9\cdot8\cdot7\cdot6\cdot5\cdot4\cdot3\cdot2\cdot1}{5\cdot4\cdot3\cdot2\cdot1\cdot8\cdot7\cdot6\cdot5\cdot4\cdot3\cdot2\cdot1}$

$$=\frac{13\cdot12\cdot11\cdot10\cdot9}{5\cdot4\cdot3\cdot2\cdot1}=13\cdot11\cdot9=1{,}287$$

c) Although there are $4\cdot1{,}287=5{,}148$ ways to choose five cards from the same suit, 40 of these ways are either straight flushes or royal flushes. So, there are $5{,}148-40=5{,}108$ ways to construct a flush.

13. We can construct a two-pair hand in stages:

Stage 1: Pick the rank of the card of which we will have a pair. This decision can be made in 13 ways.

Stage 2: Now choose these two cards from the four cards of a rank. This can be done in $C(4,2)=6$ ways

Stage 3: Now that you have chosen two cards, we select the next two for a pair. This decision can be made in 12 ways, because we don't want four-of-a-kind.

Stage 4: Now choose these two cards from the four cards of a rank. This can be done in $C(4,2)=6$ ways

Stage 5: Now the fifth card can be chosen in 44 ways since eight of the cards must be excluded.

Since there are two ways to choose the two ranks, we must divide by two.

So there are $\dfrac{13\times6\times12\times6}{2}\times44=\dfrac{5{,}616}{2}\times44=2{,}808\times44=123{,}552$ ways to choose a two-pair hand.

15. We are not simply choosing three cards from a possible 52.

17. The number of ways we can obtain the payoff for three cherries is $2 \times 5 \times 8 = 80$.
The number of ways we can obtain the payoff for three oranges is $5 \times 2 \times 2 = 20$.
Since there are fewer ways to obtain three oranges, we would expect the payoff to be higher.

19. No solution provided.

Chapter Review Exercises

1. PQ, PR, PS, QP, QR, QS, RP, RQ, RS, SP, SQ, SR

3. Without drawing the tree, you can imagine that the tree begins and there are five branches on the first shirt choice. Each of those branches split into four branches for the second choice of pants for a total of twenty branches. Each of those branches split into six branches for the choice of a tie for a total of 120 branches. Each of those branches split into two branches for the choice of a jacket for a total of 240 branches. There would therefore be **240** outfits that could be formed.

5. $4 \times 12 \times 6 = 288$

7.

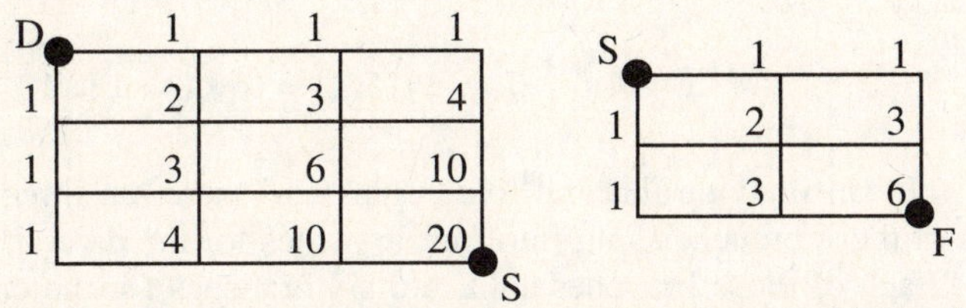

To choose direct routes, you must always be traveling down and/or to the right. The numbers indicate how many ways there are to get to the intersection below and to the right of the number. We see there are 20 ways to go directly from D to S and 6 ways to go directly from S to F. Therefore, there are $20 \times 6 = 120$ ways to go directly from D to F.

9. In a permutation, order is important; in a combination, it is not.

11. $$C(17,3) = \frac{17!}{3!(17-3)!} = \frac{17!}{3!14!} = \frac{17 \cdot 16 \cdot 15 \cdot 14 \cdot 13 \cdot 12 \cdot 11 \cdot 10 \cdot 9 \cdot 8 \cdot 7 \cdot 6 \cdot 5 \cdot 4 \cdot 3 \cdot 2 \cdot 1}{3 \cdot 2 \cdot 1 \cdot 14 \cdot 13 \cdot 12 \cdot 11 \cdot 10 \cdot 9 \cdot 8 \cdot 7 \cdot 6 \cdot 5 \cdot 4 \cdot 3 \cdot 2 \cdot 1}$$
$$= \frac{17 \cdot 16 \cdot 15}{3 \cdot 2 \cdot 1} = 17 \cdot 8 \cdot 5 = 680$$

13. $$C(n,r) = \frac{P(n,r)}{r!}$$

15. Beginning with the zeroth entry, $C(18,2)$ represents the 2nd entry of the 18th row.

17. First, we need to pick five cards from the same suit:
$$C(13,5) = \frac{13!}{5!(13-5)!} = \frac{13!}{5!8!} = \frac{13 \cdot 12 \cdot 11 \cdot 10 \cdot 9 \cdot 8 \cdot 7 \cdot 6 \cdot 5 \cdot 4 \cdot 3 \cdot 2 \cdot 1}{5 \cdot 4 \cdot 3 \cdot 2 \cdot 1 \cdot 8 \cdot 7 \cdot 6 \cdot 5 \cdot 4 \cdot 3 \cdot 2 \cdot 1}$$
$$= \frac{13 \cdot 12 \cdot 11 \cdot 10 \cdot 9}{5 \cdot 4 \cdot 3 \cdot 2 \cdot 1} = 13 \cdot 11 \cdot 9 = 1{,}287$$
Since there are four suites, there are $4 \cdot 1{,}287 = 5{,}148$ ways to choose five cards from the same suit. However, 40 of these ways are either straight flushes or royal flushes. So, there are $5{,}148 - 40 = 5{,}108$ ways to construct a flush.

Chapter Test

1. AB, AC, AD, BA, BC, BD, CA, CB, CD, DA, DB, DC

3.

Seat 1	Seat 2	Seat 3	Seat 4	Seat 5
Devaun	Emily	X	X	X
X	Devaun	Emily	X	X
X	X	Devaun	Emily	X
X	X	X	Devaun	Emily

Devaun and Emily can occupy four different groupings of two seats. There are two different arrangements for each of these four possibilities. The number of ways that the students could be arranged in these six seats is $4 \times 2 \times 3 \times 2 \times 1 = 48$.

5. $1 \times 5 \times 2 = 10$

7. $P(15,4) = 15 \times 14 \times 13 \times 12 = 32,760$

9. Without drawing the tree, you can imagine that the tree begins and there are three branches for the biology class choice. Each of those branches split into four branches for the diversity class choice for a total of twelve branches. Each of those branches split into six branches for the choice of a writing class for a total of seventy-two branches. There would therefore be **72** class schedules that could be formed.

11. $5 \times 4 \times 11 \times 6 = 1,320$

15. $3 \times 9 \times 2 \times 2 = 108$

13. $C(n,r) = \dfrac{P(n,r)}{r!}$

Chapter 14
Probability: What Are the Chances?

Section 14.1 The Basics of Probability Theory

1. We drew a tree diagram.

3. We know from Chapter 13 that there are $C(52,5)$ ways to choose five cards from 52.

5. The bread could only make one-half of a revolution before hitting the floor. With three lines, there is only a one-third chance that you are in the fastest lane.

7. $\{(1,6), (2,5), (3,4), (4,3), (5,2), (6,1)\}$

9. If we let "h" represent a heads and "t" represent tails, then the outcomes would be {hhh, hht, hth, thh}.

11. If we let "r" represent red, "b" represent blue, and "y" represent yellow, then the outcomes would be {(r,b), (r,y), (b,r), (y,r)}.

13. If we let "r" represent red, "b" represent blue, and "y" represent yellow, then the outcomes would be {(b,b,r), (b,r,b), (r,b,b), (b,b,y), (b,y,b), (y,b,b)}.

15. a) The sample space would have sixteen pairs. The sample space would be $\{(1,1), (1,2), (1,3), (1,4), (2,1), (2,2), (2,3), (2,4), (3,1), (3,2), (3,3), (3,4), (4,1), (4,2), (4,3), (4,4)\}$.

    b) The event would be $\{(1,1), (1,3), (2,2), (2,4), (3,1), (3,3), (4,2), (4,4)\}$.

    c) From (b), an even total can occur in eight ways. Since there are 16 pairs that can occur, the probability of rolling an even sum on two four-sided dice is $\frac{8}{16}=\frac{1}{2}$

    d) A total greater than six can occur in three ways. They are: $\{(3,4), (4,3), (4,4)\}$. Since there are 16 pairs that can occur, the probability of rolling a total greater than six on two four-sided dice is $\frac{3}{16}$.

17. a) $4\times3\times2\times1=24$

    b) {CKMJ, CJMK, MKCJ, MJCK, KCJM, KMJC, JCKM, JMKC}

    c) From (b), there are eight ways where the women do not perform consecutively. Since there are 24 ways for the people to perform, the probability that the women do not perform consecutively is $\frac{8}{24}=\frac{1}{3}$

    d) There are $3\times2\times1\times1=6$ ways that Carrie can perform last. Since there are 24 ways for the people to perform, the probability that Carrie performs last is $\frac{6}{24}=\frac{1}{4}$

19. a) 20; The sample space is {(s,c), (s,w), (s,d), (s,h), (c,s), (c,w), (c,d), (c,h), (w,s), (w,c), (w,d), (w,h), (d,s), (d,c), (d,w), (d,h), (h,s), (h,c), (h,w), (h,d)}. Also, we could use $P(5,2)=20$.

    b) {(s,c), (s,w), (s,d), (s,h), (c,s), (w,s), (d,s), (h,s)}.

19. (continued)

c) Since exactly one star can occur in eight ways, the probability is $\frac{8}{20}=\frac{2}{5}$.

d) No hearts can occur in twelve ways. They are: {(s,c), (s,w), (s,d), (c,s), (c,w), (c,d), (w,s), (w,c), (w,d), (d,s), (d,c), (d,w)}, thus, the probability is $\frac{12}{20}=\frac{3}{5}$.

21. Let E represent the event that a total of nine shows on the roll of two fair dice.

a) A total of nine can occur in four ways. They are: {(3,6), (4,5), (5,4), (6,3)}. Since there are 36 pairs that can occur, $P(E)=\frac{4}{36}=\frac{1}{9}\approx 0.11$

b) Since $P(E)=\frac{1}{9}$, $P(E')=1-\frac{1}{9}=\frac{9}{9}-\frac{1}{9}=\frac{8}{9}$. Thus, the odds against the event are $\frac{P(E')}{P(E)}=\frac{8/9}{1/9}=\frac{8}{1}$, or 8 to 1.

23. Let E represent the event of drawing a heart.

a) Since there are 13 hearts out of the 52 cards, $P(E)=\frac{13}{52}=\frac{1}{4}=0.25$.

b) Since $P(E)=\frac{1}{4}$, $P(E')=1-\frac{1}{4}=\frac{4}{4}-\frac{1}{4}=\frac{3}{4}$. Thus, the odds against the event are $\frac{P(E')}{P(E)}=\frac{3/4}{1/4}=\frac{3}{1}$, or 3 to 1.

25. $\frac{C(13,5)}{C(52,5)}=\frac{1,287}{2,598,960}\approx 0.000495$

27. $4\times\frac{C(13,5)}{C(52,5)}=4\times\frac{1,287}{2,598,960}\approx 0.002$

29. a) Since 5,040 people support the racetrack out of the total of 9,072 people, the probability of selecting a person that supports the racetrack is $\frac{5,040}{9,072}=\frac{5}{9}\approx 0.5556$

b) Since $P(E)=\frac{5}{9}$, $P(E')=1-\frac{5}{9}=\frac{9}{9}-\frac{5}{9}=\frac{4}{9}$. Thus, the odds for the event are $\frac{P(E)}{P(E')}=\frac{5/9}{4/9}=\frac{5}{4}$, or 5 to 4.

31. Since there are 1,951,379 females out of the 1,951,379 + 2,048,861 = 4,000,240 children, the probability of selecting a female is $\frac{1,951,379}{4,000,240}\approx 0.49$

33. Since there are 277 short plants out of the 787 + 277 = 1,064 plants, the probability that the plant will be short is $\frac{277}{1,064}\approx 0.26$. This is consistent with the theoretical results that there should be a 3 to 1 ratio in the second generation of dominant trait to recessive trait; that is, the theoretical probability of that a plant selected from the second generation will be short is 0.25.

35. The probability that the child is a carrier is $\frac{2}{4}=\frac{1}{2}$.

Second Parent

		s	n
First	s	ss	sn
Parent	s	ss	sn

37. a)

First Gen.

		r	r
First	w	wr	wr
Gen.	w	wr	wr

b) All flowers will be pink. $P(\text{pink}) = 1$; all other probabilities are 0.

39. a)

Second Parent

		N	c
First	N	NN	Nc
Parent	c	cN	cc

b) $P(\text{disease})=\frac{1}{4}$

41. There are 109 + 29 = 138 students out of the 320 surveyed that has a GPA of at least 2.5. The probability of selecting a student with a GPA of at least 2.5 is $\frac{138}{320}=\frac{69}{160}\approx 0.43$.

43. Since you are rolling two dice, there are 36 possible combinations. St. James Place, Tennessee Avenue, and New York Avenue are within 12 spaces from the Electric Company. St. James Place is 4 spaces from the Electric Company. You can land on St. James Place if you roll (1,3), (2,2), or (3,1). Tennessee Avenue is 6 spaces from the Electric Company. You can land on Tennessee Avenue if you roll (1,5), (2,4), (3,3), (4,2), or (5,1). New York Avenue is 7 spaces from the Electric Company. You can land on New York Avenue if you roll (1,6), (2,5), (3,4), (4,3), (5,2), or (6,1). Since there are 14 ways to go bankrupt, there are $36-14=22$ ways to avoid going bankrupt. Thus, the probability of avoiding bankruptcy is $\frac{22}{36}=\frac{11}{18}\approx 0.61$.

45. Since you are rolling two dice, there are 36 possible combinations. There are two railroads within 12 spaces from Virginia Avenue. The closest one is one space away. You cannot land on it, since the minimum roll would be a total of 2. The second railroad is 11 spaces from Virginia Avenue. You can land on it if you roll (5,6) or (6,5). Thus the probability of landing on a railroad is $\frac{2}{36}=\frac{1}{18}\approx 0.06$.

47. The sample space, S, consists of $10{,}818+14{,}182=25{,}000$ thousand people that are divorced. The event, call it A, is the set of 14,182 women that are divorced.

The desired probability is $P(A)=\frac{n(A)}{n(S)}=\frac{14{,}182}{25{,}000}\approx 0.5673$.

49. The sample space, S, consists of $39,435+33,377=72,812$ people that are not married. The event, call it A is the set of 39,435 people that are men.

The desired probability is $P(A)=\frac{n(A)}{n(S)}=\frac{39,435}{72,812}\approx 0.5416$.

51. Since order is important, the number of ways of choosing the horses would be a permutation. There are $P(8,3)=336$ ways to choose the horses. Since only one way is a winner, the probability of winning is $\frac{1}{336}\approx 0.002976$.

53. The odds against E are $\frac{5}{2}=\frac{5/7}{2/7}=\frac{P(E')}{P(E)}$. Thus, the probability of E is $\frac{2}{7}$.

55. The odds against W are $\frac{7}{5}=\frac{7/12}{5/12}=\frac{P(W')}{P(W)}$. Thus, the probability of winning is $\frac{5}{12}$.

57. a) The odds in favor of E (winning) are $\frac{P(E)}{P(E')}=\frac{0.30}{1-0.30}=\frac{0.30}{0.70}=\frac{3/10}{7/10}=\frac{3}{7}$, or 3 to 7.

b) The odds against E (winning) are $\frac{P(E')}{P(E)}=\frac{1-0.30}{0.30}=\frac{0.70}{0.30}=\frac{7/10}{3/10}=\frac{7}{3}$, or 7 to 3.

59. There are $C(59,6)=45,057,474$ ways to choose six numbers from the numbers 1 to 59. Since there is only one way to pick the correct numbers, the odds against winning are 45,057,474 to 1.

61. $P(E)=\frac{\text{number of times } E \text{ occurs}}{\text{number of times experiment is performed}}$

63. All plants had one yellow and one green gene, and yellow was dominant.

65. Answers may vary.
An outcome is an element in a sample space; an event is a subset of a sample space.

67. Answers may vary.
Rolling a total less than thirteen when rolling two dice.

69. – 71. No solution provided.

73. Since the hacker can enter a password every 10 seconds, he will have 6 opportunities each minute. Since the hacker is allowed three minutes, he has a total of 18 opportunities. Since the hacker is randomly trying passwords, it is conceivable that the hacker might try the same password more than once. Since there are $26\times 26\times 10\times 10\times 10=676,000$ possible passwords, the probability that the hacker would be successful in discovering a valid password is $18\times\frac{1}{676,000}\approx 0.000027$.

75. – 79. No solution provided.

Section 14.2 Complements and Unions Of Events

1. It was easier to compute $P(A')$ and then subtract it from one instead of computing $P(A)$ directly.

3. Your chances of winning a big lottery are very small compared to many other real-life events.

5. $1-0.015=0.985$

7. $1-\frac{1}{1{,}000}=\frac{1{,}000}{1{,}000}-\frac{1}{1{,}000}=\frac{999}{1{,}000}$

9. Let E be the event that either die shows a five on it. This can happen in 11 ways: {(1,5), (2,5), (3,5), (4,5), (5,5), (6,5), (5,1), (5,2), (5,3), (5,4), (5,6)}. Since there are 36 ways to roll two dice, the probability that neither die shows a five on it is $P(E')=1-P(E)=1-\frac{11}{36}=\frac{25}{36}$.

11. Let E be the event that no coin shows a head. This can happen in only one way: {TTTTT}. Since there are $2\times2\times2\times2\times2=32$ ways to flip five coins, the probability of obtaining at least one head is $P(E')=1-P(E)=1-\frac{1}{32}=\frac{31}{32}$.

13. Let F be the event that we choose a five, and let R be the event that we draw a red card. There are 4 fives in a deck, 26 red cards, and 2 red fives.

$$P(F\cup R)=P(F)+P(R)-P(F\cap R)=\frac{4}{52}+\frac{26}{52}-\frac{2}{52}=\frac{28}{52}=\frac{7}{13}$$

15. Let O be the event of rolling an odd total and E be the event of rolling a total greater than 8. There are 18 ways of rolling an odd total: {(1,2), (1,4), (1,6), (2,1), (2,3), (2,5), (3,2), (3,4), **(3,6)**, (4,1), (4,3), **(4,5)**, (5,2), **(5,4)**, **(5,6)**, (6,1), **(6,3)**, **(6,5)**}. There are 10 ways to roll a total greater than 8: {**(3,6)**, **(4,5)**, (4,6), **(5,4)**, (5,5), **(5,6)**, **(6,3)**, (6,4), **(6,5)**, (6,6)}. There are 6 ways to roll an odd total greater than 8 (bolded).

$$P(O\cup E)=P(O)+P(E)-P(O\cap E)=\frac{18}{36}+\frac{10}{36}-\frac{6}{36}=\frac{22}{36}=\frac{11}{18}$$

17.
$$\begin{aligned}P(A\cup B)&=P(A)+P(B)-P(A\cap B)\\0.85&=0.55+0.40-P(A\cap B)\\0.85&=0.95-P(A\cap B)\\-0.10&=-P(A\cap B)\\P(A\cap B)&=0.10\end{aligned}$$

19.
$$\begin{aligned}P(A\cup B)&=P(A)+P(B)-P(A\cap B)\\0.70&=0.40+P(B)-0.25\\0.70&=0.15+P(B)\\P(B)&=0.55\end{aligned}$$

21. Let H be the event that we select a heart and F be the event that we select a face card.

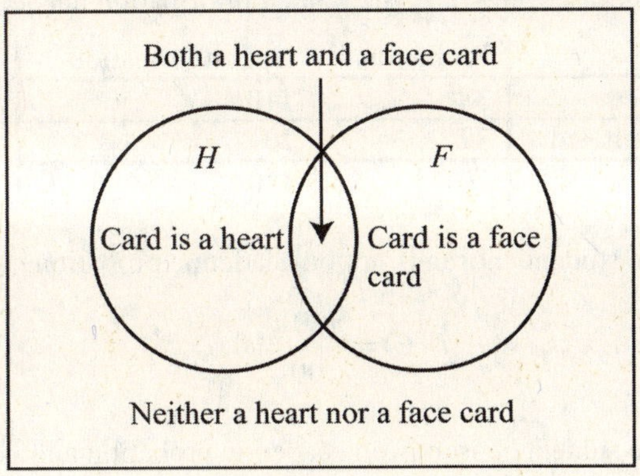

21. (continued)

$$P\left(\left(H \cup F\right)'\right) = 1 - P(H \cup F) = 1 - [P(H) + P(F) - P(H \cap F)]$$

$$= 1 - \left[\frac{13}{52} + \frac{12}{52} - \frac{3}{52}\right] = 1 - \frac{22}{52} = \frac{30}{52} = \frac{15}{26}$$

23. There are 1,483 (in thousands) surveyed. The total (in thousands) of those that are 55 or older is $61 + 53 = 114$. The probability of selecting a person less than 55 years old is $P(E) = 1 - P(E') = 1 - \frac{114}{1,483} = \frac{1,369}{1,483} \approx 0.923$.

25. Let T be the event that we select a consumer that spends 0-2 hours per month shopping on the Internet, and let A be the event that we select a consumer that has an annual income below $40,000.

$$P(T \cup A) = P(T) + P(A) - P(T \cap A) = \frac{544}{1,600} + \frac{592}{1,600} - \frac{272}{1,600} = \frac{864}{1,600} = \frac{27}{50} = 0.54$$

27. Let T be the event that we select a consumer that spends more than 2 hours per month shopping on the Internet, and let A be the event that we select a consumer that has an annual income of $60,000 or less.

$$P((T \cup A)') = 1 - P(T \cup A) = 1 - [P(T) + P(A) - P(T \cap A)]$$

$$= 1 - \left[\frac{480 + 576}{1,600} + \frac{512 + 592}{1,600} - \frac{160 + 128 + 208 + 192}{1,600}\right]$$

$$= 1 - \left[\frac{1,056}{1,600} + \frac{1,104}{1,600} - \frac{688}{1,600}\right] = 1 - \frac{1,472}{1,600} = 1 - \frac{23}{25} = \frac{2}{25} = 0.08$$

29. Let E be the event that Joanna will earn less than $1,000 in commissions.

$$P(E') = 1 - P(E) = 1 - 0.08 = 0.92$$

31. Let E be the event that Joanna will earn more than $1,999 in commissions.

$$P(E') = 1 - P(E) = 1 - (0.05 + 0.08 + 0.03) = 1 - 0.16 = 0.84$$

Use the following table for Exercises 33 – 35. The bolded information needed to be determined.

	Probation	Not on Probation	Totals
Satisfied	**38**	**110**	**148**
Not Satisfied	32	20	**52**
Totals	70	**130**	200

33. Let E be the event that the student chosen is not on academic probation.

$$P(E) = \frac{130}{200} = 0.65$$

35. Let E be the event that the student chosen is on academic probation and is satisfied with advisement.

$$P(E) = \frac{38}{200} = 0.19$$

37. Because $E \cup E' = S$, and $P(S) = 1$.

39. False, this would be true only if the intersection of events *A* and *B* was empty.

41. True, since the Rule for Computing the Probability of a Union of Two Events is $P(A \cup B) = P(A) + P(B) - P(A \cap B)$, we can solve for $P(A \cap B)$ to obtain $P(A \cap B) = P(A) + P(B) - P(A \cup B)$.

43. If *E* and *F* are disjoint sets then $P(E \cap F) = 0$.

45. Explanations will vary; $P(A) + P(B) + P(C) - P(A \cap B) - P(B \cap C)$.

47. $P(yellow') = 1 - P(yellow) = 1 - 0.04 = 0.96$

49. $$P\left((odd \cup blue)'\right) = 1 - P\left((odd \cup blue)\right) = 1 - \left(P(odd) + P(blue) - P(odd \cap blue)\right)$$
$$= 1 - (0.50 + 0.48 - 0.12) = 1 - 0.86 = 0.14$$

Section 14.3 Conditional Probability and Intersections of Events

1. The probability that the total was odd, given that the total was greater than 9.

3. We found that $P(G \mid F) \neq P(G)$, which meant that knowing the event *F* occurred had an effect on the probability of *G*.

For Exercises 5 – 7 refer to the tree diagram on the next page.

5. $P(F) = \frac{6}{36} = \frac{1}{6}$

   There are 18 times out of the 36 outcomes that yield an odd sum. Of these 18 occurrences, 6 of them yield a sum of 7.

   $P(F \mid E) = \frac{6}{18} = \frac{1}{3}$

7. $P(F) = \frac{6}{36} = \frac{1}{6}$

   There are 11 times out of the 36 outcomes that a three shows on at least one of the dice. Of these 11 occurrences, 2 of them yield a sum of less than 5.

   $P(F \mid E) = \frac{2}{11}$

9. $$P(\text{heart}|\text{red}) = \frac{P(\text{red} \cap \text{heart})}{P(\text{red})} = \frac{13/52}{26/52} = \frac{13}{26} = \frac{1}{2}$$

11. $$P(\text{seven}|\text{non-face card}) = \frac{P(\text{non-face card} \cap \text{seven})}{P(\text{non-face card})} = \frac{4/52}{40/52} = \frac{4}{40} = \frac{1}{10}$$

Use the following tree diagram for Exercises 5 – 7.

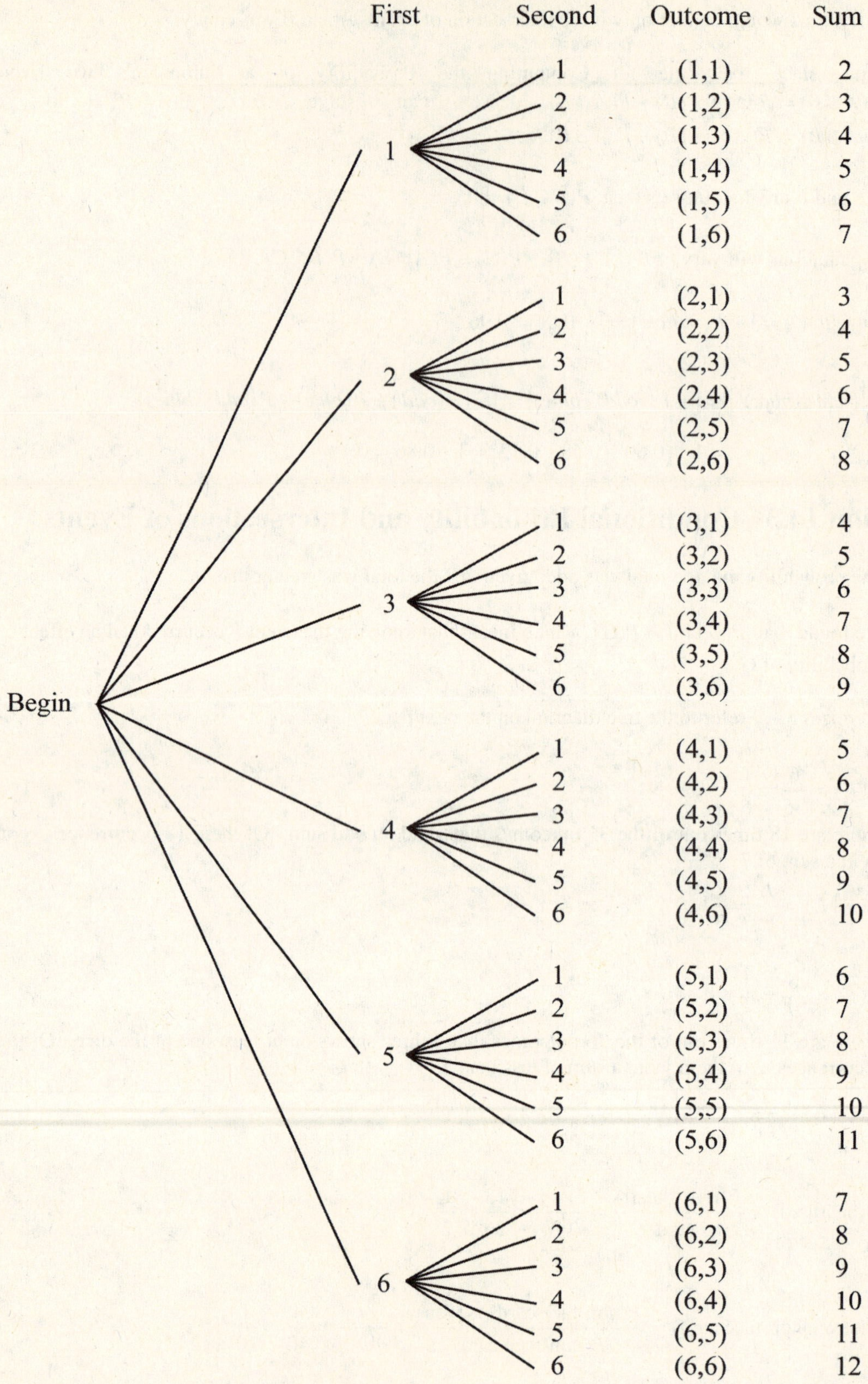

13. $P(\text{heart} \mid \text{yellow}) = \dfrac{P(\text{heart} \cap \text{yellow})}{P(\text{yellow})} = \dfrac{2/10}{4/10} = \dfrac{2}{4} = \dfrac{1}{2}$

15. $P(\text{yellow} \mid \text{heart}) = \dfrac{P(\text{yellow} \cap \text{heart})}{P(\text{heart})} = \dfrac{2/10}{3/10} = \dfrac{2}{3}$

17. $P(\text{heart} \mid \text{pink}) = \dfrac{P(\text{heart} \cap \text{pink})}{P(\text{pink})} = \dfrac{0/10}{3/10} = 0$

19. $P(F \mid E) = \dfrac{50}{61}$

21. $P(F \mid E) = \dfrac{2}{61}$

For Exercises 23 – 25 let M = the student has Mono and T = the test was positive.

23. $P(M \mid T) = \dfrac{P(M \cap T)}{P(\text{T})} = \dfrac{72/140}{76/140} = \dfrac{18}{19} \approx 0.947$

25. $P(T \mid M) = \dfrac{P(T \cap M)}{P(M)} = \dfrac{72/140}{80/140} = \dfrac{9}{10} \approx 0.90$

27. a) $P(J \cap J) = P(J) \cdot P(J \mid J) = \dfrac{4}{52} \cdot \dfrac{3}{51} = \dfrac{1}{13} \cdot \dfrac{1}{17} = \dfrac{1}{221}$

b) $P(J \cap J) = P(J) \cdot P(J) = \dfrac{4}{52} \cdot \dfrac{4}{52} = \dfrac{1}{13} \cdot \dfrac{1}{13} = \dfrac{1}{169}$

29. a) $P(F \cap N) = P(F) \cdot P(N \mid F) = \dfrac{12}{52} \cdot \dfrac{40}{51} = \dfrac{3}{13} \cdot \dfrac{40}{51} = \dfrac{120}{663} = \dfrac{40}{221}$

b) $P(F \cap N) = P(F) \cdot P(N) = \dfrac{12}{52} \cdot \dfrac{40}{52} = \dfrac{3}{13} \cdot \dfrac{10}{13} = \dfrac{30}{169}$

31. a) $P(J \cap K) = P(J) \cdot P(K \mid J) = \dfrac{4}{52} \cdot \dfrac{4}{51} = \dfrac{1}{13} \cdot \dfrac{4}{51} = \dfrac{4}{663}$

$P(K \cap J) = P(K) \cdot P(J \mid K) = \dfrac{4}{52} \cdot \dfrac{4}{51} = \dfrac{1}{13} \cdot \dfrac{4}{51} = \dfrac{4}{663}$

$\dfrac{4}{663} + \dfrac{4}{663} = \dfrac{8}{663}$

b) $P(J \cap K) = P(J) \cdot P(K) = \dfrac{4}{52} \cdot \dfrac{4}{52} = \dfrac{1}{13} \cdot \dfrac{1}{13} = \dfrac{1}{169}$

$P(K \cap J) = P(K) \cdot P(J) = \dfrac{4}{52} \cdot \dfrac{4}{52} = \dfrac{1}{13} \cdot \dfrac{1}{13} = \dfrac{1}{169}$

$\dfrac{1}{169} + \dfrac{1}{169} = \dfrac{2}{169}$

33. Let E be the event that we draw at least one face card. $\overline{E}$ is therefore the event that we draw no face cards. If N denotes a non-face card, then

$$P(E)=1-P(\overline{E})=1-P(N\cap N)=1-P(N)\cdot P(N\,|\,N)=1-\frac{40}{52}\cdot\frac{39}{51}=1-\frac{1{,}560}{2{,}652}=1-\frac{10}{17}=\frac{7}{17}$$

35. $P(F\cap F\cap F)=\frac{12}{52}\cdot\frac{11}{51}\cdot\frac{10}{50}=\frac{3}{13}\cdot\frac{11}{51}\cdot\frac{1}{5}=\frac{33}{3{,}315}=\frac{11}{1{,}105}$

37. $P(H\cap H\cap N)+P(H\cap N\cap H)+P(N\cap H\cap H)=$

$$\frac{13}{52}\cdot\frac{12}{51}\cdot\frac{39}{50}+\frac{13}{52}\cdot\frac{39}{51}\cdot\frac{12}{50}+\frac{39}{52}\cdot\frac{13}{51}\cdot\frac{12}{50}=\frac{6{,}084}{132{,}600}+\frac{6{,}084}{132{,}600}+\frac{6{,}084}{132{,}600}=\frac{18{,}252}{132{,}600}=\frac{117}{850}$$

39. A total of five can occur in four ways: (1,4), (2,3), (3,2), and (4,1). So, out of the 36 possible ways to throw a pair a dice, the probability of throwing a five is $\frac{4}{36}=\frac{1}{9}$. Therefore, the probability of a total of five on each of the three rolls is: $P(F\cap F\cap F)=\frac{1}{9}\cdot\frac{1}{9}\cdot\frac{1}{9}=\frac{1}{729}$

For Exercises 41 – 43 let O = the student preferred Obama, C = the student preferred Clinton, R = the student is a Republican, D = the student is a Democrat, and I = the student is an Independent.

41. $P(D\,|\,C)=\frac{P(D\cap C)}{P(C)}=\frac{105/240}{142/240}=\frac{105}{142}\approx 0.739$

43. $P(C\,|\,I)=\frac{P(C\cap I)}{P(I)}=\frac{25/240}{40/240}=\frac{5}{8}\approx 0.625$

For Exercises 45 – 47, refer to the tree diagram for Exercises 5 – 7.

45. dependent

$$P(F\,|\,E)=\frac{P(F\cap E)}{P(E)}=\frac{4/36}{6/36}=\frac{4}{6}=\frac{2}{3}\neq P(F)=\frac{18}{36}=\frac{1}{2}$$

47. independent

$$P(F\,|\,E)=\frac{P(F\cap E)}{P(E)}=\frac{3/36}{6/36}=\frac{3}{6}=\frac{1}{2}=P(F)=\frac{18}{36}=\frac{1}{2}$$

49. $P(F\,|\,E)=\frac{P(F\cap E)}{P(E)}=\frac{6/100}{18/100}=\frac{6}{18}=\frac{1}{3}$

51. $P(F\,|\,E)=\frac{P(F\cap E)}{P(E)}=\frac{8/100}{15/100}=\frac{8}{15}$

53. $(0.30)(0.70)(0.70)(0.70)=0.1029$

55. $6(0.30)(0.30)(0.70)(0.70)=0.2646$

57. Forty percent of the available spaces in dorm Y are rooms.

59. 0.50

61.

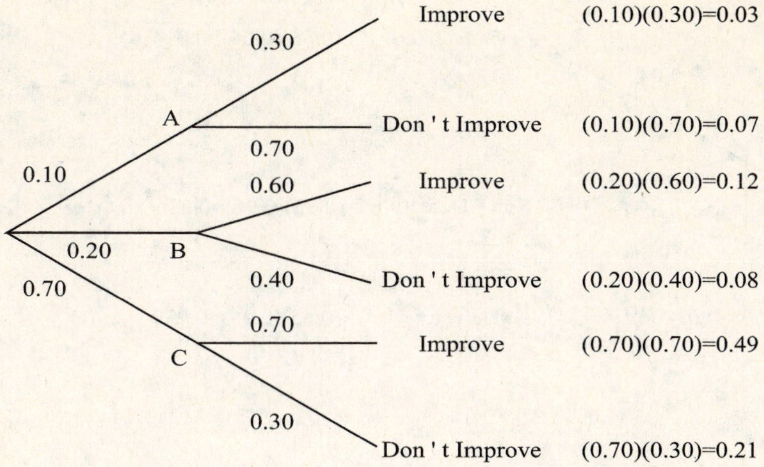

63. $0.03+0.12+0.49=0.64$

65. $\dfrac{0.18}{0.18+0.30}=\dfrac{0.18}{0.48}=0.375$

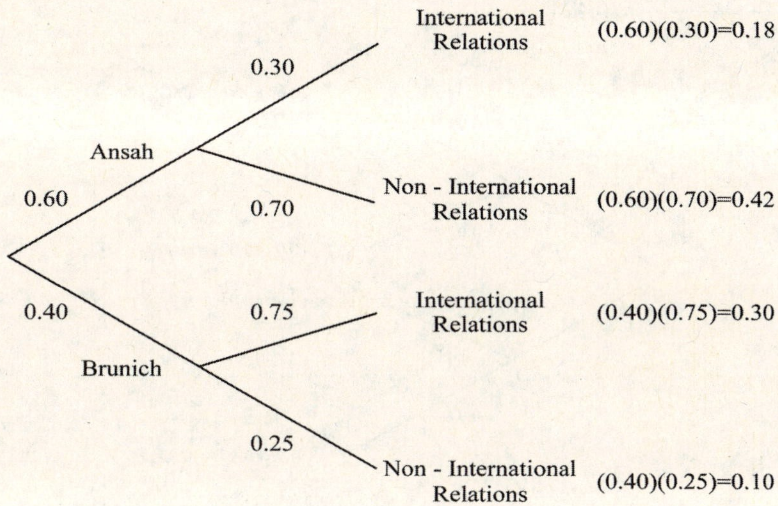

Use the following tree diagram for Exercises 67 – 69.

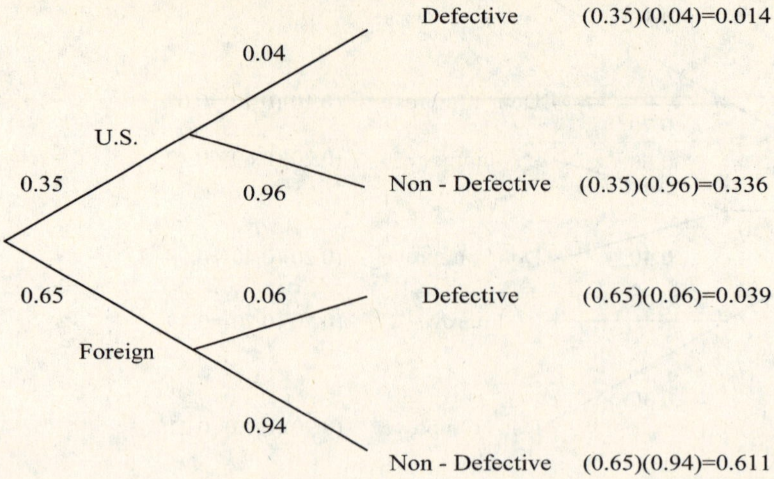

67. 0.336

69. $\dfrac{0.039}{0.014+0.039}=\dfrac{0.039}{0.053}\approx 0.736$

71. $\dfrac{0.0288}{0.0288+0.0392}=\dfrac{0.0288}{0.068}\approx 0.424$

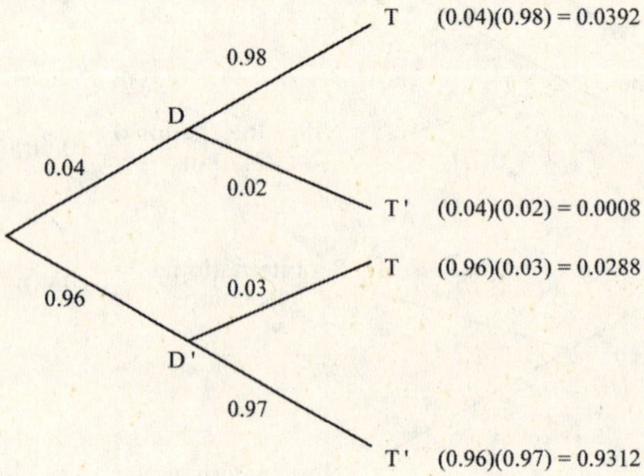

73. Multiply both sides of the equation by $P(E)$ to get $P(E\cap F)=P(F\mid E)\cdot P(E)$.

75 – 77. Answers will vary.

79. No solution provided.

81. $P(\text{duplication of birthdays})=1-P(\text{no duplication of birthdays})=$

$$1-\frac{364}{365}\cdot\frac{363}{365}\cdot\frac{362}{365}\cdot\frac{361}{365}\cdot\frac{360}{365}\cdot\frac{359}{365}\cdot\frac{358}{365}\cdot\frac{357}{365}\cdot\frac{356}{365}=1-\frac{P(364,9)}{365^9}\approx 0.117$$

83. For 21 people: $1-\frac{364}{365}\cdot\frac{363}{365}\cdot\ldots\cdot\frac{346}{365}\cdot\frac{345}{365}=1-\frac{P(364,20)}{365^{20}}\approx 0.444$

For 22 people: $1-\frac{364}{365}\cdot\frac{363}{365}\cdot\ldots\cdot\frac{345}{365}\cdot\frac{344}{365}=1-\frac{P(364,21)}{365^{21}}\approx 0.476$

For **23 people**: $1-\frac{364}{365}\cdot\frac{363}{365}\cdot\ldots\cdot\frac{344}{365}\cdot\frac{343}{365}=1-\frac{P(364,22)}{365^{22}}\approx 0.507$

Section 14.4 Expected Value

1. We multiplied the value of the laptop by the probability of it being stolen.

3. The expected value of the game is:

$$\left(\frac{1}{6}\right)\cdot(+1)+\left(\frac{1}{6}\right)\cdot(-2)+\left(\frac{1}{6}\right)\cdot(+3)+\left(\frac{1}{6}\right)\cdot(-4)+\left(\frac{1}{6}\right)\cdot(+5)+\left(\frac{1}{6}\right)\cdot(-6)$$
$$=\frac{1-2+3-4+5-6}{6}=\frac{-3}{6}=-0.50$$

This is not a fair game.

For Exercise 5, refer to the tree diagram for Exercises 5 – 7 in Section 14.3.

5. A 6, 7 or 8 will occur in 16 out of the 36 ways to roll two dice. A 2 or a 12 will occur in 2 out of the 36 ways to roll two dice.

The expected value of the game is:

$$\left(\frac{16}{36}\right)\cdot(+4)+\left(\frac{2}{36}\right)\cdot(+2)+\left(\frac{18}{36}\right)\cdot(-1)=\frac{16\cdot 4+2\cdot 2-18\cdot 1}{36}=\frac{50}{36}\approx \$1.39$$

The game should cost \$1.39 + \$1.00 = \$2.39 to be a fair game.

7. $\left(\frac{13}{52}\right)\cdot(+5)+\left(\frac{39}{52}\right)\cdot(-5)=\frac{13\cdot 5-39\cdot 5}{52}=\frac{-130}{52}=-\$2.50$

The game should cost –\$2.50 + \$5.00 = \$2.50 to be a fair game.

9. $\left(\frac{1}{1{,}000}\right)\cdot(+599)+\left(\frac{999}{1{,}000}\right)\cdot(-1)=\frac{1\cdot 599-999\cdot 1}{1{,}000}=\frac{-400}{1{,}000}=-\$0.40$

This is not a fair game. To make it a fair game, one should charge –\$0.40 + \$1.00 = \$0.60

11. $\left(\frac{1}{500}\right)\cdot(+495)+\left(\frac{2}{500}\right)\cdot(+245)+\left(\frac{5}{500}\right)\cdot(+95)+\left(\frac{492}{500}\right)\cdot(-5)=$

$$\frac{1\cdot 495+2\cdot 245+5\cdot 95-492\cdot 5}{500}=\frac{-1{,}000}{500}=-\$2.00$$

This is not a fair game. To make it a fair game, one should charge –\$2.00 + \$5.00 = \$3.00

13. The total number who earned a profit is $4+8+13+21+3+1=50$.

$$\left(\frac{4}{50}\right)\cdot(100)+\left(\frac{8}{50}\right)\cdot(200)+\left(\frac{13}{50}\right)\cdot(300)+\left(\frac{21}{50}\right)\cdot(400)+\left(\frac{3}{50}\right)\cdot(500)+\left(\frac{1}{50}\right)\cdot(600)$$
$$=\frac{400+1{,}600+3{,}900+8{,}400+1{,}500+600}{50}=\frac{16{,}400}{50}=\$328$$

15. $\left(\frac{5}{38}\right)\cdot(+6)+\left(\frac{33}{38}\right)\cdot(-1)=\frac{5\cdot 6-33\cdot 1}{38}=\frac{-3}{38}\approx -\$0.08$

17. $\left(\frac{3}{38}\right)\cdot(+8)+\left(\frac{35}{38}\right)\cdot(-1)=\frac{3\cdot 8-35\cdot 1}{38}=\frac{-11}{38}\approx -\$0.29$

19. The expected value is $\left(\frac{1}{4}\right)\cdot(+1)+\left(\frac{3}{4}\right)\cdot\left(-\frac{1}{4}\right)=\frac{4}{16}+\frac{-3}{16}=\frac{1}{16}$. Since this is positive, the student should guess.

21. The expected value when ruling out no option is: $\left(\frac{1}{5}\right)\cdot(+1)+\left(\frac{4}{5}\right)\cdot\left(-\frac{1}{2}\right)=\frac{2}{10}+\frac{-4}{10}=-\frac{2}{10}=-\frac{1}{5}$.

The expected value when ruling out one option is: $\left(\frac{1}{4}\right)\cdot(+1)+\left(\frac{3}{4}\right)\cdot\left(-\frac{1}{2}\right)=\frac{2}{8}+\frac{-3}{8}=-\frac{1}{8}$.

The expected value when ruling out **two options** is: $\left(\frac{1}{3}\right)\cdot(+1)+\left(\frac{2}{3}\right)\cdot\left(-\frac{1}{2}\right)=\frac{2}{6}+\frac{-2}{6}=0$.

23. $(0.02)\cdot(+972.50)+(0.98)\cdot(-27.50)=-\$7.50$

25. \$44.00

$$\begin{aligned}(0.02)\cdot(+2,200-x)+(0.98)\cdot(-x)&=0\\ 44-.02x-0.98x&=0\\ 44-x&=0\\ 44&=x\end{aligned}$$

27. $(0.60)\cdot(+45,000)+(0.40)\cdot(-5,000)=\$25,000$

29. – 31. No solution provided.

33. a) $\left(\frac{3}{20}\right)\cdot(+0.25)\cdot 130+\left(\frac{6}{20}\right)(+0.25)\cdot 130+\left(\frac{5}{20}\right)(+0.25)\cdot 130+\left(\frac{6}{20}\right)(+0.25\cdot 120+(-0.65)\cdot 10)=$ \$29.80

b) $\left(\frac{3}{20}\right)\cdot(+0.25)\cdot 140+\left(\frac{6}{20}\right)(+0.25)\cdot 140+\left(\frac{5}{20}\right)(+0.25\cdot 130+(-0.65)\cdot 10)+$
$\left(\frac{6}{20}\right)(+0.25\cdot 120+(-0.65)\cdot 20)=\$27.35$

35. a)

Sum of Dice

Green Die \ Red Die	1	2	3	4	5	6
1	2	3	4	5	6	7
2	3	4	5	6	7	8
3	4	5	6	7	8	9
4	5	6	7	8	9	10
5	6	7	8	9	10	11
6	7	8	9	10	11	12

$$\left(\frac{1}{36}\right)\cdot(2)+\left(\frac{2}{36}\right)\cdot(3)+\left(\frac{3}{36}\right)\cdot(4)+\left(\frac{4}{36}\right)\cdot(5)+$$
$$\left(\frac{5}{36}\right)\cdot(6)+\left(\frac{6}{36}\right)\cdot(7)+\left(\frac{5}{36}\right)\cdot(8)+\left(\frac{4}{36}\right)\cdot(9)+$$
$$\left(\frac{3}{36}\right)\cdot(10)+\left(\frac{2}{36}\right)\cdot(11)+\left(\frac{1}{36}\right)\cdot(12)=7$$

b)

Sum of Dice

Green Die \ Red Die	1	2	2	3	3	4
1	2	3	3	4	4	5
3	4	5	5	6	6	7
4	5	6	6	7	7	8
5	6	7	7	8	8	9
6	7	8	8	9	9	10
8	9	10	10	11	11	12

$$\left(\frac{1}{36}\right)\cdot(2)+\left(\frac{2}{36}\right)\cdot(3)+\left(\frac{3}{36}\right)\cdot(4)+\left(\frac{4}{36}\right)\cdot(5)+$$
$$\left(\frac{5}{36}\right)\cdot(6)+\left(\frac{6}{36}\right)\cdot(7)+\left(\frac{5}{36}\right)\cdot(8)+\left(\frac{4}{36}\right)\cdot(9)+$$
$$\left(\frac{3}{36}\right)\cdot(10)+\left(\frac{2}{36}\right)\cdot(11)+\left(\frac{1}{36}\right)\cdot(12)=7$$

37. No solution provided.

Section 14.5 Looking Deeper: Binomial Experiments

1. The dice were rolled a fixed number of times; the experiment had two outcomes (seven and non-seven); the probability of rolling a seven was constant; the rolls were independent.

3. yes

5. No; the number of trials is not fixed.

7. yes

9. A total of five can be made in 4 ways, (1,4), (2,3), (3,2), (4,1). The probability of obtaining a total of five on a single roll is $\frac{4}{36}=\frac{1}{9}$. The probability of rolling a total of five exactly once is

$$B\left(3,1,\frac{1}{9}\right)=C(3,1)\cdot\left(\frac{1}{9}\right)^{1}\cdot\left(1-\frac{1}{9}\right)^{3-1}=\frac{3!}{1!2!}\cdot\frac{1}{9}\cdot\left(\frac{8}{9}\right)^{2}=3\cdot\frac{1}{9}\cdot\frac{64}{81}=\frac{1}{3}\cdot\frac{64}{81}=\frac{64}{243}\approx 0.2634.$$

11. An experiment is performed 5 times. $B\left(5,3,\frac{1}{4}\right)$ represents the probability of obtaining a certain outcome 3 times, where the probability that an individual trial results in that certain outcome is $\frac{1}{4}$.

$$B\left(5,3,\frac{1}{4}\right)=C(5,3)\cdot\left(\frac{1}{4}\right)^{3}\cdot\left(1-\frac{1}{4}\right)^{5-3}=\frac{5!}{3!2!}\cdot\frac{1}{64}\cdot\left(\frac{3}{4}\right)^{2}=\frac{5\cdot 4}{2}\cdot\frac{1}{64}\cdot\frac{9}{16}=\frac{180}{2048}=\frac{45}{512}\approx 0.0879$$

13. k is larger than n

15. $$B\left(12,9,\frac{1}{2}\right)=C(12,9)\cdot\left(\frac{1}{2}\right)^{9}\cdot\left(1-\frac{1}{2}\right)^{12-9}=\frac{12!}{9!3!}\cdot\frac{1}{512}\cdot\left(\frac{1}{2}\right)^{3}=\frac{12\cdot 11\cdot 10}{3\cdot 2}\cdot\frac{1}{512}\cdot\frac{1}{8}=\frac{1,320}{24,576}$$
$$=\frac{55}{1,024}\approx 0.0537$$

17. $B\left(12,9,\frac{1}{2}\right)=C(12,9)\cdot\left(\frac{1}{2}\right)^{9}\cdot\left(1-\frac{1}{2}\right)^{12-9}=\frac{12!}{9!3!}\cdot\frac{1}{512}\cdot\left(\frac{1}{2}\right)^{3}=\frac{12\cdot 11\cdot 10}{3\cdot 2}\cdot\frac{1}{512}\cdot\frac{1}{8}=\frac{1{,}320}{24{,}576}$

$=\frac{55}{1{,}024}=0.0537109375$

$B\left(12,10,\frac{1}{2}\right)=C(12,10)\cdot\left(\frac{1}{2}\right)^{10}\cdot\left(1-\frac{1}{2}\right)^{12-10}=\frac{12!}{10!2!}\cdot\frac{1}{1{,}024}\cdot\left(\frac{1}{2}\right)^{2}=\frac{12\cdot 11}{2}\cdot\frac{1}{1{,}024}\cdot\frac{1}{4}$

$=\frac{132}{8{,}192}\approx 0.01611328125$

$B\left(12,11,\frac{1}{2}\right)=C(12,11)\cdot\left(\frac{1}{2}\right)^{11}\cdot\left(1-\frac{1}{2}\right)^{12-11}=\frac{12!}{11!1!}\cdot\frac{1}{2{,}048}\cdot\left(\frac{1}{2}\right)^{1}=12\cdot\frac{1}{2{,}048}\cdot\frac{1}{2}=\frac{12}{4{,}096}$

$=\frac{3}{1{,}024}=0.0029296875$

$B\left(12,12,\frac{1}{2}\right)=C(12,12)\cdot\left(\frac{1}{2}\right)^{12}\cdot\left(1-\frac{1}{2}\right)^{12-12}=1\cdot\frac{1}{4{,}096}\cdot\left(\frac{1}{2}\right)^{0}=\frac{1}{4{,}096}\approx 0.0002441406$

$0.0537109375+0.01611328125+0.0029296875+0.0002441406=0.07299804685\approx 0.0730$

19. $B(10,8,0.80)=C(10,8)\cdot(0.80)^{8}\cdot(1-0.80)^{10-8}=\frac{10!}{8!2!}\cdot 0.16777216\cdot(0.20)^{2}$

$=\frac{10\cdot 9}{2}\cdot 0.16777216\cdot 0.04=0.301989888$

$B(10,9,0.80)=C(10,9)\cdot(0.80)^{9}\cdot(1-0.80)^{10-9}=\frac{10!}{9!1!}\cdot 0.1342217728\cdot(0.20)^{1}$

$=10\cdot 0.134217728\cdot 0.20=0.268435456$

$B(10,10,0.80)=C(10,10)\cdot(0.80)^{10}\cdot(1-0.80)^{10-10}=\frac{10!}{10!0!}\cdot 0.1073741824\cdot(0.20)^{0}$

$=1\cdot 0.1073741824\cdot 1=0.1073741824$

$0.301989888+0.268435456+0.1073741824=0.6777995264\approx 0.6778$

21. $B(5,2,0.250)=C(5,2)\cdot(0.250)^{2}\cdot(1-0.250)^{5-2}=\frac{5!}{2!3!}\cdot 0.0625\cdot(0.750)^{3}$

$=\frac{5\cdot 4}{2}\cdot 0.0625\cdot 0.421875=0.263671875\approx 0.2637$

23. We need to determine the probability that the hospital does not meet the need so we need to determine the probability that the hospital has 0, 1, or 2 donors with type A+ blood waiting.

$B(12,0,0.30)=C(12,0)\cdot(0.30)^{0}\cdot(1-0.30)^{12-0}=\frac{12!}{0!12!}\cdot 1\cdot(0.70)^{12}$

$\approx 1\cdot 1\cdot 0.013841287201=0.013841287201$

23. (continued)

$$B(12,1,0.30) = C(12,1)\cdot(0.30)^1\cdot(1-0.30)^{12-1} = \frac{12!}{1!11!}\cdot 0.30\cdot(0.70)^{11}$$

$$= 12\cdot 0.30\cdot 0.01977326743 = 0.071183762748$$

$$B(12,2,0.30) = C(12,2)\cdot(0.30)^2\cdot(1-0.30)^{12-2} = \frac{12!}{2!10!}\cdot 0.09\cdot(0.70)^{10}$$

$$\approx \frac{12\cdot 11}{2}\cdot 0.09\cdot 0.0282475249 \approx 0.167790297906$$

0.013841287201 + 0.071183762748 + 0.167790297906 = 0.252815347855

1 – 0.252815347855=0.747184652145 $\approx$ 0.7472

25. $\frac{3}{4}$

27. The total number of purchases to obtain a complete set T is: the number of purchases to obtain the first new figure + the number of purchases to obtain the second new figure + the number of purchases to obtain the third new figure + the number of purchases to obtain the fourth new figure + the number of purchases to obtain the fifth new figure + the number of purchases to obtain the sixth new figure =

$$1+\frac{1}{5/6}+\frac{1}{4/6}+\frac{1}{3/6}+\frac{1}{2/6}+\frac{1}{1/6}=1+\frac{6}{5}+\frac{6}{4}+\frac{6}{3}+\frac{6}{2}+\frac{6}{1}=\frac{147}{10}=14.7$$

The child should expect to purchase 15 boxes.

29. The experiment has a fixed number of trials. The experiment has two outcomes: "success" and "failure". The probability of success is the same from trial to trial. The trials are independent of each other.

31. No solution provided.

33. We need to determine the probability that 4 or fewer people will come down with the cold without the vaccine.

$$B(10,4,0.50) = C(10,4)\cdot(0.50)^4\cdot(1-0.50)^{10-4} = \frac{10!}{4!6!}\cdot 0.0625\cdot(0.50)^6 = \frac{10\cdot 9\cdot 8\cdot 7}{4\cdot 3\cdot 2}\cdot 0.0625\cdot 0.015625$$

$$= 0.205078125$$

$$B(10,3,0.50) = C(10,3)\cdot(0.50)^3\cdot(1-0.50)^{10-3} = \frac{10!}{3!7!}\cdot 0.125\cdot(0.50)^7 = \frac{10\cdot 9\cdot 8}{3\cdot 2}\cdot 0.125\cdot 0.0078125$$

$$= 0.1171875$$

$$B(10,2,0.50) = C(10,2)\cdot(0.50)^2\cdot(1-0.50)^{10-2} = \frac{10!}{2!8!}\cdot 0.25\cdot(0.50)^8 = \frac{10\cdot 9}{2}\cdot 0.25\cdot 0.00390625$$

$$= 0.0439453125$$

$$B(10,1,0.50) = C(10,1)\cdot(0.50)^1\cdot(1-0.50)^{10-1} = \frac{10!}{1!9!}\cdot 0.50\cdot(0.50)^9 = 10\cdot 0.50\cdot 0.001953125$$

$$= 0.009765625$$

$$B(10,0,0.50) = C(10,0)\cdot(0.50)^0\cdot(1-0.50)^{10-0} = \frac{10!}{0!10!}\cdot 1\cdot(0.50)^{10} = 1\cdot 1\cdot 0.0009765625$$

$$= 0.0009765625$$

0.205078125 + 0.1171875 + 0.0439453125 + 0.009765625 + 0.0009765625 = 0.376953125 $\approx$ 0.3770

Chapter Review Exercises

1. a) If we let "H" represent a heads and "T" represent tails then we would have three outcomes. The set of outcomes would be {HHT, HTH, THH}.

   b) We would have five outcomes. The set of outcomes would be {(2,6), (3,5), (4,4), (5,3), (6,2)}.

3. No solution provided.

5. a) Let D be the event that the Dodgers win the World Series.

   The odds against D are $\frac{17}{2}=\frac{17/19}{2/19}=\frac{P(D')}{P(D)}$. Thus, the probability of D is $\frac{2}{19}$.

   b) Let R be the event that it rains.

   The odds against R are $\frac{P(E')}{P(E)}=\frac{1-0.55}{0.55}=\frac{0.45}{0.55}=\frac{0.45\times 100}{0.55\times 100}=\frac{45}{55}=\frac{9}{11}$, or 9 to 11.

7. Let F be the event that we select a face card and R be the event that we select a red card.

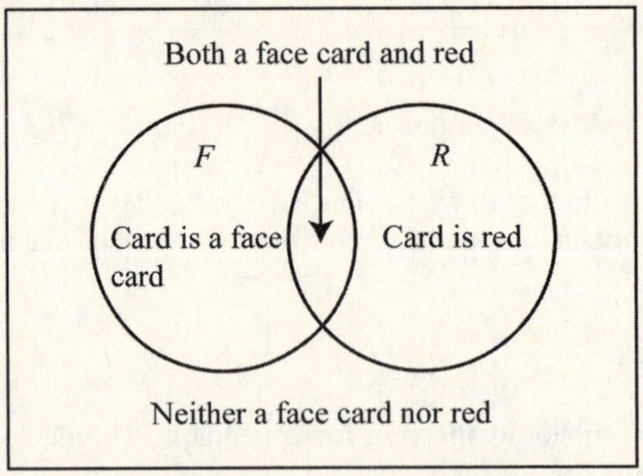

$$P(F\cup R)=P(F)+P(R)-P(F\cap R)=\frac{12}{52}+\frac{26}{52}-\frac{6}{52}=\frac{32}{52}=\frac{8}{13}$$

9. No solution provided.

11. a) $P(H\cap H)=P(H)\cdot P(H\mid H)=\frac{13}{52}\cdot\frac{12}{51}=\frac{1}{4}\cdot\frac{12}{51}=\frac{12}{204}=\frac{1}{17}$

    b) $P(Q\cap A)=P(Q)\cdot P(A\mid Q)=\frac{4}{52}\cdot\frac{4}{51}=\frac{1}{13}\cdot\frac{4}{51}=\frac{4}{663}$

13. $\dfrac{0.036}{0.036+0.0576}=\dfrac{0.036}{0.0936}\approx 0.385$

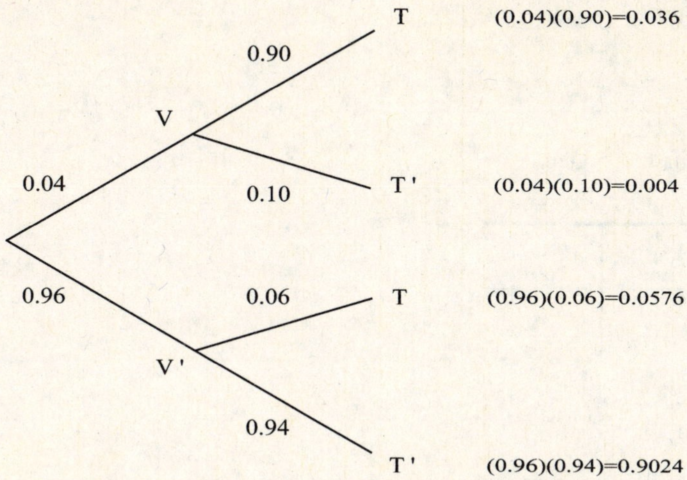

15. $\left(\dfrac{12}{52}\right)\cdot(+15)+\left(\dfrac{40}{52}\right)\cdot(-4)=\dfrac{12\cdot 15-40\cdot 4}{52}=\dfrac{20}{52}\approx \$0.38$

17. $B\left(8,3,\dfrac{1}{2}\right)=C(8,3)\cdot\left(\dfrac{1}{2}\right)^3\cdot\left(1-\dfrac{1}{2}\right)^{8-3}=\dfrac{8!}{3!5!}\cdot\dfrac{1}{8}\cdot\left(\dfrac{1}{2}\right)^5=\dfrac{8\cdot 7\cdot 6}{3\cdot 2}\cdot\dfrac{1}{8}\cdot\dfrac{1}{32}=\dfrac{336}{1{,}536}$

$=\dfrac{7}{32}\approx 0.2199$

Chapter Test

1. a) There would be six outcomes. The set of outcomes would be {(4,6), (6,4), (5,5), (5,6), (6,5), (6,6)}.
   b) If "H" represents heads and "T" represent tails, there would be four outcomes. The set of outcomes would be {HHHT, HHTH, HTHH, THHH}.

3. a) Let D be the event that the Dolphins win the Super Bowl.

   The odds against D are $\dfrac{28}{3}=\dfrac{28/31}{3/31}=\dfrac{P(D')}{P(D)}$. Thus, the probability of D is $\dfrac{3}{31}$.

   b) Let R be the event that it rains.

   The odds against R are $\dfrac{P(R')}{P(R)}=\dfrac{1-0.15}{0.15}=\dfrac{0.85}{0.15}=\dfrac{0.85\times 100}{0.15\times 100}=\dfrac{85}{15}=\dfrac{17}{3}$, or 17 to 3.

5. $P(B\,|\,A)$ is the probability of B assuming that A has occurred. $P(A\,|\,B)$ is the probability of A assuming that B has occurred.

7. Since it costs \$1 to play, the expected value of this game is $\left(\dfrac{6}{36}\right)\cdot(+4)+\left(\dfrac{30}{36}\right)\cdot(-1)=\dfrac{6\cdot 4-30\cdot 1}{36}$

$=-\dfrac{30}{36}=-\dfrac{1}{6}\approx -\$0.17.$

9. $B\left(5,3,\frac{1}{4}\right)=C(5,3)\cdot\left(\frac{1}{4}\right)^3\cdot\left(1-\frac{1}{4}\right)^{5-3}=\frac{5!}{3!2!}\cdot\frac{1}{64}\cdot\left(\frac{3}{4}\right)^2=\frac{5\cdot 4}{2}\cdot\frac{1}{64}\cdot\frac{9}{16}=\frac{180}{2{,}048}$

$$=\frac{45}{512}\approx 0.0879$$

11. The Punnett square is

		First Gen.	
		w	r
First	w	ww	wr
Gen.	r	rw	rr

$P(\text{pink})=\frac{1}{2}$

13. Since is costs $2 to play, the expected value of the raffle is

$\left(\frac{1}{500}\right)\cdot(+248)+\left(\frac{3}{500}\right)\cdot(98)+\left(\frac{5}{500}\right)\cdot(48)+\left(\frac{491}{500}\right)\cdot(-2)=\frac{1\cdot 248+3\cdot 98+5\cdot 48-2\cdot 491}{500}=$

$-\frac{200}{500}=-\$0.40.$

15. a) $P(F\cap F)=P(F)\cdot P(F\,|\,F)=\frac{12}{52}\cdot\frac{11}{51}=\frac{3}{13}\cdot\frac{11}{51}=\frac{1}{13}\cdot\frac{11}{17}=\frac{11}{221}.$

b) $P(K\,\&\,A)=P(K\cap A)+P(A\cap K)=P(K)\cdot P(A\,|\,K)+P(A)\cdot P(K\,|\,A)=$

$\frac{4}{52}\cdot\frac{4}{51}+\frac{4}{52}\cdot\frac{4}{51}=\frac{1}{13}\cdot\frac{4}{51}+\frac{1}{13}\cdot\frac{4}{51}=\frac{8}{663}$

Chapter 15
Descriptive Statistics: What a Data Set Tells Us

Section 15.1 Organizing and Visualizing Data

1. selection bias and leading-question bias

3. Example 4 dealt with weight, which is a continuous variable, so a bar graph would not be appropriate.

5. Gallup's sampling method was better.

7.

x	Frequency
2	1
3	0
4	0
5	3
6	2
7	4
8	4
9	3
10	3

x	Relative Frequency
2	0.05
3	0.00
4	0.00
5	0.15
6	0.10
7	0.20
8	0.20
9	0.15
10	0.15

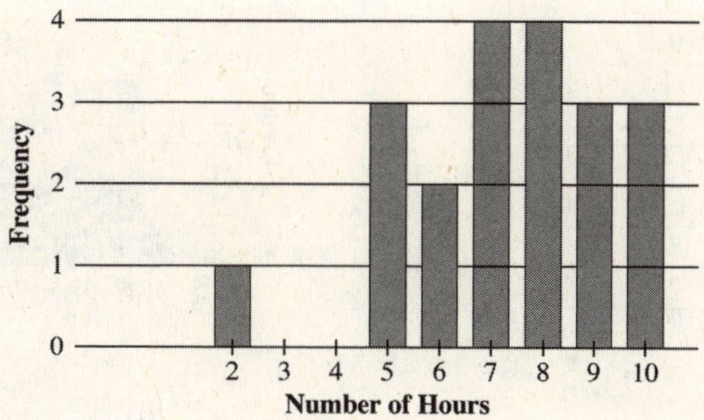

9.

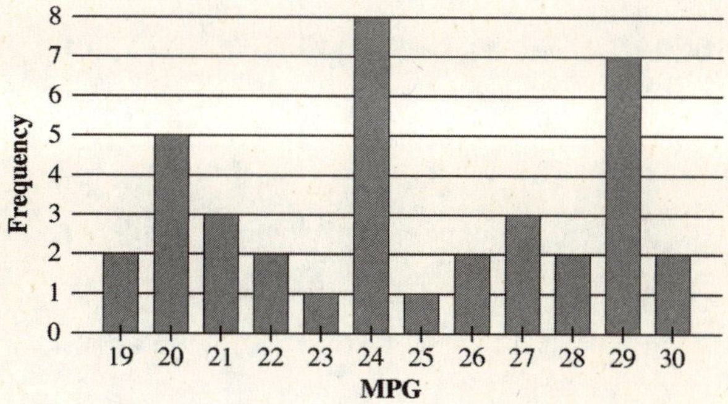

11.

x	Frequency
21	4
22	6
23	10
24	9
25	5
26	8
27	10
28	5
29	3

x	Relative Frequency
21	$4/60 \approx 0.067 = 6.7\%$
22	$6/60 = 0.10 = 10\%$
23	$10/60 \approx 0.167 = 16.7\%$
24	$9/60 = 0.15 = 15\%$
25	$5/60 \approx 0.083 = 8.3\%$
26	$8/60 \approx 0.133 = 13.3\%$
27	$10/60 \approx 0.167 = 16.7\%$
28	$5/60 \approx 0.083 = 8.3\%$
29	$3/60 = 0.05 = 5\%$

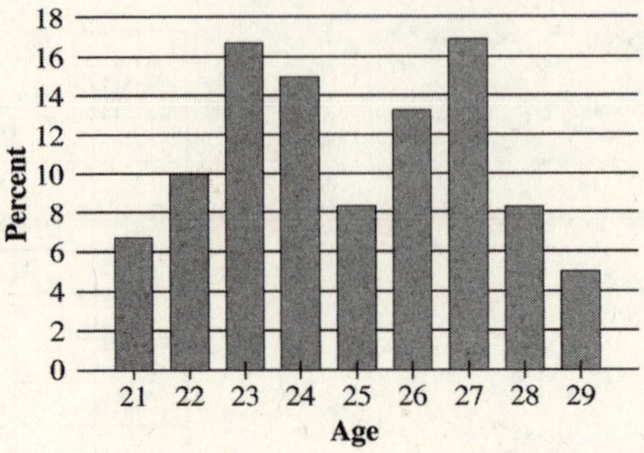

13.

x	Frequency
64.5+ to 66.5	2
66.5+ to 68.5	9
68.5+ to 70.5	11
70.5+ to 72.5	8
72.5+ to 74.5	15
74.5+ to 76.5	12
76.5+ to 78.5	5
78.5+ to 80.5	2

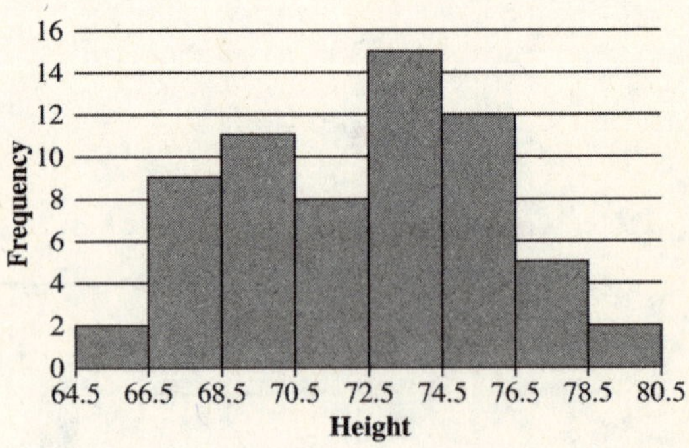

15.

A		B
8 8	1	3 6 8
9 6 2 2 1	2	0 1 1 2 2 2 3 4 6 8 9
9 9 8 7 4 3 2	3	2 3 3 8 9
7 3 3 3 2 2	4	7

17.

x	Frequency
1.80+ to 2.00	2
2.00+ to 2.20	3
2.20+ to 2.40	11
2.40+ to 2.60	4
2.60+ to 2.80	5
2.80+ to 3.00	7
3.00+ to 3.20	4

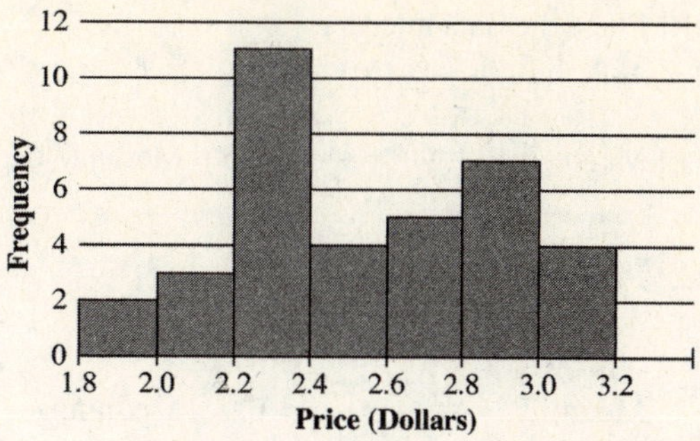

19. a) 0 occurred thirteen times

b) 12 occurred two times

c) 6

d) $13 + 11 + 5 + 8 + 12 + 9 + 6 + 5 + 2 = 71$

e) $\frac{9+6+5+2}{71} = \frac{22}{71}$

21. a) 5+3=8

b) 3

c) 5+3+4=12

d) $3 + 2 + 3 + 5 + 3 = 16$

23. Answers may vary.
This exercise was based on real–world data. An approximation of 10% would be reasonable. The exact value is 10.2%.

25. 25 and older, earning at least $10.00 per hour

27. 16 to 19, earning less than $7.51 per hour

29. The distribution is shifted to higher numbers, so it appears that the orientation program is succeeding.

No Training		With Training
9	1	
9 8 8 7 6 6 3 2	2	1 8 8 9 9
4 2	3	2 2 3 6 6 7 9
3 3 2 1 0	4	1 3 4 5

31. The population is the set of all objects being studied; a sample is a subset of the population.

33. A continuous variable can take on arbitrary values; a discrete variable cannot.

35. – 45. No solution provided.

Section 15.2 Measures of Central Tendency

1. They ignore the frequencies and divided by 6, which is the number of different scores.

3. A median is the middle strip between a divided highway. This helps us remember that the median is the middle score after the scores have been arranged in order.

5. Place the data in order.
3, 4, 4, 5, 6, 7, 8, 9, 11

Mean: $\bar{x} = \dfrac{\sum x}{n} = \dfrac{57}{9} \approx 6.3$; Median: 6; Mode: 4

7. Place the data in order.
3, 4, 4, 6, 6, 7, 9, 9, 10, 11

Mean: $\bar{x} = \dfrac{\sum x}{n} = \dfrac{69}{10} = 6.9$; Median: $\dfrac{6+7}{2} = \dfrac{13}{2} = 6.5$; Mode: no mode

9. Place the data in order.
1, 2, 3, 4, 5, 5, 6, 7, 8, 9

Mean: $\bar{x} = \dfrac{\sum x}{n} = \dfrac{50}{10} = 5$; Median: $\dfrac{5+5}{2} = \dfrac{10}{2} = 5$; Mode: 5

11. Place the data in order.
2, 5, 5, 5, 6, 6, 7, 7, 7, 7, 8, 8, 8, 8, 9, 9, 9, 10, 10, 10

Mean: $\bar{x} = \dfrac{\sum x}{n} = \dfrac{146}{20} = 7.3$; Median: $\dfrac{7+8}{2} = \dfrac{15}{2} = 7.5$; Mode: 7, 8

13. $\sum(x \cdot f) = 5 \cdot 4 + 6 \cdot 3 + 7 \cdot 5 + 8 \cdot 2 + 9 \cdot 6 = 20 + 18 + 35 + 16 + 54 = 143$

$n = \sum f = 4 + 3 + 5 + 2 + 6 = 20$; Mean: $\bar{x} = \dfrac{\sum(x \cdot f)}{\sum f} = \dfrac{143}{20} = 7.45$

10th piece of data: 7; 11th piece of data: 7; Median: $\dfrac{7+7}{2} = \dfrac{14}{2} = 7$; Mode: 9

15. $\sum(x \cdot f) = 2 \cdot 1 + 6 \cdot 3 + 11 \cdot 5 + 14 \cdot 3 + 15 \cdot 5 = 2 + 18 + 55 + 42 + 75 = 192$

$n = \sum f = 1 + 3 + 5 + 3 + 5 = 17$; Mean: $\bar{x} = \dfrac{\sum(x \cdot f)}{\sum f} = \dfrac{192}{17} \approx 11.29$

Median: 9th piece of data: 11; Modes: 11 and 15

17. $\sum(x \cdot f) = 3 \cdot 2 + 4 \cdot 5 + 6 \cdot 3 + 7 \cdot 1 + 8 \cdot 4 + 10 \cdot 3 + 11 \cdot 2 = 6 + 20 + 18 + 7 + 32 + 30 + 22 = 135$

$n = \sum f = 2 + 5 + 3 + 1 + 4 + 3 + 2 = 20$; Mean: $\bar{x} = \dfrac{\sum(x \cdot f)}{\sum f} = \dfrac{135}{20} = 6.75$

10th piece of data: 6; 11th piece of data: 7; Median: $\dfrac{6+7}{2} = \dfrac{13}{2} = 6.5$; Mode: 4

19. $\sum(x \cdot f) = 5 \cdot 1 + 8 \cdot 3 + 11 \cdot 6 + 12 \cdot 2 + 14 \cdot 5 + 15 \cdot 4 + 18 \cdot 3 = 5 + 24 + 66 + 24 + 70 + 60 + 54 = 303$

$n = \sum f = 1 + 3 + 6 + 2 + 5 + 4 + 3 = 24$; Mean: $\bar{x} = \dfrac{\sum(x \cdot f)}{\sum f} = \dfrac{303}{24} \approx 12.6$

12th piece of data: 12; 13th piece of data: 14; Median: $\dfrac{12+14}{2} = \dfrac{26}{2} = 13$; Mode: 11

21. a) Place the data in order.

11, 17, 18, 23, 25, 26, 26, 31, 31, 33, 41, 44, 45, 48, 53

lower half: 11, 17, 18, 23, 25, 26, 26

upper half: 31, 33, 41, 44, 45, 48, 53

minimum: 11

first quartile is the median of the lower half: $Q_1 = 23$

median: 31

third quartile is the median of the upper half: $Q_3 = 44$

maximum: 53

b)

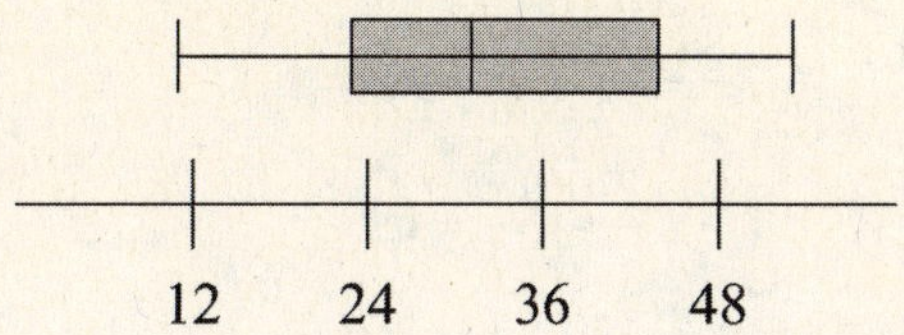

23. a) Place the data in order.

25, 28, 29, 30, 31, 32, 33, 33, 34, 37, 38, 41, 41, 45, 49

lower half: 25, 28, 29, 30, 31, 32, 33

upper half: 34, 37, 38, 41, 41, 45, 49

minimum: 25

first quartile is the median of the lower half: $Q_1 = 30$

median: 33

third quartile is the median of the upper half: $Q_3 = 41$

maximum: 49

b)

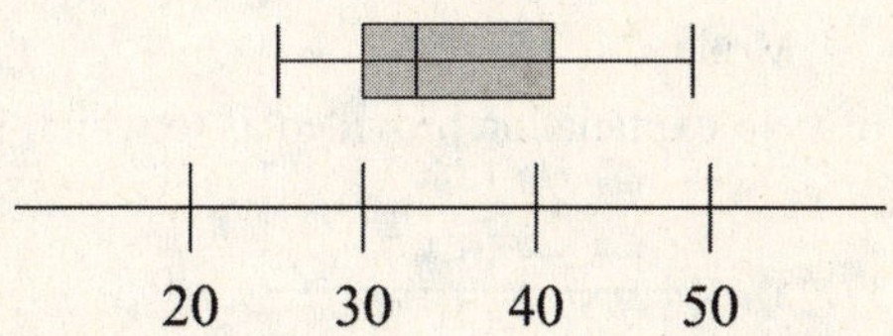

25. Place the data in order.

18, 19, 20, 21, 21, 22, 23, 25, 25, 25, 26, 26, 26

Mean: $\bar{x} = \frac{\sum x}{n} = \frac{297}{13} \approx 22.85$; Median: 23; Modes: 25 and 26

No explanation provided as to which measure is typical

27. Place this data in order.

6, 7, 9, 13, 15, 18, 22, 35, 44, 54

Mean: $\bar{x} = \frac{\sum x}{n} = \frac{223}{10} = 22.3$; Median: $\frac{15+18}{2} = \frac{33}{2} = 16.5$; Mode: no mode

No explanation provided as to which measure is typical

29. Place the data in order.

3, 21, 30, 31, 37, 43, 44, 66, 78, 181, 250, 635

Mean: $\bar{x} = \frac{\sum x}{n} = \frac{1{,}419}{12} = 118.25$; Median: $\frac{43+44}{2} = 43.5$; Mode: no mode

No explanation provided as to which measure is typical

31. $\frac{3\times4+2\times3+3\times4+3\times0+4\times3}{15} = \frac{42}{15} \approx 2.8$

33. Let x be the incorrect final exam score. Assuming that the three exams are equally weighted we have

$$\frac{84+86+x}{3}=69$$
$$\frac{170+x}{3}=69$$
$$170+x=207$$
$$x=37.$$

The correct grade was 73.

35. Mean: $\overline{x}=\frac{\sum x\cdot f}{\sum f}=\frac{1,435}{58}\approx 24.74$

Median: 25; You would need to add together the 29th and the 30th values and divide by two. Both of these numbers are 25.

Mode: 24, 25

No explanation provided as to which measure is typical

37. a)
$$\frac{78+82+56+72+x}{5}=70$$
$$\frac{288+x}{5}=70$$
$$288+x=350$$
$$x=62$$

b)
$$\frac{78+82+56+72+x}{5}=80$$
$$\frac{288+x}{5}=80$$
$$288+x=400$$
$$x=112$$

39. $\frac{9\cdot 50+15\cdot 125+7\cdot 245}{31}=\frac{4,040}{31}\approx \$130.32$

41. 1975–1989: Place the data in order.

31, 36, 37, 37, 37, 38, 38, 39, 40, 40, 47, 48, 48, 49, 52

lower half: 31, 36, 37, 37, 37, 38, 38

upper half: 40, 40, 47, 48, 48, 49, 52

minimum: 31

first quartile is the median of the lower half: $Q_1 = 37$

median: 39

third quartile is the median of the upper half: $Q_3 = 48$

maximum: 52

41. (continued)

1990–2004: Place the data in order.

35, 38, 40, 40, 43, 46, 47, 47, 48, 49, 49, 50, 65, 70, 73

lower half: 35, 38, 40, 40, 43, 46, 47

upper half: 48, 49, 49, 50, 65, 70, 73

minimum: 35

first quartile is the median of the lower half: $Q_1 = 40$

median: 47

third quartile is the median of the upper half: $Q_3 = 50$

maximum: 73

The top box-and-whisker plot is for 1975–1989, the bottom is for 1990–2004

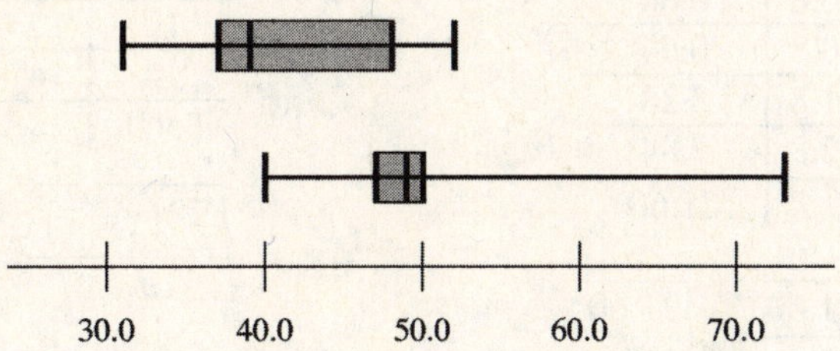

43. liberal arts

45. education; no; No discussion provided

47. The maximum and minimum salaries are not too far from the middle 50% of the salaries.

49. With $\sum x$, we are adding the individual scores in the distribution; with $\sum x \cdot f$, we are multiplying each scores by its frequency before adding.

51. a box-and-whisker plot

53. mean in both cases

55. – 67. No solution provided.

Section 15.3 Measures of Dispersion

1. One outlier can make the range meaningless.

3. It allows us to compare variation between different sets of data.

5. Place data in order.

17, 18, 18, 18, 19, 19, 20, 20, 21, 22

Range: largest – smallest = 22 – 17 = 5

Mean: $\bar{x} = \frac{\sum x}{n} = \frac{192}{10} = 19.2$

Standard deviation:

Data Value	$x - \bar{x}$	$(x - \bar{x})^2$
17	17 – 19.2 = – 2.2	4.84
18	18 – 19.2 = – 1.2	1.44
18	18 – 19.2 = – 1.2	1.44
18	18 – 19.2 = – 1.2	1.44
19	19 – 19.2 = – 0.2	0.04
19	19 – 19.2 = – 0.2	0.04
20	20 – 19.2 = 0.8	0.64
20	20 – 19.2 = 0.8	0.64
21	21 – 19.2 = 1.8	3.24
22	22 – 19.2 = 2.8	7.84
Total	0	21.60

$$s = \sqrt{\frac{\sum(x-\bar{x})^2}{n-1}} = \sqrt{\frac{21.60}{10-1}} = \sqrt{\frac{21.60}{9}}$$

≈ 1.55

7. Place data in order.

4, 5, 6, 7, 7, 8, 9, 10

Range: largest – smallest = 10 – 4 = 6

Mean: $\bar{x} = \frac{\sum x}{n} = \frac{56}{8} = 7$

Standard deviation:

Data Value	$x - \bar{x}$	$(x - \bar{x})^2$
4	4 – 7 = – 3	9
5	5 – 7 = – 2	4
6	6 – 7 = – 1	1
7	7 – 7 = 0	0
7	7 – 7 = 0	0
8	8 – 7 = 1	1
9	9 – 7 = 2	4
10	10 – 7 = 3	9
Total	0	28

$$s = \sqrt{\frac{\sum(x-\bar{x})^2}{n-1}} = \sqrt{\frac{28}{8-1}} = \sqrt{\frac{28}{7}} = 2$$

9. Place data in order.

2, 3, 5, 8, 9, 11, 12, 14

Range: largest – smallest = 14 – 2 = 12

Mean: $\bar{x} = \frac{\sum x}{n} = \frac{64}{8} = 8$

Standard deviation:

Data Value	$x - \bar{x}$	$(x - \bar{x})^2$
2	2 – 8 = – 6	36
3	3 – 8 = – 5	25
5	5 – 8 = – 3	9
8	8 – 8 = 0	0
9	9 – 8 = 1	1
11	11 – 8 = 3	9
12	12 – 8 = 4	16
14	14 – 8 = 6	36
Total	0	132

$$s = \sqrt{\frac{\sum(x-\bar{x})^2}{n-1}} = \sqrt{\frac{132}{8-1}} = \sqrt{\frac{132}{7}} \approx 4.34$$

11. Place data in order.

3, 7, 7, 7, 8, 9, 13, 18

Range: largest – smallest = 18 – 3 = 15

Mean: $\bar{x} = \frac{\sum x}{n} = \frac{72}{8} = 9$

Standard deviation:

Data Value	$x - \bar{x}$	$(x - \bar{x})^2$
3	3 – 9 = – 6	36
7	7 – 9 = – 2	4
7	7 – 9 = – 2	4
7	7 – 9 = – 2	4
8	8 – 9 = – 1	1
9	9 – 9 = 0	0
13	13 – 9 = 4	16
18	18 – 9 = 9	81
Total	0	146

$$s = \sqrt{\frac{\sum(x-\bar{x})^2}{n-1}} = \sqrt{\frac{146}{8-1}} = \sqrt{\frac{146}{7}} \approx 4.57$$

13. Place data in order.

3, 3, 3, 3, 3, 3, 3

Range: largest – smallest = 3 – 3 = 0

Mean: $\bar{x} = \frac{\sum x}{n} = \frac{21}{7} = 3$

Standard deviation:

Data Value	$x-\bar{x}$	$(x-\bar{x})^2$
3	3 – 3 = 0	0
3	3 – 3 = 0	0
3	3 – 3 = 0	0
3	3 – 3 = 0	0
3	3 – 3 = 0	0
3	3 – 3 = 0	0
3	3 – 3 = 0	0
Total	0	0

$$s = \sqrt{\frac{\sum (x-\bar{x})^2}{n-1}} = \sqrt{\frac{0}{7-1}} = \sqrt{\frac{0}{6}} = 0$$

15. $\bar{x} = \frac{\sum x \cdot f}{\sum f} = \frac{140}{20} = 7$ and $s = \sqrt{\frac{\sum (x-\bar{x})^2 \cdot f}{n-1}} = \sqrt{\frac{86}{20-1}} = \sqrt{\frac{86}{19}} \approx 2.13$

Number x	Frequency f	Product $x \cdot f$	Deviation $x-\bar{x}$	Deviation² $(x-\bar{x})^2$	Product $(x-\bar{x})^2 \cdot f$
2	1	2	2 – 7 = – 5	25	25 · 1 = 25
3	1	3	3 – 7 = – 4	16	16 · 1 = 16
4	0	0	4 – 7 = – 3	9	9 · 0 = 0
5	2	10	5 – 7 = – 2	4	4 · 2 = 8
6	3	18	6 – 7 = – 1	1	1 · 3 = 3
7	4	28	7 – 7 = 0	0	0 · 4 = 0
8	4	32	8 – 7 = 1	1	1 · 4 = 4
9	3	27	9 – 7 = 2	4	4 · 3 = 12
10	2	20	10 – 7 = 3	9	9 · 2 = 18
	$\sum f = 20$	$\sum x \cdot f = 140$			$\sum (x-\bar{x})^2 \cdot f = 86$

17. $\bar{x} = \frac{\sum x \cdot f}{\sum f} = \frac{48}{12} = 4$ and $s = \sqrt{\frac{\sum (x-\bar{x})^2 \cdot f}{n-1}} = \sqrt{\frac{28}{12-1}} = \sqrt{\frac{28}{11}} \approx 1.60$

Number x	Frequency f	Product $x \cdot f$	Deviation $x-\bar{x}$	Deviation² $(x-\bar{x})^2$	Product $(x-\bar{x})^2 \cdot f$
2	2	4	2 – 4 = – 2	4	4 · 2 = 8
3	4	12	3 – 4 = – 1	1	1 · 4 = 4
4	2	8	4 – 4 = 0	0	0 · 2 = 0
5	0	0	5 – 4 = 1	1	1 · 0 = 0
6	4	24	6 – 4 = 2	4	4 · 4 = 16
	$\sum f = 12$	$\sum x \cdot f = 48$			$\sum (x-\bar{x})^2 \cdot f = 28$

19. $\overline{x}=\dfrac{\sum x\cdot f}{\sum f}=\dfrac{105}{15}=7$ and $s=\sqrt{\dfrac{\sum(x-\overline{x})^2\cdot f}{n-1}}=\sqrt{\dfrac{42}{15-1}}=\sqrt{\dfrac{42}{14}}\approx 1.73$

Number x	Frequency f	Product $x\cdot f$	Deviation $x-\overline{x}$	Deviation² $(x-\overline{x})^2$	Product $(x-\overline{x})^2\cdot f$
3	1	3	3 − 7 = − 4	16	16·1 = 16
4	1	4	4 − 7 = − 3	9	9·1 = 9
5	0	0	5 − 7 = − 2	4	4·0 = 0
6	2	12	6 − 7 = − 1	1	1·2 = 2
7	5	35	7 − 7 = 0	0	0·5 = 0
8	3	24	8 − 7 = 1	1	1·3 = 3
9	3	27	9 − 7 = 2	4	4·3 = 12
	$\sum f=15$	$\sum x\cdot f=105$			$\sum(x-\overline{x})^2\cdot f=42$

21. a) set a b) set c

23. $\overline{x}=\dfrac{\sum x\cdot f}{\sum f}=\dfrac{1,580}{20}=79$ and $s=\sqrt{\dfrac{\sum(x-\overline{x})^2\cdot f}{n-1}}=\sqrt{\dfrac{826}{20-1}}=\sqrt{\dfrac{826}{19}}\approx 6.59$

Number x	Frequency f	Product $x\cdot f$	Deviation $x-\overline{x}$	Deviation² $(x-\overline{x})^2$	Product $(x-\overline{x})^2\cdot f$
72	7	504	72 − 79 = − 7	49	49·7 = 343
73	1	73	73 − 79 = − 6	36	36·1 = 36
78	3	234	78 − 79 = − 1	1	1·3 = 3
84	6	504	84 − 79 = 5	25	25·6 = 150
86	2	172	86 − 79 = 7	49	49·2 = 98
93	1	93	93 − 79 = 14	196	196·1 = 196
	$\sum f=20$	$\sum x\cdot f=1,580$			$\sum(x-\overline{x})^2\cdot f=826$

25. $\overline{x}=\dfrac{\sum x}{n}=\dfrac{43.66}{8}\approx 5.46$ and $s=\sqrt{\dfrac{\sum(x-\overline{x})^2}{n-1}}=\sqrt{\dfrac{21.617}{8-1}}=\sqrt{\dfrac{21.617}{7}}\approx 1.76$

Number x	Deviation $x-\overline{x}$	Deviation² $(x-\overline{x})^2$
3.36	3.36 – 5.46 = –2.1	4.4100
4.63	4.63 – 5.46 = –0.83	0.6889
8.25	8.25 – 5.46 = 2.79	7.7841
6.45	6.45 – 5.46 = 0.99	0.9801
5.30	5.30 – 5.46 = –0.16	0.0256
6.85	6.85 – 5.46 = 1.39	1.9321
3.07	3.07 – 5.46 = –2.39	5.7121
5.75	5.75 – 5.46 = 0.29	0.0841
$\sum x=43.66$		$\sum(x-\overline{x})^2=21.617$

27. $\bar{x} = \dfrac{\sum x \cdot f}{\sum f} = \dfrac{313}{71} \approx 4.41$ and $s = \sqrt{\dfrac{\sum (x - \bar{x})^2 \cdot f}{n-1}} = \sqrt{\dfrac{801.1551}{71-1}} = \sqrt{\dfrac{801.1551}{70}} \approx 3.38$

Number x	Frequency f	Product $x \cdot f$	Deviation $x - \bar{x}$	Deviation[2] $(x-\bar{x})^2$	Product $(x-\bar{x})^2 \cdot f$
0	13	0	0 – 4.41 = – 4.41	19.4481	252.8253
1	11	11	1 – 4.41 = – 3.41	11.6281	127.9091
2	5	10	2 – 4.41 = – 2.41	5.8081	29.0405
3	0	0	3 – 4.41 = – 1.41	1.9881	0
4	0	0	4 – 4.41 = – 0.41	0.1681	0
5	8	40	5 – 4.41 = 0.59	0.3481	2.7848
6	12	72	6 – 4.41 = 1.59	2.5281	30.3372
7	9	63	7 – 4.41 = 2.59	6.7081	60.3729
8	6	48	8 – 4.41 = 3.59	12.8881	77.3286
9	5	45	9 – 4.41 = 4.59	21.0681	105.3405
10	0	0	10 – 4.41 = 5.59	31.2481	0
11	0	0	11 – 4.41 = 6.59	43.4281	0
12	2	24	12 – 4.41 = 7.59	57.6081	115.2162
	$\sum f = 71$	$\sum x \cdot f =$ 313			$\sum (x-\bar{x})^2 \cdot f =$ 801.1551

29. $\bar{x} = \dfrac{\sum x}{n} = \dfrac{392}{8} = 49$ and $s = \sqrt{\dfrac{\sum (x-\bar{x})^2}{n-1}} = \sqrt{\dfrac{18}{8-1}} = \sqrt{\dfrac{18}{7}} \approx 1.60$

Number x	Deviation $x - \bar{x}$	Deviation[2] $(x-\bar{x})^2$
47	47 – 49 = – 2	4
48	48 – 49 = – 1	1
50	50 – 49 = 1	1
49	49 – 49 = 0	0
51	51 – 49 = 2	4
47	47 – 49 = – 2	4
49	49 – 49 = 0	0
51	51 – 49 = 2	4
$\sum x = 392$		$\sum (x-\bar{x})^2 = 18$

$\dfrac{51-49}{1.6} = 1.25$; Family H's income is 1.25 standard deviations above the mean.

31. $\overline{x} = \frac{\sum x}{n} = \frac{800}{10} = 80$ and $s = \sqrt{\frac{\sum (x-\overline{x})^2}{n-1}} = \sqrt{\frac{184}{10-1}} = \sqrt{\frac{184}{9}} \approx 4.52$

Number x	Deviation $x-\overline{x}$	Deviation^2 $(x-\overline{x})^2$
80	80 – 80 = 0	0
76	76 – 80 = – 4	16
81	81 – 80 = 1	1
84	84 – 80 = 4	16
79	79 – 80 = – 1	1
80	80 – 80 = 0	0
90	90 – 80 = 10	100
75	75 – 80 = – 5	25
75	75 – 80 = – 5	25
80	80 – 80 = 0	0
$\sum x = 800$		$\sum (x-\overline{x})^2 = 184$

$\frac{76-80}{4.52} \approx -0.88$; A grade of 76 is 0.88 standard deviations below the mean. This would be a grade of "D".

33. Apple Computer is more volatile.

Apple Computer: $CV = \frac{s}{\overline{x}} \cdot 100\% = \frac{12.3}{123.76} \cdot 100\% \approx 9.9\%$

Dell: $CV = \frac{s}{\overline{x}} \cdot 100\% = \frac{7.2}{78.6} \cdot 100\% \approx 9.2\%$

35. WebMaster is more volatile.

DJIA: $CV = \frac{s}{\overline{x}} \cdot 100\% = \frac{72.17}{11261.12} \cdot 100\% \approx 0.6\%$;

WebMaster: $CV = \frac{s}{\overline{x}} \cdot 100\% = \frac{1.7}{37.6} \cdot 100\% \approx 4.5\%$

37. Coffee

$$\overline{x}=\frac{\sum x}{n}=\frac{41.8}{12}\approx 3.48 \text{ and } s=\sqrt{\frac{\sum(x-\overline{x})^2}{n-1}}=\sqrt{\frac{0.147577}{12-1}}=\sqrt{\frac{0.147577}{11}}\approx 0.116$$

Price x	Deviation $x-\overline{x}$					Deviation² $(x-\overline{x})^2$
3.288	3.288	–	3.484	=	0.196	0.038416
3.456	3.456	–	3.484	=	0.028	0.000784
3.475	3.475	–	3.484	=	0.009	0.000081
3.437	3.437	–	3.484	=	0.047	0.002209
3.308	3.308	–	3.484	=	0.176	0.030976
3.407	3.407	–	3.484	=	0.077	0.005929
3.529	3.529	–	3.484	=	–0.045	0.002025
3.497	3.497	–	3.484	=	–0.013	0.000169
3.537	3.537	–	3.484	=	–0.053	0.002809
3.577	3.577	–	3.484	=	–0.093	0.008649
3.607	3.607	–	3.484	=	–0.123	0.015129
3.685	3.685	–	3.484	=	–0.201	0.040401
$\sum x = 41.8$						$\sum(x-\overline{x})^2 = 0.147577$

Gasoline

$$\overline{x}=\frac{\sum x}{n}=\frac{33.607}{12}\approx 2.801 \text{ and } s=\sqrt{\frac{\sum(x-\overline{x})^2}{n-1}}=\sqrt{\frac{0.908343}{12-1}}=\sqrt{\frac{0.908343}{11}}\approx 0.287$$

Price x	Deviation $x-\overline{x}$					Deviation² $(x-\overline{x})^2$
2.274	2.274	–	2.801	=	0.527	0.277729
2.285	2.285	–	2.801	=	0.516	0.266256
2.592	2.592	–	2.801	=	0.209	0.043681
2.860	2.860	–	2.801	=	–0.059	0.003481
3.130	3.130	–	2.801	=	–0.329	0.108241
3.052	3.052	–	2.801	=	–0.251	0.063001
2.961	2.961	–	2.801	=	–0.16	0.025600
2.782	2.782	–	2.801	=	0.019	0.000361
2.789	2.789	–	2.801	=	0.012	0.000144
2.793	2.793	–	2.801	=	0.008	0.000064
3.069	3.069	–	2.801	=	–0.268	0.071824
3.020	3.020	–	2.801	=	–0.219	0.047961
$\sum x = 33.607$						$\sum(x-\overline{x})^2 = 0.908343$

39. G went from the mean income to 0.935 (0.94 rounded) standard deviations above the mean.

Family	Year One $\frac{x-\bar{x}}{s}$	Year Two $\frac{x-\bar{x}}{s}$	Change in Standard Deviations
A	$\frac{47-49}{1.60}=-1.25$	$\frac{49-51}{1.07}\approx-1.87$	Down 0.62
B	$\frac{48-49}{1.60}=-0.625$	$\frac{51-51}{1.07}=0$	Up 0.625
C	$\frac{50-49}{1.60}=0.625$	$\frac{52-51}{1.07}\approx 0.935$	Up 0.31
D	$\frac{49-49}{1.60}=0$	$\frac{51-51}{1.07}=0$	No Change
E	$\frac{51-49}{1.60}=1.25$	$\frac{51-51}{1.07}=0$	Down 1.25
F	$\frac{47-49}{1.60}=-1.25$	$\frac{50-51}{1.07}\approx-0.935$	Up 0.315
G	$\frac{49-49}{1.60}=0$	$\frac{52-51}{1.07}\approx 0.935$	**Up 0.935**
H	$\frac{51-49}{1.60}=1.25$	$\frac{52-51}{1.07}\approx 0.935$	Down 0.315

41. In calculating a sample's standard deviation, we divide by $n-1$; for a population, we divide by n.

43. true; Explanations/examples may vary.

45. – 47. No solution provided.

Section 15.4 The Normal Distribution

1. See box on page 752.

3. The normal curve is symmetric about the mean.

5. Since 12 is one standard deviation above the mean, we would expect $\frac{68\%}{2}=34\%$ of the values to lie between 10 and 12.

7. Since 14 is two standard deviations above the mean, we would expect $\frac{95\%}{2}=47.5\%$ of the values to lie between 10 and 14. Therefore, 50% – 47.5%=2.5% of the values should fall above 14.

9. Since 12 is one standard deviation above the mean, we would expect $\frac{68\%}{2}=34\%$ of the values to lie between 10 and 12. Therefore, 50% – 34%=16% of the values should fall above 12.

11. Since 9 is one standard deviation below the mean, we would expect $\frac{68\%}{2}=34\%$ of the values to lie between 9 and 12. Therefore, 50% – 34%=16% of the values should fall below 9.

13. Since 6 is two standard deviations below the mean, we would expect $\frac{95\%}{2}=47.5\%$ of the values to lie between 6 and 12. Therefore, 50% + 47.5%=97.5% of the values should fall above 6.

15. Since 18 is two standard deviations above the mean, we would expect $\frac{95\%}{2} = 47.5\%$ of the values to lie between 12 and 18. Similarly, since 15 is one standard deviation above the mean, we would expect $\frac{68\%}{2} = 34\%$ of the values to lie between 12 and 15. Therefore, we would expect 47.5% – 34%=13.5% of the values to lie between 15 and 18.

17. 39.1%

19. 42.4% – 31.3% = 11.1%

21. 27.3%

23. 47.4% – 39.4% = 8%

25. 27.6% –14.8% = 12.8%

27. 50% – 42.7% = 7.3%

29. 50% – 41.9% = 8.1%

31. 50% + 40.8% = 90.8%

33. 50% + 30.0% = 80%

35. We need to find a z–score that corresponds to 0.5 – 0.1 = 0.4. The closest z–score is 1.28.

37. We need to find a z–score that corresponds to 0.5 – 0.12 = 0.38. Since 0.38 is halfway between 0.379 and 0.381, the z–score would be 1.175. This z–score needs to be –1.175, however, since it indicates an area less than 50% to the left of the mean.

39. We need to find a z–score that corresponds to 0.6 – 0.5 = 0.1. Since 0.1 is one-fourth of the way between 0.099 and 0.103, the z–score would be 0.2525.

41. We need to find a z–score that corresponds to 0.75 – 0.5 = 0.25. Since 0.25 is one-third of the way between 0.249 and 0.252, the z–score would be about 0.6733. This z–score needs to be – 0.6733 however, since it indicates an area more than 50%.

43. $z = \frac{87-80}{5} = 1.40$

45. $z = \frac{14-21}{4} = -1.75$

47. $z = \frac{48-38}{10.3} \approx 0.97$

49. $0.84 = \frac{x-60}{5}$

$4.2 = x - 60$

$64.2 = x$

51. $-0.45 = \frac{x-35}{3}$

$-1.35 = x - 35$

$33.65 = x$

53. $1.64 = \frac{x-28}{2.25}$

$3.69 = x - 28$

$31.69 = x$

55. a) 50%

b) Since 7.5 is one standard deviation below the mean, we would expect $\frac{68\%}{2} = 34\%$ of the values to lie between 7.5 and 8. Therefore, 50% – 34%=16% of the cups should have less that 7.5 ounces.

57. a) Since 202 is one standard deviation above the mean, we would expect $\frac{68\%}{2} = 34\%$ of the bags to have between 200 and 202 pieces in them.

b) 50% – 34%=16%

59. $z = \frac{140-120}{12} \approx 1.67$

61. 6 feet 8 inches corresponds to 80 inches and 7 feet corresponds to 84 inches. $z = \frac{84-80}{3} \approx 1.33$, and the percent of area that corresponds to players between 80 inches and 84 inches is 40.8%. Therefore, 50% – 40.8% = 9.2% have a height over 7 feet. Since $0.092 \cdot 324 = 29.808$, we would expect about 29 or 30 players out of the 324 to be over 7 feet.

63. $z = \frac{70-68}{4} = 0.50$, and this corresponds to an area of 19.2% + 50% = 69.2% for women with heart rates of less than 70 beats per minute. Since $0.692 \cdot 200 = 138.4$, we would expect 138 or 139 out of the 200 women to have heart rates of less that 70 beats per minute.

65. Since roughly 95% of data values are within 2 standard deviations of the mean, the standard deviation should be 9 since 250 is 18 below 268 and 286 is 18 above 268. We now need to find the z–score associated with 275 days. $z = \frac{275-268}{9} \approx 0.78$ and therefore, the percentage of pregnancies that would be expected to last at least 275 days would be 50% – 28.2% = 21.8%.

67. $1.5 = \frac{x-37}{11}$

$16.5 = x - 37$

$x = 53.5$

69. Carew was more dominant.

Jackie Robinson: $z = \frac{0.342-0.267}{0.0326} \approx 2.30$ above the mean

Rod Carew: $z = \frac{0.350-0.261}{0.0317} \approx 2.81$ above the mean

71. The z–score that corresponds to 50% – 4%= 46% = 0.46 is 1.75. Moreover, this z–score should be – 1.75, since the accumulated area is less than 50%.

$-1.75 = \frac{x-2{,}000}{800} \Rightarrow -1{,}400 = x - 2{,}000 \Rightarrow x = 600$ hours of play. Since a game is played for about 2 hours a day, the number of days the game is played is $\frac{600}{2} = 300$. This implies that the warranty should be for about 10 months.

73. a) $z = \frac{9-7.8}{1.3} \approx 0.92$ and this corresponds to an area of 50% – 32.1% = 17.9%. So of the past 15 years, you would expect to receive at least 9% on the investments $0.179 \cdot 15 = 2.685$ or 2 or 3 years.

b) $z = \frac{6-7.8}{1.3} \approx -1.38$, and this corresponds to an area of 50% – 41.6% = 8.4%. So of the past 15 years, you would expect to receive less than 6% on the investments $0.084 \cdot 15 = 1.26$ or 1 or 2 years.

75. $z = \frac{480-500}{100} = -0.20$, and this corresponds to 50% – 7.9% = 42.1%.

77. The score is almost two standard deviations above the mean.

79. By looking at the graph of the normal curve, there is clearly more area under the curve between 0.0 and 0.7 than there is between 1.3 and 2.0.

81. Answers may vary.
The mean is 20 in both distributions; standard deviation is about 2 or 3 in distribution 1; standard deviation is about 4 or 5 in distribution 2. The estimation for the means comes from determining for what value the curves are symmetric about. The estimation for the standard deviations comes from applying the properties that govern how much data will fall within 1, 2, or 3 standard deviations from the mean.

83. – 85. No solution provided.

87. The z–score that corresponds to 80% – 50% = 30% = 0.300 is 0.84.

89. The z–score that approximately corresponds to 75% – 50%= 25% = 0.250 is 0.673.
$0.673 = \frac{x-40}{4} \Rightarrow 2.692 = x - 40 \Rightarrow x = 42.692$

91. $z = \frac{75-68}{4} = 1.75$ corresponds to an accumulated area of 50% + 46% = 96%. This implies the 96th percentile.

Section 15.5 Looking Deeper: Linear Correlation

1. Because the grades generally increased as the number of tutoring sessions increased, there was some relationship between the two variables.

3. positive correlation

5. a)

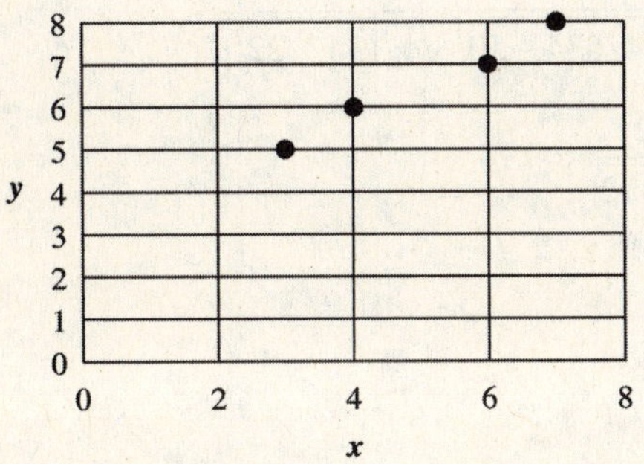

b) No solution provided.

c)

x	y	x^2	y^2	xy
3	5	9	25	15
7	8	49	64	56
4	6	16	36	24
6	7	36	49	42
$\sum x = 20$	$\sum y = 26$	$\sum x^2 = 110$	$\sum y^2 = 174$	$\sum xy = 137$

5. (continued)

$$r = \frac{n\sum xy - \left(\sum x\right)\left(\sum y\right)}{\sqrt{n\left(\sum x^2\right) - \left(\sum x\right)^2}\sqrt{n\left(\sum y^2\right) - \left(\sum y\right)^2}} = \frac{4\cdot 137 - 20\cdot 26}{\sqrt{4\cdot 110 - 20^2}\sqrt{4\cdot 174 - 26^2}}$$

$$= \frac{548 - 520}{\sqrt{440 - 400}\sqrt{696 - 676}} = \frac{28}{\sqrt{40}\sqrt{20}} = \frac{28}{\sqrt{800}} \approx 0.99$$

7. a)

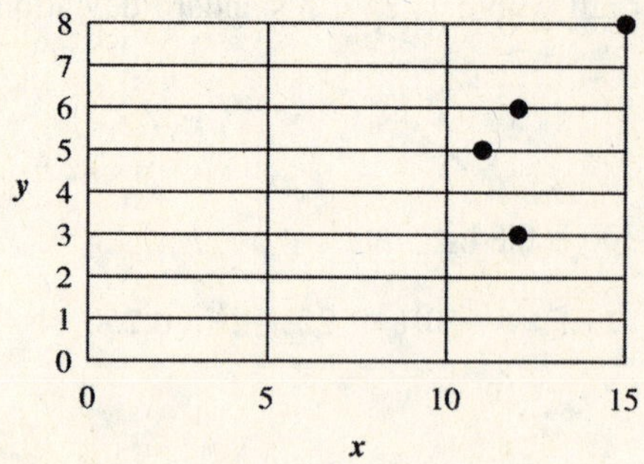

b) No solution provided.

c)

x	y	x^2	y^2	xy
11	5	121	25	55
15	8	225	64	120
12	3	144	9	36
12	6	144	36	72
$\sum x = 50$	$\sum y = 22$	$\sum x^2 = 634$	$\sum y^2 = 134$	$\sum xy = 283$

$$r = \frac{n\sum xy - \left(\sum x\right)\left(\sum y\right)}{\sqrt{n\left(\sum x^2\right) - \left(\sum x\right)^2}\sqrt{n\left(\sum y^2\right) - \left(\sum y\right)^2}} = \frac{4\cdot 283 - 50\cdot 22}{\sqrt{4\cdot 634 - 50^2}\sqrt{4\cdot 134 - 22^2}}$$

$$= \frac{1{,}132 - 1{,}100}{\sqrt{2{,}536 - 2{,}500}\sqrt{536 - 484}} = \frac{32}{\sqrt{36}\sqrt{52}} = \frac{32}{\sqrt{1{,}872}} \approx 0.74$$

9. We can be 95% confident, but not 99% confident.

11. neither

13. $y = 0.7x + 3$

x	y	x^2	xy
3	5	9	15
7	8	49	56
4	6	16	24
6	7	36	42
$\sum x = 20$	$\sum y = 26$	$\sum x^2 = 110$	$\sum xy = 137$

13. (continued)

$$m=\frac{n\sum xy-\left(\sum x\right)\left(\sum y\right)}{n\left(\sum x^2\right)-\left(\sum x\right)^2}=\frac{4\cdot 137-20\cdot 26}{4\cdot 110-20^2}=\frac{548-520}{440-400}=\frac{28}{40}=0.7$$

$$b=\frac{\sum y-m\left(\sum x\right)}{n}=\frac{26-0.7\cdot 20}{4}=\frac{26-14}{4}=\frac{12}{4}=3$$

15. $y=0.89x-5.61$

x	y	x^2	xy
11	5	121	55
15	8	225	120
12	3	144	36
12	6	144	72
$\sum x=50$	$\sum y=22$	$\sum x^2=634$	$\sum xy=283$

$$m=\frac{n\sum xy-\left(\sum x\right)\left(\sum y\right)}{n\left(\sum x^2\right)-\left(\sum x\right)^2}=\frac{4\cdot 283-50\cdot 22}{4\cdot 634-50^2}=\frac{1{,}132-1{,}100}{2{,}536-2{,}500}=\frac{32}{36}=\frac{8}{9}\approx 0.89$$

$$b=\frac{\sum y-m\left(\sum x\right)}{n}=\frac{22-\frac{8}{9}\cdot 50}{4}=\frac{198-8\cdot 50}{36}=\frac{198-400}{36}=\frac{-202}{36}=-\frac{101}{18}\approx -5.61$$

17. We can be 99% confident that there is positive linear correlation.

x	y	x^2	y^2	xy
0	23	0	529	0
1	22	1	484	22
2	27	4	729	54
2	28	4	784	56
5	35	25	1,225	175
$\sum x=10$	$\sum y=135$	$\sum x^2=34$	$\sum y^2=3{,}751$	$\sum xy=307$

$$r=\frac{n\sum xy-\left(\sum x\right)\left(\sum y\right)}{\sqrt{n\left(\sum x^2\right)-\left(\sum x\right)^2}\sqrt{n\left(\sum y^2\right)-\left(\sum y\right)^2}}=\frac{5\cdot 307-10\cdot 135}{\sqrt{5\cdot 34-10^2}\sqrt{5\cdot 3{,}751-135^2}}$$

$$=\frac{1{,}535-1{,}350}{\sqrt{170-100}\sqrt{18{,}755-18{,}225}}=\frac{185}{\sqrt{70}\sqrt{530}}=\frac{185}{\sqrt{37{,}100}}\approx 0.96$$

19. neither

x	y	x^2	y^2	xy
30	26	900	676	780
40	31	1,600	961	1,240
50	33	2,500	1,089	1,650
60	31	3,600	961	1,860
70	26	4,900	676	1,820
$\sum x=250$	$\sum y=147$	$\sum x^2=13{,}500$	$\sum y^2=4{,}363$	$\sum xy=7{,}350$

19. (continued)

$$r = \frac{n\sum xy - \left(\sum x\right)\left(\sum y\right)}{\sqrt{n\left(\sum x^2\right) - \left(\sum x\right)^2}\sqrt{n\left(\sum y^2\right) - \left(\sum y\right)^2}} = \frac{5\cdot 7{,}350 - 250\cdot 147}{\sqrt{5\cdot 13{,}500 - 250^2}\sqrt{5\cdot 4{,}363 - 147^2}}$$

$$= \frac{36{,}750 - 36{,}750}{\sqrt{67{,}500 - 62{,}500}\sqrt{21{,}815 - 21{,}609}} = \frac{0}{\sqrt{5{,}000}\sqrt{206}} = \frac{0}{\sqrt{1{,}030{,}000}} = 0.00$$

21. $y = 2.64x + 21.71$

x	y	x^2	xy
0	23	0	0
1	22	1	22
2	27	4	54
2	28	4	56
5	35	25	175
$\sum x = 10$	$\sum y = 135$	$\sum x^2 = 34$	$\sum xy = 307$

$$m = \frac{n\sum xy - \left(\sum x\right)\left(\sum y\right)}{n\left(\sum x^2\right) - \left(\sum x\right)^2} = \frac{5\cdot 307 - 10\cdot 135}{5\cdot 34 - 10^2} = \frac{1{,}535 - 1{,}350}{170 - 100} = \frac{185}{70} = \frac{37}{14} \approx 2.64$$

$$b = \frac{\sum y - m\left(\sum x\right)}{n} = \frac{135 - \frac{37}{14}\cdot 10}{5} = \frac{1{,}890 - 370}{70} = \frac{1{,}520}{70} = \frac{152}{7} \approx 21.71$$

23. $y = 0\cdot x + 29.4$

x	y	x^2	xy
30	26	900	780
40	31	1600	1240
50	33	2500	1650
60	31	3600	1860
70	26	4900	1820
$\sum x = 250$	$\sum y = 147$	$\sum x^2 = 13{,}500$	$\sum xy = 7{,}350$

$$m = \frac{n\sum xy - \left(\sum x\right)\left(\sum y\right)}{n\left(\sum x^2\right) - \left(\sum x\right)^2} = \frac{5\cdot 7{,}350 - 250\cdot 147}{5\cdot 13{,}500 - 250^2} = \frac{36{,}750 - 36{,}750}{67{,}500 - 62{,}500} = \frac{0}{5{,}000} = 0$$

$$b = \frac{\sum y - m\left(\sum x\right)}{n} = \frac{147 - 0\cdot 250}{5} = \frac{147 - 0}{5} = \frac{147}{5} = 29.4$$

25. We can be 99% confident that there is a significant linear correlation between the variables of car weight and mileage.

27. – 29. No solution provided.

31. 1

Chapter Review Exercises

1.

Number of Accidents	Frequency
4	1
5	3
6	6
7	2
8	10
9	4
10	5

3.

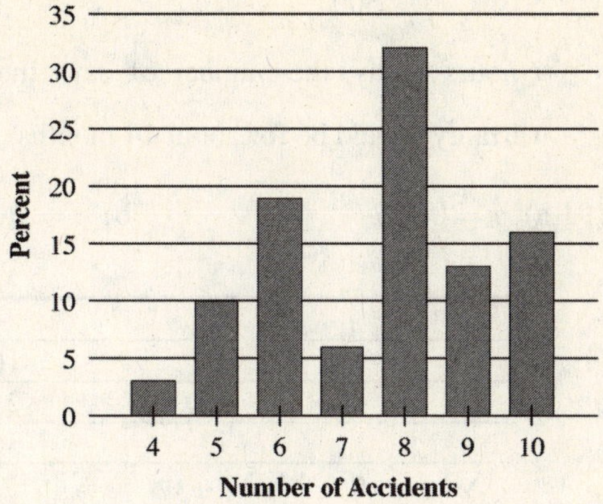

5. a)

F		M
9 9 8 6 6 5 1	2	9
9 8 5 5 5 4 3 3 3 2 1 0 0	3	1 2 5 6 7 7 7 7 8 8 9
9 9 5 2 1	4	0 2 3 3 5 5 5 6 7
	5	0 1 2 2 4
1 1	6	0 1
4	7	6
0	8	

b) The actors generally seem older than the actresses.

7. The mean is the arithmetic average; the median is the middle score; the mode is the most frequent score.

9.

minimum: 3
first quartile: $Q_1 = 6$
median: 8
third quartile: $Q_3 = 12$
maximum: 20

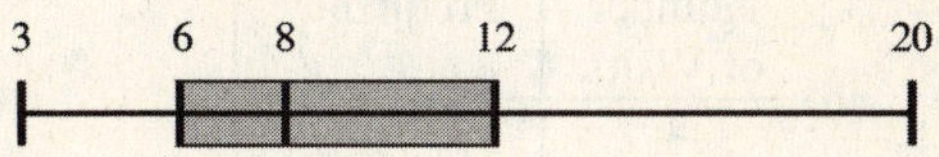

11. the spread of the distribution

13. $z = \dfrac{85-80}{7} \approx 0.71$

15. For 75, $z = \dfrac{75-72}{4} = 0.75$, so 2.73% of the scores are between 75 and the mean of 72.

For 82 $z = \dfrac{82-72}{4} = 2.5$, so 4.94% of the scores are between the mean of 72 and 82.

So, 4.94% – 2.73% =2.21% of the scores are expected to lie between 75 and 82, so $1{,}000(.0.221) = 221$ scores should lie between 75 and 82.

17. The z–score that corresponds to 50% – 4%= 46% = 0.46 is 1.75. Moreover, this z–score should be –1.75, since the accumulated area is less than 50%.

$$-1.75 = \frac{x-2{,}500}{500} \Rightarrow -875 = x - 2{,}500 \Rightarrow x = 1{,}625$$ hours of use. Since an iTouch is used for about 4 hours a day, the number of days the iTouch is used is $\frac{1{,}625}{4} = 406.25$. This implies that the warranty should be for about 14 months.

19. a) $r = 0.977$ b) $y = 2.5x - 2$

x	y	x^2	y^2	xy
3	5	9	25	15
4	9	16	81	36
5	10	25	100	50
6	13	36	169	78
$\sum x = 18$	$\sum y = 37$	$\sum x^2 = 86$	$\sum y^2 = 375$	$\sum xy = 179$

$$r = \frac{n\sum xy - \left(\sum x\right)\left(\sum y\right)}{\sqrt{n\left(\sum x^2\right) - \left(\sum x\right)^2}\sqrt{n\left(\sum y^2\right) - \left(\sum y\right)^2}} = \frac{4\cdot 179 - 18\cdot 37}{\sqrt{4\cdot 86 - 18^2}\sqrt{4\cdot 375 - 37^2}}$$

$$= \frac{716-666}{\sqrt{344-324}\sqrt{1{,}500-1{,}369}} = \frac{50}{\sqrt{20}\sqrt{131}} = \frac{50}{\sqrt{2{,}620}} \approx 0.977$$

$$m = \frac{n\sum xy - \left(\sum x\right)\left(\sum y\right)}{n\left(\sum x^2\right) - \left(\sum x\right)^2} = \frac{4\cdot 179 - 18\cdot 37}{4\cdot 86 - 18^2} = \frac{716-666}{344-324} = \frac{50}{20} = 2.5$$

$$b = \frac{\sum y - m\left(\sum x\right)}{n} = \frac{37 - 2.5\cdot 18}{4} = \frac{37-45}{4} = \frac{-8}{4} = -2$$

Chapter Test

1.

Number of Visits	Frequency
4	1
5	4
6	8
7	9
8	7
9	6
10	2
11	2
12	1

3.

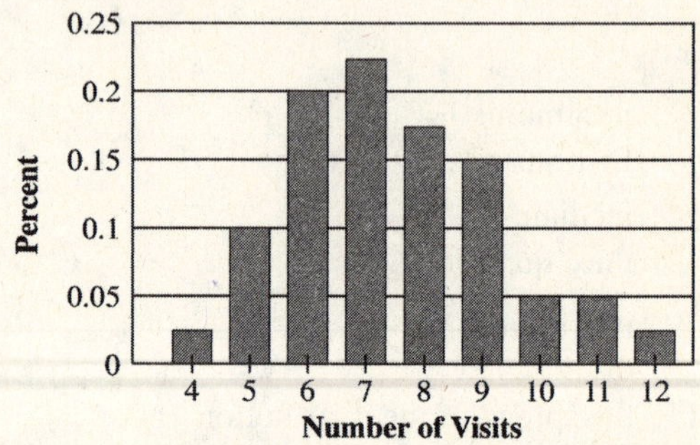

5. $\bar{x} = \dfrac{\sum x}{n} = \dfrac{54}{9} = 6$ and $s = \sqrt{\dfrac{\sum (x-\bar{x})^2}{n-1}} = \sqrt{\dfrac{32}{9-1}} = \sqrt{\dfrac{32}{8}} = 2$

Number x	Deviation $x-\bar{x}$	Deviation² $(x-\bar{x})^2$
3	$3-2=-3$	9
4	$4-2=-2$	4
5	$5-2=-1$	1
6	$6-2=0$	0
5	$5-2=-1$	1
6	$6-2=0$	0
8	$8-2=2$	4
9	$9-2=3$	9
8	$8-2=2$	4
$\sum x = 54$		$\sum (x-\bar{x})^2 = 32$

7.

Ruth		Aaron
5 2	2	
5 4	3	0 2 4 4 8 9 9
9 7 6 6 6 1 1	4	0 0 4 4 4 4 5 7
9 4 4	5	
0	6	

9. The statistics exam score is relatively better, since it has a higher z–score.

Statistics: $z = \dfrac{85-78}{4} = 1.75$

Sociology: $z = \dfrac{88-80}{5} = 1.60$

11. a) no linear correlation
b) negative linear correlation

13. Place the data in order.
3, 5, 5, 6, 6, 6, 7, 7, 8, 8, 9, 9, 9, 9, 9, 9, 10, 10, 11, 11

Mean: $\bar{x} = \dfrac{\sum x}{n} = \dfrac{157}{20} = 7.85$

Median: $\dfrac{8+9}{2} = \dfrac{17}{2} = 8.5$

Mode: 9

15. $z = \dfrac{82-75}{5} = 1.4$

17. The z–score that corresponds to 50% – 6%= 44% = 0.44 is 1.555, since 0.44 lies between 1.55 and 1.56. Moreover, this z–score should be –1.555, since the accumulated area is less than 50%.

$-1.555 = \frac{x - 2,800}{400} \Rightarrow -622 = x - 2,800 \Rightarrow x = 2,178$ hours of use. Since the iTouch is used for about 4 hours a day, the number of days the iTouch is used is $\frac{2,178}{4} = 544.5$. This implies that the warranty should be for about 18 months.